Satellite
Communications

Other McGraw-Hill Communications Books of Interest

Satellite Communications

Dennis Roddy

Second Edition

McGraw-Hill

New York   San Francisco   Washington, D.C.   Auckland   Bogotá
Caracas   Lisbon   London   Madrid   Mexico City   Milan
Montreal   New Delhi   San Juan   Singapore
Sydney   Tokyo   Toronto

Library of Congress Cataloging-in-Publication Data

Roddy, Dennis, (date).
   Satellite communications / Dennis Roddy. — 2nd ed.
      p.   cm.
   Includes index.
   ISBN 0-07-053370-9 (HC)
   1. Artificial satellites in telecommunication. I. Title.
TK5104.R627   1995
621.382'5—dc20   95-22077
                                                 CIP

McGraw-Hill

A Division of The **McGraw·Hill** *Companies*

  3 4 5 6 7 8 9 0 DOC/DOC 9 0 0 9 8 7

ISBN 0-07-053370-9

The sponsoring editor for this book was Stephen S. Chapman, the editing supervisor was Peggy Lamb, and the production supervisor was Suzanne W. B. Rapcavage. It was set in Century Schoolbook by Victoria Khavkina of McGraw-Hill's Professional Book Group composition unit.

Printed and bound by R. R. Donnelley & Sons Company.

This book is printed on acid-free paper.

McGraw-Hill books are available at special quantity discounts to use as premiums and sales promotions, or for use in corporate training programs. For more information, please write to the Director of Special Sales, McGraw-Hill, 11 West 19th Street, New York, NY 10011. Or contact your local bookstore.

Contents

Preface

The overall objectives of the second edition remain the same as those of the first edition, namely, to provide a broad coverage of satellite communication technology. The material has been considerably expanded and includes the fundamentals of orbital mechanics, propagation effects, descriptions of earth and space systems employed in satellite links, the analog and digital type signals carried by satellites, interference peculiar to satellite networks, multiple access methods, and descriptions of selected satellite services.

The book should be suitable for senior students in engineering-technology programs and as first course material in engineering-degree programs. Lengthy mathematical derivations are avoided, but references are given and mathematical results used and explained as required. Although many of the examples and problems given in the text may be worked using a good calculator or even better one of the several commercial computer packages available, it has been the author's experience that Mathcad[1] is an excellent tool for this purpose. Consequently, many of the solutions are given using Mathcad and an appendix is included to explain some of the basics of Mathcad.

It is assumed that the student is familiar with basic communication circuits and systems, but material which is particularly applicable to satellite communications is reviewed at the appropriate places. Because of the widespread use of decibels and decilogs in link budget calculations these logarithmic units are reviewed in an appendix.

As acknowledged in the first edition, my thanks go to the staff at the Communications Research Centre, Department of Communications, Ottawa, and to Lakehead University for providing the opportunity to gather and study material for the book, to the students at Lakehead University who suffered through the unedited version of the manu-

---

[1]Mathcad is a registered trademark of Mathsoft Inc.

script used for class notes, and who suggested improvements and corrections; to Dr. Henry Driver of Computer Sciences Corporation who sent in corrections and references for calculation of geodetic position. Finally my thanks go to Peggy Lamb, Editing Manager, Professional Book Group, McGraw-Hill for the courteous manner in which she managed to keep the production on schedule.

Dennis Roddy
Thunder Bay, Ontario

Preface to the First Edition

Although satellite communications would seem to be a straightforward extension of terrestrial radio systems, the use of satellites brings in new operational features not found in terrestrial systems. At the same time, new technology is required to exploit the unique features of satellite systems.

The inquiring student will also wish to understand something about the mechanics of orbital motion: why, for example, there is only one geostationary orbit, and how one might track nonstationary satellites.

This book, which covers these fundamental aspects, is intended for senior students in engineering-technology programs; it should also provide suitble material for an introductory course in undergraduate engineering-degree programs. It is assumed that the reader is familiar with, or is concurrently studying, basic communication circuits and systems, including modulation, noise, and microwave propagation, although these topics are briefly reviewed. Lengthy mathematical derivations are avoided, but references are given and mathematical results are used and explained as required.

Most of the calculations for satellite system performance are carried out in decibels or related units. These units and calculations are explained in detail, but as a prerequisite the student should have a good working knowledge of the decibel.

I would like to thank the many people and organizations who freely provided photographs, figures, and data. These have been acknowledged at the appropriate places in the text. Much of the material for the book was gathered while I was on sabbatical leave at the Communications Research Centre, Department of Communications, Ottawa, Canada. I am grateful for the generous assistance provided by the department, both financially and by way of access to resources. My thanks go to the staff at the Communications Research Centre for providing much technical information and guidance, and to the library staff for all their help.

My thanks also go to Lakehead University for providing me with a sabbatical leave, and to the students at the university who suggested improvements and corrections to the text while working through the first few drafts of the manuscript.

Finally, I would like to thank production editor John Fleming for his courteous and helpful editing of the manuscript, and the Prentice-Hall reviewers for their positive comments and suggestions for improvements, most of which have been incorporated in the book.

Dennis Roddy
Thunder Bay, Ontario

Overview of Satellite Systems

1.1 Introduction

The use of satellites in communications systems is very much a fact of everyday life, as is evidenced by the many homes which are equipped with antennas, or "dishes," used for reception of satellite television. What may not be so well known is that satellites form an essential part of telecommunications systems worldwide, carrying large amounts of data and telephone traffic in addition to television signals.

Satellites offer a number of features not readily available with other means of communications. Because very large areas of the earth are visible from a satellite, the satellite can form the star point of a communications net linking together many users simultaneously, users who may be widely separated geographically. The same feature enables satellites to provide communications links to remote communities in sparsely populated areas which are difficult to access by other means. Of course, satellite signals ignore political boundaries as well as geographical ones, which may or may not be a desirable feature.

To give some idea of cost, the construction and launch costs of the Canadian Anik-E1 satellite (in 1994 Canadian dollars) were $281.2 million, and the Anik-E2, $290.5 million. The combined launch insurance for both satellites was $95.5 million. A feature of any satellite system is that the cost is *distance insensitive,* meaning that it costs about the same to provide a satellite communications link over a short distance as it does over a large distance. Thus, a satellite communications system is economical only where the system is in continuous use and the costs can be reasonably spread over a large number of users.

Satellites are also used for remote sensing, examples being the detection of water pollution and the monitoring and reporting of weather conditions. Some of these remote sensing satellites also form a vital link in search and rescue operations for downed aircraft and the like.

A good overview of the role of satellites is given by Pritchard, 1984, and Brown, 1981. To provide a general overview of satellite systems here, three different types of applications are briefly described in this chapter: (1) the largest international system, Intelsat; (2) the domestic satellite system in the United States, Domsat; and (3) U.S. National Oceanographic and Atmospheric Administration (NOAA) series of polar orbiting satellites used for environmental monitoring and search and rescue.

1.2   Frequency Allocations for Satellite Services

Allocating frequencies to satellite services is a complicated process which requires international coordination and planning. This is carried out under the auspices of the International Telecommunication Union. To facilitate frequency planning, the world is divided into three regions:

Region 1: Europe, Africa, what was formerly the Soviet Union, and Mongolia

Region 2: North and South America and Greenland

Region 3: Asia (excluding region 1 areas), Australia, and the southwest Pacific

Within these regions, frequency bands are allocated to various satellite services, although a given service may be allocated different frequency bands in different regions. Some of the services provided by satellites are

Fixed satellite service (FSS)

Broadcasting satellite service (BSS)

Mobile satellite services

Navigational satellite services

Meteorological satellite services

There are many subdivisions within these broad classifications; for example, the fixed satellite service provides links for existing telephone networks as well as for transmitting television signals to cable companies for distribution over cable systems. Broadcasting satellite services are intended mainly for direct broadcast to the home, sometimes referred to as *direct broadcast satellite* (DBS) service [in Europe it may be known as *direct-to-home* (DTH) service]. Mobile satellite services would include land mobile, maritime mobile, and aeronautical mobile. Navigational satellite services include global positioning systems, and satellites intended for the meteorological services often provide a search and rescue service.

Table 1.1 lists the frequency band designations in common use for satellite services. The Ku band signifies the band *u*nder the K band, and the Ka band is the band *a*bove the K band. The Ku band is the one used at present for direct broadcast satellites, and it is also used for certain fixed satellite services. The C band is used for fixed satellite services, and no direct broadcast services are allowed in this band. The VHF band is used for certain mobile and navigational services and for data transfer from weather satellites. The L band is used for mobile satellite services and navigation systems. For the fixed satellite service in the C band, the most widely used subrange is approximately 4 to 6 GHz. The higher frequency is nearly always used for the uplink to the satellite, for reasons which will be explained later, and common practice is to denote the C band by 6/4 GHz, giving the uplink frequency first. For the direct broadcast service in the Ku band the most widely used range is approximately 12 to 14 GHz, which is denoted by 14/12 GHz. Although frequency assignments are made much more precisely, and they may lie somewhat outside the values quoted here (an example of assigned frequencies in the Ku band is 14,030 and 11,730 MHz), the approximate values stated above are quite satisfactory for use in calculations involving frequency, as will be shown later in the text.

Care must be exercised when using published references to frequency bands, as the designations have developed somewhat differently for radar and communications applications; in addition not all countries use the same designations. New designations have been

TABLE 1.1   Frequency Band Designations

Frequency range, GHz	Band designation
0.1–0.3	VHF
0.3–1.0	UHF
1.0–2.0	L
2.0–4.0	S
4.0–8.0	C
8.0–12.0	X
12.0–18.0	Ku
18.0–24.0	K
24.0–40.0	Ka
40.0–100.0	mm

TABLE 1.2   Modern Frequency Band Designations

Frequency range, GHz	Band designation
0.1–0.25	A
0.25–0.5	B
0.5–1.0	C
1.0–2.0	D
2.0–3.0	E
3.0–4.0	F
4.0–5.5	G
5.5–8.0	H
8.0–10.0	I
10.0–20.0	J
20.0–40.0	K
40.0–60.0	L
60.0–100.0	M

proposed, and these appear in Table 1.2 for completeness. However, in this text the designations given in Table 1.1 will be used, along with 6/4 GHz for the C band, and 14/12 GHz for the Ku band.

1.3   Intelsat

Intelsat is an acronym derived from *International Telecommunications Satellite*. The Intelsat Organization was established in 1964 to handle the myriad of technical and administrative problems associated with a worldwide telecommunications system. A listing of the 119 member countries and signatories is given in Khan, 1992.

The early development of Intelsat is covered in Colino, 1985. Starting with a single satellite, Early Bird (Intelsat I), which was launched in 1965 and provided 480 voice channels, a series of satellites, designated Intelsat I, II, III, IV, V, and VI were launched. Figure 1.1 shows the evolution of the Intelsat satellites. It will be seen that up to the Intelsat VI series there was a progressive increase in size and capacity, Intelsat VI being capable of providing 80,000 voice channels. The latest in the series, Intelsat VII, is smaller than its predecessor as shown in Table 1.3.

The Intelsat VII is similar in construction to the V and VA/VB series shown in Fig. 1.1, in that it has solar sails rather than a cylindrical body. This type of construction is described more fully in Chap. 6. The design lifetime for the Intelsat VII is 15 years (Lilly, 1990), and the lifetimes given in Fig. 1.1 show that there is some overlap in the available service time. The Intelsat VII series of satellites was planned for service in the Pacific Ocean region and also for some of the less demanding services in the Atlantic Ocean region (the traffic carried by the Atlantic Ocean region satellites is much greater than

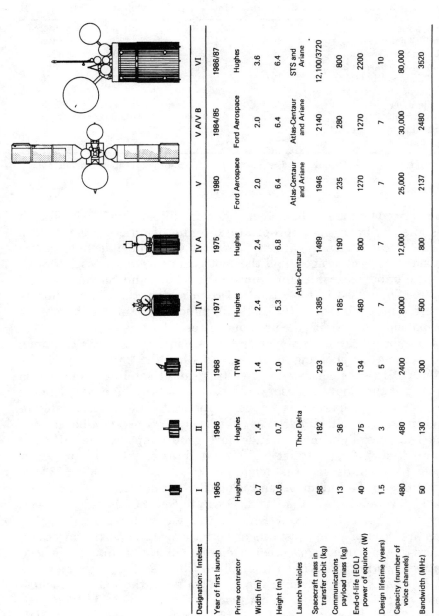

Designation: Intelsat	I	II	III	IV	IV A	V	V A/V B	VI
Year of first launch	1965	1966	1968	1971	1975	1980	1984/85	1986/87
Prime contractor	Hughes	Hughes	TRW	Hughes	Hughes	Ford Aerospace	Ford Aerospace	Hughes
Width (m)	0.7	1.4	1.4	2.4	2.4	2.0	2.0	3.6
Height (m)	0.6	0.7	1.0	5.3	6.8	6.4	6.4	6.4
Launch vehicles	Thor Delta			Atlas-Centaur		Atlas-Centaur and Ariane	Atlas-Centaur and Ariane	STS and Ariane
Spacecraft mass in transfer orbit (kg)	68	182	293	1385	1489	1946	2140	12,100/3720
Communications payload mass (kg)	13	36	56	185	190	235	280	800
End-of-life (EOL) power of equinox (W)	40	75	134	480	800	1270	1270	2200
Design lifetime (years)	1.5	3	5	7	7	7	7	10
Capacity (number of voice channels)	480	480	2400	8000	12,000	25,000	30,000	80,000
Bandwidth (MHz)	50	130	300	500	800	2137	2480	3520

Figure 1.1 Evolution of Intelsat Satellites *(From Colino, 1985. Courtesy "ITU Telecommunication Journal.")*

TABLE 1.3   Comparison of Earliest and More Recent Generations of Intelsat
Satellite

	Earlybird	Intelsat VA	Intelsat VI	Intelsat VII
First launch	1965	1980	1989	1992
Satellite mass in orbit, kg	38.5	900	1870	1425
Prime power, W	40	1200	2200	3900
Number of transponders	2	30	48	36
Total bandwidth, MHz	50	2160	3030	2300
Telephone channel capacity:				
All analog	480	3300	48,000	38,000
All digital		180,000	270,000	200,000
Suitability for small earth terminals		*	**	***

SOURCE: Lilly, 1990. (Asterisks indicate increased suitability, not mentioned in original reference.)

that carried by the Pacific Ocean region satellites). The VII satellite is designed to have the appropriate antenna beam coverage for the Pacific Ocean region. Figure 1.2 shows the C-band hemispheric coverages and zone coverages, and the spot beam coverages possible with the Ku-band steerable antennas (Lilly, 1990, and Sachdev et al., 1990). When used in the Atlantic Ocean region, the VII satellite is inverted north for south (Lilly, 1990), minor adjustments then being needed only to optimize the antenna patterns for this region.

A number of satellites of a given type are in service at any one time. For example, in planning for the next decade, the "initial buy" of the Intelsat VI series consisted of five Intelsat VI satellites, and it was envisaged that four of these would be successfully launched, leaving one as a ground spare. It is standard practice to have a spare satellite in orbit on high-reliability routes (which can carry preemptible traffic), and thus a system may include both ground spares and orbiting spares. As of 1992, Intelsat had six satellites in the Atlantic Ocean region (AOR) at degrees east longitude 307, 325.5, 332.5, 338.5, 341.5, 359; three in the Indian Ocean region (IOR) at degrees east longitude 60, 63, 66; and three in the Pacific Ocean region (POR) at degrees east longitude 174, 177, and 180 (Khan, 1992). For each region, the satellites are positioned in geostationary orbit above the particular ocean, where they provide a transoceanic telecommunications route. The geostationary orbit is described in detail in Chap. 2. The coverage areas for Intelsat VI are shown in Fig. 1.3. Traffic in the AOR is about 3 times that of the IOR, and about twice that of the IOR and POR combined. Thus the system design is tailored mainly around the AOR requirements (Thompson and Johnston, 1983).

In addition to providing transoceanic routes, the Intelsat satellites are also used for domestic services within any given country and

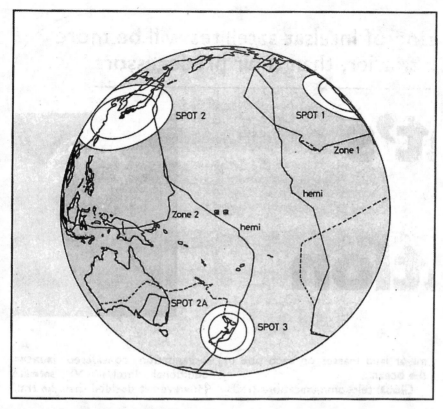

Figure 1.2 Intelsat VII coverage (Pacific Ocean region; global, hemispheric, and spot beams). (*From Lilly, 1990, with permission.*)

regional services between countries. Two such services are Vista for telephony and Intelnet for data exchange. Figure 1.4 shows typical Vista applications.

1.4 U.S. Domsats

Domsat is an abbreviation for *domestic satellite*. Domestic satellites are used to provide various telecommunications services, such as voice, data, and video transmissions, within a country. In the United States, all domsats are situated in geostationary orbit. As is well-known, they make available a wide selection of TV channels for the home entertainment market, in addition to carrying a large amount of commercial telecommunications traffic.

U.S. Domsats which provide a direct-to-home television service can be classified broadly as high power, medium power, and low power

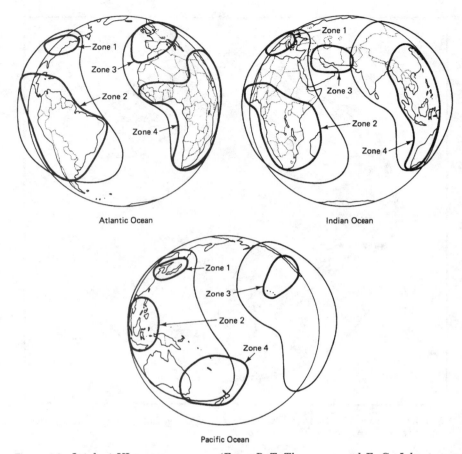

Figure 1.3 Intelsat VI coverage areas. (*From P. T. Thompson and E. C. Johnston, INTELSAT VI: A New Satellite Generation for 1986–2000, "International Journal of Satellite Communications," vol. 1, 3–14, 1983. © John Wiley & Sons, Ltd.*)

(Reinhart, 1990). The defining characteristics of these categories are shown in Table 1.4.

The main distinguishing feature of these categories is the equivalent isotropic radiated power (EIRP). This is explained in more detail in Chap. 10, but for present purposes it should be noted that the upper limit of EIRP is 60 dBW for the high-power category and 37 dBW for the low-power, a difference of 23 dB. This represents an increase in received power of $10^{2.3}$ or about 200:1 in the high-power category, which allows much smaller antennas to be used with the receiver. As noted in the table, the primary purpose of satellites in the high-power category is to provide a DBS service. In the medium-

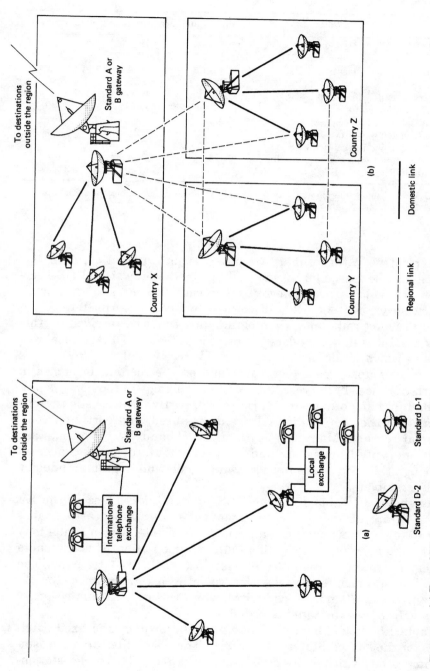

Figure 1.4 (a) Typical Vista application; (b) domestic/regional Vista network with standard A or B gateway. (From Colino, 1985. Courtesy "ITU Telecommunication Journal.")

TABLE 1.4   Defining Characteristics of Three Categories of United States DBS Systems

	High power	Medium power	Low power
Band	Ku	Ku	C
Downlink frequency allocation, GHz	12.2–12.7	11.7–12.2	3.7–4.2
Uplink frequency allocation, GHz	17.3–17.8	14–14.5	5.925–6.425
Space service	BSS	FSS	FSS
Primary intended use	DBS	Point to point	Point to point
Allowed additional use	Point to point	DBS	DBS
Terrestrial interference possible	No	No	Yes
Satellite spacing, degrees	9	2	2–3
Satellite spacing determined by	ITU	FCC	FCC
Adjacent satellite interference possible?	No	Yes	Yes
Satellite EIRP range, dBW	51–60	40–48	33–37

ITU: International Telecommunication Union; FCC: Federal Communications Commission.
SOURCE: Reinhart, 1990.

power category, the primary purpose is point-to-point services, but space may be leased on these satellites for the provision of DBS services. In the low-power category, no official DBS services are provided. However, it was quickly discovered by home experimeters that a wide range of radio and TV programming could be received on this band, and it is now considered to provide a de facto DBS service, witness to which is the large number of TV receive-only (TVRO) dishes which have appeared in the yards and on the rooftops of homes in North America. TVRO reception of C-band signals in the home is prohibited in many other parts of the world, partly for aesthetic reasons because of the comparatively large dishes used, and partly for commercial reasons. Many North American C-band TV broadcasts are now encrypted, or scrambled, to prevent unauthorized access, although this also seems to be spawning a new underground industry in descramblers.

As shown in Table 1.4, true DBS service takes place in the Ku band. Figure 1.5 shows the components of a direct broadcasting satellite system (Government of Canada, 1983). The television signal may be relayed over a terrestrial link to the uplink station. This transmits a very narrowbeam signal to the satellite in the 14-GHz band. The satellite retransmits the television signal in a wide beam in the 12-GHz frequency band. Individual receivers within the beam coverage area will receive the satellite signal.

Table 1.5 shows the orbital assignments for domestic fixed satellites for the United States (FCC, 1994). These satellites are in geostationary orbit, which is discussed further in Sec. 2.11. The geostationary orbit is located directly above the equator, and the positions of the

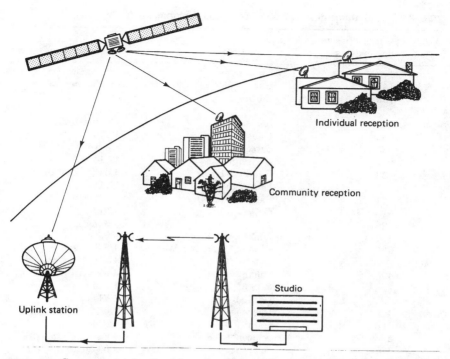

Figure 1.5 Components of a direct broadcasting satellite system. (*From Government of Canada, 1983. With permission.*)

satellites are given in degrees longitude, as shown. Figure 1.6 gives a pictorial representation of the satellites in orbit for North America (Mexican satellites are not shown). In 1983, the U.S. Federal Communications Commission (FCC) adopted a policy objective setting 2° as the minimum orbital spacing for satellites operating in the 6/4-GHz band, and 1.5° for those operating in the 14/12-GHz band (FCC, 1983). It is clear that interference between satellite circuits is likely to increase as satellites are positioned closer together. These spacings represent the minimum presently achievable in each band at acceptable interference levels. In fact, it seems likely that in some cases home satellite receivers in the 6/4-GHz band may be subject to excessive interference where 2° spacing is employed.

1.5   Polar Orbiting Satellites

Polar orbiting satellites orbit the earth in such a way as to cover the north and south polar regions. (Note that the term *polar orbiting* does *not* mean that the satellite orbits around one or the other of the

TABLE 1.5   U.S. Domestic Fixed Satellite Orbital Assignments (as of 1994)

Orbital positions (west longitude), degrees	User	Frequency bands, GHz
143	Unassigned	4/6 (vertical)
141	Unassigned	4/6 (horizontal)
139	Aurora 2	4/6 (vertical)
	AMSC	12/14
137	Satcom C-1	4/6 (horizontal)
	Unassigned	12/14
135	Satcom C-4	4/6 (vertical)
	Unassigned	12/14
133	Galaxy 1/1-R	4/6 (horizontal)
	Unassigned	12/14
131	Satcom I-R/C-3	4/6 (vertical)
	Galaxy B-R	12/14
129	ASCI	4/6 (horizontal) 12/14
127	Unassigned	4/6 (vertical) 12/14
125	Westar 5/Galaxy 5-W	4/6 (horizontal)
	Gstar 4	12/14
123	Telstar 303	4/6 (vertical)
	SBS 5	12/14
121	Gstar 1/1-R	12/14
120	Spacenet 1	4/6/12/14 (horizontal)
105	Gstar 2	12/14
103	Spacenet 1-R	4/6 (horizontal) / 12/14
101	Spacenet 4-n	4/6 (vertical) / 12/14
99	Westar 4/Galaxy 4-H	4/6 (horizontal) / 12/14
97	Telstar 301/401	4/6 (vertical) 12/14
95	Galaxy 3/3-R	4/6 (horizontal)
	SBS 3	12/14
93	Unassigned	4/6 (vertical) 12/14
91	SBS 4/Galaxy 7-H	4/6 (horizontal) / 12/14
89	Telstar 402	4/6 (vertical) / 12/14
87	Spacenet 3	4/6 (horizontal) / 12/14
85	Telstar 302/Satcom H-1	4/6 (vertical)
	Satcom K-1/Satcom H-1	12/14
83	Unassigned	4/6 (horizontal) 12/14
81	Satcom 4	4/6 (vertical)
	Satcom K-2	12/14
79	Unassigned	4/6 (horizontal) / 12/14
76	Constar D-2/D-4	4/6 (vertical)
	Unassigned	12/14
74	Galaxy 2/2-R	4/6 (horizontal)
	SBS-2	12/14
72	Satcom 2-R	4/6 (vertical)
	SBS 6	12/14
69	Spacenet 2/2-R	4/6 (horizontal) / 12/14
67	Unassigned	4/6 (vertical)
	Unassigned	12/14
64	Unassigned	4/6 (horizontal)
	Unassigned	12/14
62	Unassigned	4/6 (vertical)
	AMSC	12/14

SOURCE: FCC, 1994.

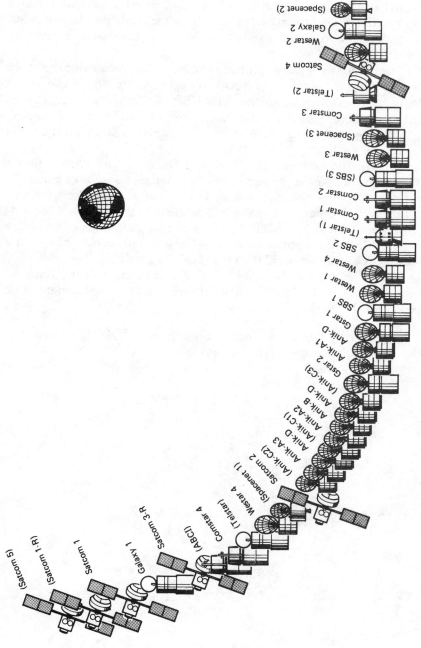

Figure 1.6 Satellite Orbital Positions for North America, operating and (authorized for launch). *(From "Scientific Atlanta," 1985/6. With permission.)*

poles). Several such satellites are in orbit, for example, the Russian communications satellites which fly in highly elliptical orbits to cover the northern regions of Russia and the U.S. weather satellites which fly in relatively low, nearly circular orbits, passing close to the poles. A brief description of the services provided by the latter is given in this section.

Figure 1.7 shows a polar orbit in relation to the geostationary orbit. Whereas there is only one geostationary orbit, there are, in theory, an infinite number of polar orbits. The United States experience with weather satellites has led to the use of relatively low orbits, ranging in altitude between 800 and 900 km, compared to 36,000 km for the geostationary orbit.

In the United States, the National Oceanic and Atmospheric Administration manages the program for a series of satellites generally known as the Tiros-N series, where Tiros is an acronym for *television and infrared observational satellite*. The current program is known as the Advanced Tiros-N (or ATN) program, and the series of satellites carry the NOAA designations shown in Table 1.6 (data obtained from a NASA-NOAA undated booklet on the NOAA-J satellite).

The main mission of the NOAA series of spacecraft is environmental monitoring as typified by the following list of instrumentation planned for the NOAA-J spacecraft:

Advanced very high resolution radiometer (AVHRR) determines cloud cover and surface temperature.

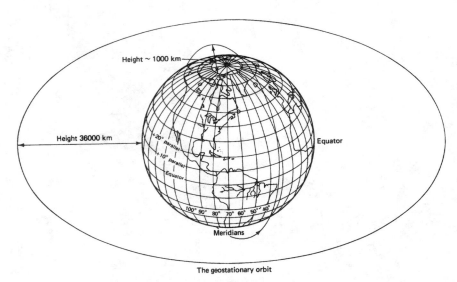

The geostationary orbit

Figure 1.7 Geostationary orbit and one possible polar orbit.

TABLE 1.6   Tiros-N (ATN) Spacecraft Designations and Launch Dates

Prelaunch designation	Date of launch	In-orbit designation	Status
NOAA-A	June 27, 1979	NOAA-6	Deactivated March 31, 1987
NOAA-B	May 29, 1980		Failed to reach orbit
NOAA-C	June 23, 1981	NOAA-7	Deactivated June 1986
NOAA-E	March 28, 1983	NOAA-8	Lost December 1985
NOAA-F	December 12, 1984	NOAA-9	On standby operation, some equipment failure
NOAA-G	September 17, 1986	NOAA-10	Currently in operation, some equipment failure
NOAA-H	September 24, 1988	NOAA-11	In operation
NOAA-D	May 14, 1991	NOAA-12	In operation
NOAA-I	August 9, 1993	NOAA-13	Failed after launching
NOAA-J	December 1994	NOAA-14	
NOAA-K			
NOAA-L	To be determined		
NOAA-M			
NOAA-N			

Solar backscatter ultraviolet spectral radiometer (SBUV) mod 2 measures solar irradiance and backscattered solar energy to determine ozone concentration.

Tiros operational vertical sounder system (TOVS) measures radiant energy from various altitudes of the atmosphere, which in turn enables the atmosphere's temperature profile to be determined.

Space environmental monitor (SEM) measures the population of the earth's radiation belts, which provides information on solar particle energy and warning of solar storms.

For weather forecasting it is particularly important to gather such information for the polar regions of the earth, which is the main reason for having these satellites in polar orbits. Environmental data are also transmitted to the NOAA satellites from platforms consisting of buoys, free-floating balloons, and remote weather stations. This part of the mission is known as the Argos data collection system (DCS).

In addition to providing environmental data services, the NOAA satellites help locate ships and aircraft in distress. A satellite offering this service is known as Sarsat for *search and rescue satellite*. In 1979, Canada, the United States, and France agreed to joint testing of the concept, and in 1980, an agreement was signed with what was then the Soviet Union for a joint demonstration project utilizing Russian and NOAA satellites, the combined system being known as Cospas-Sarsat. Presently there are 23 participating countries (Cospas-Sarsat, 1994a).

The nominal space segment of the Cospas-Sarsat system consists of two Cospas and two Sarsat satellites. In operation, the satellite receives a signal from an emergency beacon set off automatically at the distress site. The beacon transmits in the VHF/UHF range, at a precisely controlled frequency. The satellite moves at some velocity relative to the beacon, and this results in a doppler shift in frequency received at the satellite. As the satellite approaches the beacon, the received frequency appears to be higher than the transmitted value. As the satellite recedes from the beacon, the received frequency appears to be lower than the transmitted value. Figure 1.8 shows how the beacon frequency, as received at the satellite, varies for different passes. In all cases, the received frequency goes from being higher to being lower than the transmitted value, as the satellite approaches and then recedes from, the beacon. The longest record and the greatest change in frequency are obtained if the satellite passes over the site, as shown for pass no. 2. This is because the satellite is visible for the longest period during this pass. Knowing the orbital parameters for the satellite, the beacon frequency, and the doppler shift for any one pass, the distance of the beacon relative to the projection of the orbit on the earth can be determined. However, whether the beacon is east or west of the orbit cannot easily be determined from a single pass. For two successive passes, the effect of the earth's rotation on the doppler shift can be estimated more accurately, and from this it

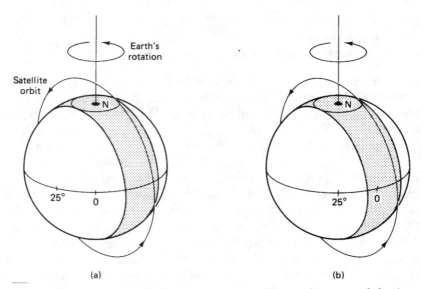

Figure 1.8 Polar orbiting satellite: (*a*) first pass; (*b*) second pass, earth having rotated 25°. Satellite period is 102 min.

can be determined whether the beacon is approaching or receding from the orbital path. In this way, the ambiguity in east/west positioning is resolved. Figure 1.9 illustrates the doppler shifts for successive passes.

The satellite must of course get the information back to an earth station so that the search and rescue operation can be completed, successfully one hopes. The Sarsat communicates on a down-link frequency of 1544.5 MHz to one of several local user terminals (LUTs) established at various locations throughout the world. There are 23 LUTs in operation at present with another seven under test, and two planned for the future (Cospas-Sarsat, 1994b). From the information received, the LUT determines the distress location and passes the information onto a mission control center (MCC). At present there are 13 MCCs in operation worldwide, with three more under test, and two planned for the future (Cospas-Sarsat, 1994). The MCC alerts the rescue coordination center (RCC) nearest the location where the distress signal originated and the RCC takes the appropriate action to effect a rescue.

In the original Sarsat system the beacons operated on frequencies of 121.5 and 243 MHz, these being standard frequencies for distress

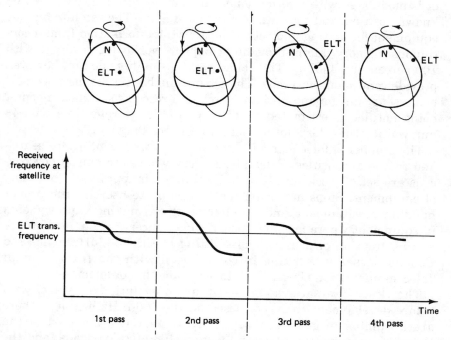

Figure 1.9 Showing the doppler shift in received frequency on successive passes of the satellite. ELT = emergency locator transmitter.

beacons. The beacon transmitters are known as *emergency locator transmitters* (ELTs). It was found that the original Sarsat system was very sensitive to voice communication channels operating on nearby frequencies, with the result that it was often difficult to identify distress signals. Also, these beacons operate at low power, typically a few tenths of a watt, which limits the accuracy of location to about 10 to 20 km. Another limitation with the ELT system is that there are no storage facilities aboard the satellite for the 121.5/243-MHz beacons so that the ELT location and a LUT must be in view simultaneously from the satellite for the distress signal to get through.

The current estimate for the number of distress beacons at 121.5 MHz is 550,000 worldwide with a forecast of 600,000 for the year 2000. The current estimate for newer beacons working at 406 MHz as described below is 85,000, with 120,000 being forecast for the year 1995 (Cospas-Sarsat, 1994).

New beacons operating at a frequency of 406 MHz are now being introduced and power has been increased to 5 W, so it should be possible to resolve position to within 3 to 5 km (Scales and Swanson, 1984). These are known as *emergency position-indicating radio beacons* (EPIRBs) [406-MHz units are also available for personnel use, these being known as *personal locator beacons* (PLBs)]. The 406-MHz carrier is modulated with information such as an identifying code, the last known position, and the nature of the emergency. The satellite has the equipment for storing and retransmitting the information from a continuous memory dump, providing complete worldwide coverage with 100 percent availability. The polar orbiting satellites do not, however, provide continuous coverage. The mean time between a distress alert being sent and the appropriate search and rescue coordination center being notified is estimated at 27 min satellite storage time plus 44 min waiting time, for a total of 71 min (Cospas-Sarsat, 1994b).

The number of false alerts is a matter of growing concern for search and rescue authorities. False alerts occur where the emergency beacons are set off accidentally, usually through inappropriate handling of equipment. Steps are being taken to educate users in the proper handling of equipment, and manufacturers are making improvements in equipment where feasible. Also of some concern is the interference arising from other sources transmitting in the 406-MHz band, and Cospas-Sarsat is working in cooperation with the International Telecommunication Union (ITU) to overcome this problem.

The Cospas-Sarsat system is still at an evolutionary stage, with future developments being focussed on the 406-MHz beacons. There are a number of geostationary satellite systems, not part of the Cospas-Sarsat system, which may carry 406-MHz payloads, and the operational and technical aspects of these are currently under study by the Cospas-Sarsat organization. It should be noted that geostationary satellites do not give good coverage of the polar regions. Also being considered is the encoding of position information onto the 406-MHz beacon. Low-cost receiving equipment is now available for use with the Global Positioning Satellite (GPS) system (see Chap. 13), and it should be possible to incorporate these receivers into the EPIRBs.

As mentioned previously, the NOAA satellites are placed in a low earth orbit typified by the NOAA-J satellite. The NOAA-J satellite will orbit the earth in approximately 102.12 min. The orbit is arranged to rotate eastward at a rate of 0.9856°/day, to make it *sun-synchronous*. Sun-synchronous orbits are discussed more fully in Chap. 2, but, very briefly, in a sun-synchronous orbit the satellite crosses the same spot on the earth at the same local time each day. One advantage of a sun-synchronous orbit is that the same area of the earth can be viewed under approximately the same lighting conditions each day. By definition, an orbital pass from south to north is referred to as an *ascending pass,* and from north to south, as a *descending pass.* The NOAA-J orbit crosses the equator at about 1:40 p.m. local solar time on its ascending pass and at about 1:40 a.m. local solar time on its descending pass.

Because of the eastward rotation of the satellite orbit, the earth rotates approximately 359° relative to it in 24 h of mean solar time (ordinary clock time), and therefore in 102.12 min the earth will have rotated about 25.59° relative to the orbit. The satellite "footprint" is displaced each time by this amount, as shown in Fig. 1.7. At the equator, 25.59° corresponds to a distance of about 2848 km. The width of ground seen by the satellite sensors is about 5000 km, which means that some overlap occurs between passes. The overlap is greatest at the poles.

1.6 Problems

1.1 Describe briefly the main advantages offered by satellite communications. Explain what is meant by a *distance-insensitive communications system.*

1.2 Comparisons are sometimes made between satellite and optical fiber communications systems. State briefly the areas of application for which you feel each system is best suited.

1.3 (*a*) Explain what the name *Intelsat* stands for. (*b*) The Intelsat organization has geostationary satellites positioned at longitudes 335.5°E, 174°E, and 61.4°E. Given that the satellites are at a height of 36,000 km above the equator, indicate the intercontinental coverage possible from each position.

1.4 From Table 1.5, determine typical orbital spacings, in degrees, for (a) the 6/4-GHz band and (b) the 14/12-GHz band.

1.5 Suggest reasons why the 14/12-GHz band has been selected for direct-to-home satellite broadcasting.

1.6 Explain briefly what is meant by a *polar orbiting satellite*. A NOAA polar orbiting satellite completes one revolution around the earth in 102 min. The satellite makes a north to south equatorial crossing at longitude 90°W. Assuming that the orbit is circular and crosses exactly over the poles, estimate the position of the subsatellite point at the following times after the equatorial crossing: (a) 0 h 10 min; (b) 1 h 42 min; (c) 2 h 0 min.

1.7 Explain how polar orbiting satellites are used in the Sarsat search and rescue system.

1.8 Discuss briefly why the emergency locator transmitters used in the Sarsat system transmit in the VHF range. State the frequencies used by the ELTs, and the frequency at which the satellite retransmits the emergency signal received by it. Is the retransmission made immediately after it is received? Give reasons for your answer.

1.9 The following satellites are listed in App. D. State, with reasons, whether these satellites are in the geostationary orbit or in inclined orbits: (a) Solidaridad; (b) Orion-1; (c) DFH-3; (d) Molyna-1; (e) USA-107; (f) Faisat.

1.10 The following acronyms and abbreviations apply to a few of the many satellite systems currently in use: (a) Immarsat; (b) Arabsat; (c) GPS. Research the literature on these and write brief descriptions of each, highlighting the main features.

2

Orbits and Launching Methods

2.1 Introduction

Satellites (spacecraft) which orbit the earth follow the same laws that govern the motion of the planets around the sun. From early times much has been learned about planetary motion through careful observations. From these observations Johannes Kepler (1571–1630) was able to derive empirically three laws describing planetary motion. Later, in 1665, Sir Isaac Newton (1642–1727) was able to derive Kepler's laws from his own laws of mechanics and develop the theory of gravitation (for very readable accounts of much of the work of these two great men see Arons, 1965, and Bate et al., 1971).

Kepler's laws apply quite generally to any two bodies in space which interact through gravitation. The more massive of the two bodies is referred to as the *primary,* the other the *secondary,* or *satellite.*

2.2 Kepler's First Law

Kepler's first law states that the path followed by the satellite around the primary will be an ellipse. An ellipse has two focal points shown as F_1 and F_2 in Fig. 2.1. The center of mass of the two-body system, termed the *barycenter,* is always centered on one of the foci. In our specific case, because of the enormous difference between the masses of the earth and the satellite, the center of mass coincides with the center of the earth, which is therefore always at one of the foci.

The semimajor axis of the ellipse is denoted by a, and the semiminor axis by b. The eccentricity e is given by

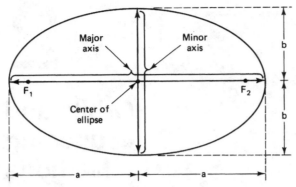

Figure 2.1 Showing the foci F_1 and F_2, the semimajor axis a, and the semiminor axis b, of an ellipse.

$$e = \frac{\sqrt{a^2 - b^2}}{a} \tag{2.1}$$

The eccentricity and the semimajor axis are two of the orbital parameters specified for satellites (spacecraft) orbiting the earth. For an elliptical orbit $0 < e < 1$. When $e = 0$, the orbit becomes circular. The geometrical significance of eccentricity, along with some of the other geometrical properties of the ellipse, is developed in App. B.

2.3   Kepler's Second Law

Kepler's second law states that, for equal time intervals, the satellite will sweep out equal areas in its orbital plane, focused at the barycenter. Referring to Fig. 2.2, assuming the satellite travels distances S_1 and S_2 meters in 1 s, then the areas A_1 and A_2 will be equal. The average velocity in each case is S_1 and S_2 meters per second, and because

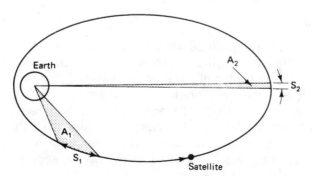

Figure 2.2 Illustrating Kepler's second law. The areas A_1 and A_2 swept out in unit time are equal.

of the equal area law, it follows that the velocity at S_2 is less than that at S_1. An important consequence of this is that the satellite takes longer to travel a given distance when it is farther away from earth. Use is made of this property to increase the length of time a satellite can be seen from particular geographic regions of the earth.

2.4 Kepler's Third Law

Kepler's third law states that the square of the periodic time of orbit is proportional to the cube of the mean distance between the two bodies. The mean distance is equal to the semimajor axis a. For the artificial satellites orbiting the earth, Kepler's third law can be written in the form

$$a^3 = \frac{\mu}{n^2} \tag{2.2}$$

where n is the mean motion of the satellite in radians per second and μ is the earth's geocentric gravitational constant. With a in meters its value is (see Wertz, 1984, Table L3):

$$\mu = 3.986005 \times 10^{14} \text{ m}^3/\text{s}^2 \tag{2.3}$$

Eq. (2.2) applies only to the ideal situation of a satellite orbiting a perfectly spherical earth of uniform mass, with no perturbing forces acting, such as atmospheric drag. Later, in Sec. 2.8, the effects of the earth's oblateness and atmospheric drag will be taken into account.

With n in radians per second the orbital period in seconds is given by

$$P = \frac{2\pi}{n} \tag{2.4}$$

The importance of Kepler's third law is that it shows there is a fixed relationship between period and size. One very important orbit in particular, known as the *geostationary orbit,* is determined by the rotational period of the earth and is described in Sec. 2.10. In anticipation of this, the approximate radius of the geostationary orbit is determined in the following example.

Example 2.1

Calculate the radius of a circular orbit for which the period is 1-day.

solution

The mean motion, in rad/day, is:

$$n := \frac{2 \cdot \pi}{1 \text{ day}}$$

Note that in Mathcad this will be automatically recorded in rad/sec. Thus, for the record:

$$n = 7.272 \cdot 10^{-5} \cdot \frac{\text{rad}}{\text{sec}}$$

The earth's gravitational constant is:

$$\mu : = 3.986005 \cdot 10^{14} \cdot m^3 \cdot sec^{-2}$$

Kepler's 3rd law gives:

$$a : = \left(\frac{\mu}{n^2} \right)$$

$$a = 42241 \cdot km$$

$$= = = = = = = = = = = = =$$

Since the orbit is circular the semimajor axis is also the radius.

2.5 Definitions of Terms for Earth-Orbiting Satellites

As previously mentioned, Kepler's laws apply in general to satellite motion around a primary body. For the particular case of earth-orbiting satellites, certain terms are used to describe the position of the orbit with respect to the earth.

Apogee   The point farthest from earth. Apogee height is shown as h_a in Fig. 2.3.

Perigee   The point of closest approach to earth. The perigee height is shown as h_p in Fig. 2.3.

Line of apsides   The line joining the perigee and apogee through the center of the earth.

Ascending node   The point where the orbit crosses the equatorial plane going from south to north.

Descending node   The point where the orbit crosses the equatorial plane going from north to south.

Line of nodes   The line joining the ascending and descending nodes through the center of the earth.

Inclination   The angle between the orbital plane and the earth's equatorial plane. It is measured at the ascending node from the equator to the orbit, going from east to north. The inclination is shown as i in Fig. 2.3. It will be seen that the greatest latitude, north or south, is equal to the inclination.

Prograde orbit   An orbit in which the satellite moves in the same direction as the earth's rotation, as shown in Fig. 2.4. The prograde orbit is also known as a *direct orbit*. The inclination of a prograde orbit always lies between 0 and

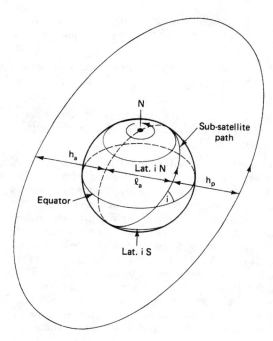

Figure 2.3 Illustrating apogee height h_a, perigee height h_p, and inclination i. l_a is the line of apsides.

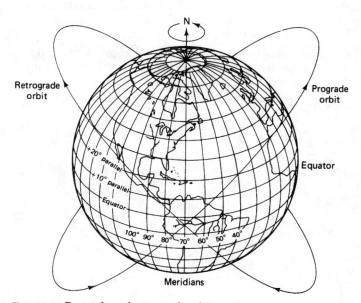

Figure 2.4 Prograde and retrograde orbits.

90°. Most satellites are launched in a prograde orbit because the earth's rotational velocity provides part of the orbital velocity with a consequent saving in launch energy.

Retrograde orbit An orbit in which the satellite moves in a direction counter to the earth's rotation as shown in Fig. 2.4. The inclination of a retrograde orbit always lies between 90 and 180°.

Argument of perigee The angle from ascending node to perigee, measured in the orbital plane at the earth's center, in the direction of satellite motion. The argument of perigee is shown as ω in Fig. 2.5.

Right ascension of the ascending node To define completely the position of the orbit in space, the position of the ascending node is specified. However, because the earth spins, while the orbital plane remains stationary (slow drifts which do occur are discussed later), the longitude of the ascending node is not fixed, and it cannot be used as an absolute reference. For the practical determination of an orbit, the longitude and time of crossing of the ascending node are frequently used. However, for an absolute measurement, a fixed reference in space is required. The reference chosen is the *first point of Aries,* otherwise known as the vernal, or spring, equinox. The vernal equinox occurs when the sun crosses the equator going from south to north, and an imaginary line drawn from this equatorial crossing, through the center of the sun, points to the first point of Aries (symbol ϒ). This is the *line of Aries.* The right ascension of the ascending node is then the angle measured eastward, in the equatorial plane, from the ϒ line to the ascending node, shown as Ω in Fig. 2.5.

Mean anomaly Mean anomaly M gives an average value of the angular position of the satellite with reference to the perigee. For a circular orbit, M gives the angular position of the satellite in the orbit. For an elliptical orbit, the position is much more difficult to calculate, and M is used as an intermediate step in the calculation as described in Sec. 2.9.5.

True anomaly The true anomaly is the angle from perigee to the satellite position, measured at the earth's center. This gives the true angular position

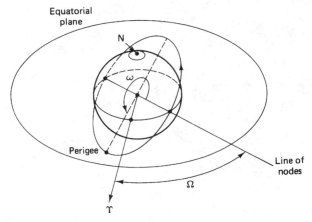

Figure 2.5 Showing the argument of perigee ω and the right ascension of the ascending node Ω

of the satellite in the orbit as a function of time. A method of determining the true anomaly is described in Sec. 2.9.5.

2.6 Orbital Elements

Earth-orbiting artificial satellites are defined by six orbital elements referred to as the *keplerian element set*. Two of these, the semimajor axis a and the eccentricity e described in Sec. 2.2, give the shape of the ellipse. A third, the mean anomaly M_0 gives the position of the satellite in its orbit at a reference time known as the *epoch*. A fourth, the argument of perigee ω, gives the rotation of the orbit's perigee point relative to the orbit's line of nodes in the earth's equatorial plane. The remaining two elements, the inclination i and the right ascension of the ascending node Ω, relate the orbital plane's position to the earth. These four elements are described in Sec. 2.5.

Because the equatorial bulge causes slow variations in ω and Ω, and because other perturbing forces may alter the orbital elements slightly, the values are specified for the reference time or epoch, and thus the epoch must also be specified.

Appendix C lists the two-line elements provided to users by the U.S. National Aeronautics and Space Administration (NASA). The two-line elements may be downloaded from the NASA electronic bulletin board. Originally the data were provided on hard-copy bulletin sheets which also gave equatorial crossings. The hard-copy bulletins were discontinued on August 16, 1994. Figure 2.6 shows two-line elements taken from one of these, and used as example data in Table 2.1.

It will be seen that the semimajor axis is not specified, but this can be calculated from the data given. An example calculation is presented in Example 2.2.

Example 2.2

Calculate the semimajor axis for the satellite parameters given in Table 2.1.

solution

The mean motion is given in Table 2.1 as:

$$NN := 14.22296917 \cdot \text{day}^{-1}$$

This can be converted to rad/sec as:

$$n_0 := NN \cdot 2 \cdot \pi$$

(Note that Mathcad automatically converts time to the fundamental unit of second.)
Eq. (2.3) gives:

$$\mu := 3.986005 \cdot 10^{14} \cdot \text{m}^3 \cdot \text{sec}^{-2}$$

NASA PREDICTION BULLETIN

NASA 51004

NASA GODDARD SPACE FLIGHT CENTER, CODE 513 GREENBELT, MD. 20771
ISSUE DATE: AUGUST 9, 1993

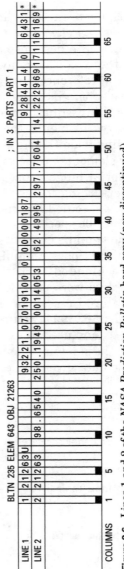

Figure 2.6   Lines 1 and 2 of the *NASA Prediction Bulletin* hard copy (now discontinued).

Kepler's 3rd law gives:

$$a := \left(\frac{\mu}{n_o^2}\right)^{\frac{1}{3}}$$

$$a = 7195.7 \cdot km$$

$$=========$$

TABLE 2.1   Details from the NASA Bulletins

Line no.	Columns	Description
1	3–7	*Satellite number:* 21263
1	19–20	*Epoch year* (last two digits of the year): 93
1	21–32	*Epoch day* (day and fractional day of the year): 221.07019100 (this is discussed further in Sec. 2.9.2). This gives a S-N equatorial crossing time (see Prob. 2.29).
1	34–43	*First time derivative of the mean motion* (rev/day^2): 0.00000187
2	9–16	*Inclination* (degrees): 98.6540
2	18–25	*Right ascension of the ascending node* (degrees): 250.1949
2	27–33	*Eccentricity* (leading decimal point assumed): 0014053
2	35–42	*Argument of perigee* (degrees): 62.4995
2	44–51	*Mean anomaly* (degrees): 297.7604
2	53–63	*Mean motion* (rev/day): 14.22296917
2	64–68	*Revolution number at epoch* (rev/day): 11,616

2.7   Apogee and Perigee Heights

Although not specified as orbital elements, the apogee height and perigee height are often required. As shown in App. B, the length of the radius vectors at apogee and perigee can be obtained from the geometry of the ellipse

$$r_a = a(1 + e) \tag{2.5}$$

$$r_p = a(1 - e) \tag{2.6}$$

In order to find the apogee and perigee heights, the radius of the earth must be subtracted from the radii lengths, as shown in the following example.

Example 2.3

Calculate the apogee and perigee heights for the orbital parameters given in Table 2.1. The earth's polar radius may be taken as 6356.755 km.

solution

The required data from Table 2.1 are: e : = .0014053   a : = 7195.7·km

(Note, that value for a was determined in Example 2.2)

Given data:

$$R_p : = 6356.755 \cdot km$$

$r_a := a \cdot (1 + e)$	...eq.(2.5)	$r_a = 7205.8 \cdot km$
$r_p := a \cdot (1 - e)$	...eq.(2.6)	$r_p = 7185.6 \cdot km$
$h_a := r_a - R_p$	$h_a = 849.1 \cdot km$	

=========

$h_p := r_p - R_p$	$h_p = 828.8 \cdot km$

=========

2.8   Orbit Perturbations

The type of orbit described so far, referred to as a *keplerian orbit,* is elliptical for the special case of an artificial satellite orbiting the earth. However, the keplerian orbit is ideal in the sense that it assumes that the earth is a uniform spherical mass, and that the only force acting is the centrifugal force, resulting from satellite motion balancing the gravitational pull of the earth. In practice, other forces which can be significant are the gravitational forces of the sun and the moon and atmospheric drag. The gravitational pulls of sun and moon have negligible effect on low-orbiting satellites, but they do affect satellites in the geostationary orbit as described in Sec. 2.11. Atmospheric drag, on the other hand, has negligible effect on geostationary satellites, but does affect low-orbiting earth satellites below about 1000 km.

2.8.1   Effects of a nonspherical earth

For a spherical earth of uniform mass, Kepler's third law, Eq. (2.2) gives the nominal mean motion n_0 as

$$n_0 = \sqrt{\frac{\mu}{a^3}} \qquad (2.7)$$

The 0 subscript is included as a reminder that this result applies for a perfectly spherical earth of uniform mass. However, it is known that the earth is not perfectly spherical, there being an equatorial bulge and a flattening at the poles, a shape described as an *oblate spheroid.* When the earth's oblateness is taken into account the mean motion, denoted in this case by symbol n, is modified to (Wertz, 1984):

$$n = n_0 \left[1 + \frac{K_1(1 - 1.5 \sin^2 i)}{a^2(1 - e^2)^{1.5}} \right] \tag{2.8}$$

K_1 is a constant which evaluates to 66,063.1704 km^2. The earth's oblateness has negligible effect on the semimajor axis a, and if a is known the mean motion is readily calculated. The orbital period taking into account the earth's oblateness is termed the *anomalistic period* (e.g., from perigee to perigee). The mean motion specified in the NASA bulletins is the reciprocal of the anomalistic period. The anomalistic period is

$$P_A = \frac{2\pi}{n} \quad s \tag{2.9}$$

where n is in radians per second.

If the known quantity is n (as is given in the NASA bulletins, for example), one can solve Eq. (2.8) for a, keeping in mind that n_0 is also a function of a. Equation (2.8) may be solved for a by finding the root of the following equation:

$$n - \sqrt{\frac{\mu}{a^3}} \left[1 + \frac{K_1(1 - 1.5 \sin^2 i)}{a^2(1 - e^2)^{1.5}} \right] = 0 \tag{2.10}$$

This is illustrated in the following example.

Example 2.4

A satellite is orbiting in the equatorial plane with a period from perigee to perigee of 12 hr. Given that the eccentricity is .002 calculate the semimajor axis. The earth's equatorial radius is 6378.1414 km.

solution

Given data:

$$e := .002 \qquad i := 0 \cdot \deg \qquad P := 12 \cdot hr$$

$$K_1 := 66063.1704 \cdot km^2 \qquad a_E := 6378.1414 \cdot km \qquad \mu := 3.986005 \cdot 10^{14} \cdot m^3 \cdot sec^{-2}$$

The mean motion is:

$$n := \frac{2 \cdot \pi}{P}$$

Kepler's 3rd law gives:

$$a := \left(\frac{\mu}{n^2} \right)^{\frac{1}{3}}$$

$a = 26597 \cdot km$ ...this is the non-perturbed value which can be used as a guess
======== value for the root function.

Perturbed value:

$$a := \text{root}\left[n - \left(\sqrt{\frac{\mu}{a^3}}\right) \cdot \left[1 + \qquad\qquad\qquad\right], a\right]$$

a = 26598.6·km

==========

The oblateness of the earth also produces two rotations of the orbital plane. The first of these, known as *regression of the nodes,* is where the nodes appear to slide along the equator. In effect, the line of nodes, which is in the equatorial plane, rotates about the center of the earth. Thus Ω, the right ascension of the ascending node, shifts its position.

If the orbit is prograde (see Fig. 2.4), the nodes slide westward, and if retrograde, they slide eastward. As seen from the ascending node, a satellite in prograde orbit moves eastward, and in a retrograde orbit, westward. The nodes therefore move in a direction opposite to the direction of satellite motion, hence the term *regression of the nodes.* For a polar orbit ($i = 90°$) the regression is zero.

The second effect is rotation of apsides in the orbital plane, described below. Both effects depend on the mean motion n, the semimajor axis a, and the eccentricity e. These factors can be grouped into one factor K given by

$$K = \qquad\qquad\qquad\qquad\qquad (2.11)$$

K will have the same units as n. Thus with n in *rad/day*, K will be in *rad/day*, and with n in *°/day*, K will be in *°/day*. An approximate expression for the rate of change of Ω with respect to time is (Wertz, 1984)

$$\frac{d\Omega}{dt} = -K \cos i \qquad\qquad\qquad (2.12)$$

where i is the inclination.

The rate of regression of the nodes will have the same units as n.

When the rate of change given by Eq. (2.12) is negative the regression is westward, and when the rate is positive, the regression is eastward. It will be seen therefore that for eastward regression, i must be greater than 90°, or the orbit must be retrograde. It is possible to choose values of a, e, and i such that the rate of rotation is 0.9856°/day eastward. Such an orbit is said to be *sun-synchronous* and is described further in Sec. 2.10.

In the other major effect produced by the equatorial bulge, rotation of the line of apsides in the orbital plane, the argument of perigee changes with time, in effect, the rate of change being given by (Wertz, 1984).

$$\frac{d\omega}{dt} = K(2 - 2.5 \sin^2 i) \tag{2.13}$$

Again, the units for the rate of rotation of the line of apsides will be the same as those for n.

When the inclination i is equal to $63.435°$, the term within the brackets is equal to zero and hence no rotation takes place. Use is made of this fact in the orbit chosen for the Russian Molniya satellites (see Probs. 2.23 and 2.24).

Denoting the epoch time by t_0, the right ascension of the ascending node by Ω_0, and the argument of perigee by ω_0 at epoch gives the new values for Ω and ω at time t as

$$\Omega = \Omega_0 + \frac{d\Omega}{dt}(t - t_0) \tag{2.14}$$

$$\omega = \omega_0 + \frac{d\omega}{dt}(t - t_0) \tag{2.15}$$

Keep in mind that the orbit is not a physical entity, and it is the forces resulting from an oblate earth which act on the satellite to produce the changes in the orbital parameters. Thus, rather than follow a closed elliptical path in a fixed plane, the satellite drifts as a result of the regression of the nodes, and the latitude of the point of closest approach (the perigee) changes as a result of the rotation of the line of apsides. With this in mind it is permissible to visualize the satellite as following a closed elliptical orbit but with the orbit itself moving relative to the earth as a result of the changes in Ω and ω. Thus, as stated above, the period P_A is the time required to go around the orbital path from perigee to perigee, even though the perigee has moved relative to the earth.

Suppose, for example, that the inclination is $90°$ so that the regression of the nodes is zero [from Eq. (2.12)], and the rate of rotation of the line of apsides is $-K/2$ [from Eq. (2.13)], and further, imagine the situation where the perigee at the start of observations is exactly over the ascending node. One period later the perigee would be at an angle $-KP_A/2$ relative to the ascending node, or in other words, would be south of the equator. The time between crossings *at the ascending node* would be $P_A(1 + K/2n)$, which would be the period observed from the earth. Recall that K will have the same units as n, for example, radians per second.

Example 2.5

Determine the rate of regression of the nodes, and the rate of rotation of the line of apsides for the satellite parameters specified in Table 2.1. The value for a obtained in Example 2.2 may be used.

solution

Data from Table 2.1 and Example 2.2:

$$i := 98.6540 \cdot \text{deg} \qquad e := .0014053$$

$$n := 14.22296917 \cdot \text{day}^{-1} \qquad a := 7195.7 \cdot \text{km}$$

Known constant: $K_1 := 66063.1704 \cdot \text{km}^2$

$n := 2 \cdot \pi \cdot n$ ...converts n to SI units of rad/sec.

$$K := \frac{n \cdot K_1}{a^2 \cdot (1 - e^2)^2} \qquad\qquad K = 6.533 \cdot \frac{\text{deg}}{\text{day}}$$

$$\Omega' := -K \cdot \cos(i) \qquad\qquad \Omega' = 0.983 \cdot$$

$$=========$$

$$\omega' := K \cdot \left(2 - 2.5 \cdot \sin(i)^2\right) \qquad\qquad \omega' = -2.897 \cdot$$

$$=========$$

Example 2.6

Calculate, for the satellite in Example 2.5, the new values for ω and Ω one period after epoch.

solution

From Example 2.5:

$$\Omega' := .983 \cdot \qquad\qquad \omega' := -2.897 \cdot$$

From Table 2.1:

$$n := 14.22296917 \cdot \text{day}^{-1} \qquad \omega_0 := 62.4995 \cdot \text{deg} \qquad \Omega_0 := 250.1949 \cdot \text{deg}$$

The period is:

$$P_A =$$

$$\Omega := \Omega_O + \Omega' \cdot P_A \qquad\qquad \Omega = 250.264 \cdot \text{deg}$$
$$\qquad\qquad\qquad\qquad\qquad\qquad ==========$$

$$\omega := \omega_O + \omega' \cdot P_A \qquad\qquad \omega = 62.296 \cdot \text{deg}$$
$$\qquad\qquad\qquad\qquad\qquad\qquad ==========$$

In addition to the equatorial bulge, the earth is not perfectly circular in the equatorial plane; it has a small eccentricity of the order of 10^{-5}. This is referred to as the *equatorial ellipticity*. The effect of the equatorial ellipticity is to set up a gravity gradient which has a pronounced effect on satellites in geostationary orbit (Sec. 2.11). Very briefly, a satellite in geostationary orbit ideally should remain fixed relative to the earth. The gravity gradient resulting from the equatorial ellipticity causes the satellites in geostationary orbit to drift to one of two stable points, which coincide with the minor axis of the equatorial ellipse. These two points are separated by 180° on the equator, and are at approximately 75° E longitude and 105° W longitude. Satellites in service are prevented from drifting to these points through station-keeping maneuvers, described in Sec. 6.4. Because old, out-of-service satellites eventually do drift to these points, they are referred to as "satellite graveyards."

It may be noted that the effect of equatorial ellipticity is negligible on most other satellite orbits.

2.8.2 Atmospheric Drag

For near-earth satellites, below about 1000 km, the effects of atmospheric drag are significant. Because the drag is greatest at the perigee, the drag acts to reduce the velocity at this point, with the result that the satellite does not reach the same apogee height on successive revolutions. The result is that the semimajor axis and the eccentricity are both reduced. Drag does not noticeably change the other orbital parameters, including perigee height. In the program used for generating the orbital elements given in the NASA bulletins, a "pseudo-drag" term is generated which is equal to one-half the rate of change of mean motion (ADC USAF, 1980). An approximate expression for the change of major axis is

$$a \cong a_0 \left(\qquad\qquad\qquad \right)^{2/3} \tag{2.16}$$

The mean anomaly is also changed. An approximate expression for the amount by which it changes is

$$\delta M = \quad (t - t_0)^2 \tag{2.17}$$

From Table 2.1 it is seen that the first time derivative of the mean motion is listed in columns 34–43 of line 1 of the NASA bulletin. For the example shown in Fig. 2.6, this is 0.00000187 rev/day^2. Thus the changes resulting from the drag term will be significant only for long time intervals, and for present purposes will be ignored. For a more accurate analysis, suitable for long-term predictions, the reader is referred to ADC USAF, 1980.

2.9 Inclined Orbits

A study of the general situation of a satellite in an inclined elliptical orbit is complicated by the fact that different parameters are referred to different reference frames. The orbital elements are known with reference to the plane of the orbit, the position of which is fixed (or slowly varying) in space, while the location of the earth station is usually given in terms of the local geographic coordinates which rotate with the earth. Rectangular coordinate systems are generally used in calculations of satellite position and velocity in space, while the earth station quantities of interest may be the azimuth and elevation angles and range. Transformations between coordinate systems are therefore required.

Here, in order to illustrate the method of calculation for elliptical inclined orbits, the problem of finding the earth station look angles and range will be considered. It should be kept in mind that with inclined orbits the satellites are not geostationary and therefore the required look angles and range will change with time. Detailed and very readable treatments of orbital properties in general will be found for example in Bate et al., 1971, and Wertz, 1984. Much of the explanation and the notation in this section is based on these two references.

Determination of the look angles and range involves the following quantities and concepts:

1. The *orbital elements,* as published in the NASA bulletins and described in Sec. 2.6

2. Various measures of *time*

3. The *perifocal coordinate system,* which is based on the orbital plane

4. The *geocentric-equatorial coordinate system,* which is based on the earth's equatorial plane

5. The *topocentric-horizon coordinate system,* which is based on the observer's horizon plane

The two major coordinate transformations which are needed are

- The satellite position measured in the perifocal system is transformed to the geocentric-horizon system in which the earth's rotation is measured, thus enabling the satellite position and the earth station location to be coordinated.
- The satellite-to-earth station position vector is transformed to the topocentric-horizon system, which enables the look angles and range to be calculated.

2.9.1 Calendars

A calendar is a timekeeping device in which the year is divided into months, weeks, and days. Calendar days are units of time based on the earth's motion relative to the sun. Of course it is more convenient to think of the sun moving relative to the earth. This motion is not uniform, and so a fictitious sun, termed the *mean sun,* is introduced.

The mean sun does move at a uniform speed, but otherwise requires the same time as the real sun to complete one orbit of the earth, this time being the *tropical year.* A day measured relative to this mean sun is termed a *mean solar day.* Calendar days are mean solar days, and generally they are just referred to as days.

A tropical year contains 365.2422 days. In order to make the calendar year, also referred to as the civil year, more easily usable, it is normally divided into 365 days. The extra 0.2422 of a day is significant, and for example after 100 years there would be a discrepancy of 24 days between the calendar year and the tropical year. Julius Caesar made the first attempt to correct for the discrepancy by introducing the *leap year,* in which an extra day is added to February whenever the year number is divisible by four. This gave the *Julian calendar* in which the civil year was 365.25 days on average, a reasonable approximation to the tropical year.

By the year 1582 an appreciable discrepancy once again existed between the civil and tropical years. Pope Gregory XIII took matters in hand by abolishing the days October 5 through October 14, 1582, to bring the civil and tropical years into line, and by placing an additional constraint on the leap year in that years ending in two zeros must be divisible by 400 to be reckoned as leap years. This dodge was used to miss out three days every four hundred years. The resulting calendar is the *Gregorian calendar,* which is the one in use today.

Example 2.7   Calculate the average length of the civil year in the Gregorian calendar.

solution   The nominal number of days in 400 years is 400 × 365 = 146,000. The nominal number of leap years is 400/4 = 100, but this must be reduced by 3 and therefore the number of days in 400 years of the Gregorian calendar is 146,000 + 100 − 3 = 146,097. This gives a yearly average of 146,097/400 = 365.2425.

In calculations requiring satellite predictions it is necessary to determine whether a year is a leap year or not, and the simple rule is: If the year number ends in two zeros and is divisible by 400 it is a leap year. Otherwise, if the year number is divisible by 4 it is a leap year.

Example 2.8   Determine which of the following years are leap years: (*a*) 1987, (*b*) 1988, (*c*) 2000, (*d*) 2100.

solution

(*a*) 1987/4 = 496.75 (therefore 1987 is not a leap year)
(*b*) 1988/4 = 497 (therefore 1988 is a leap year)
(*c*) 2000/400 = 5 (therefore 2000 is a leap year)
(*d*) 2100/400 = 5.25 (therefore 2100 is not a leap year)

2.9.2   Universal time

Universal time coordinated (UTC) is the time used for all civil time-keeping purposes, and it is the time reference which is broadcast by the National Bureau of Standards as a standard for setting clocks. It is based on an atomic time-frequency standard. The fundamental unit for UTC is the *mean solar day* (see App. J in Wertz, 1984). In terms of "clock time" the mean solar day is divided into 24 hours, an hour into 60 minutes, and a minute into 60 seconds. Thus there are 86,400 "clock seconds" in a mean solar day. Satellite-orbit epoch time is given in terms of UTC.

Example 2.9   Calculate the time in days, hours, minutes, and seconds for the epoch day 324.95616765.

solution   This represents the 324th day of the year plus 0.95616765 mean solar day. The decimal fraction in hours is 24 × 0.95616765 = 22.948022; the decimal fraction of this, 0.948022, in minutes is 60 × 0.948022 = 56.881344; the decimal fraction of this in seconds is 60 × 0.881344 = 52.88064. The epoch is at 22 h, 56 min, 52.88 s on the 324th day of the year.

Universal time coordinated is equivalent to *Greenwich mean time* (GMT), also *Zulu* (Z) *time*. There are a number of other "universal time" systems, all interrelated (see Wertz, 1984) , and all with the mean solar day as the fundamental unit. For present purposes the distinction between these systems is not critical, and the term *universal time,* abbreviation UT, will be used from now on.

For computations, UT will be required in two forms: as a fraction of a day and in degrees. Given UT in the normal form of hours, minutes, and seconds, it is converted to fractional days as

$$\mathrm{UT}_{\mathrm{day}} = \left(\mathrm{hours} + \qquad + \qquad \right) \qquad (2.18)$$

In turn, this may be converted to degrees as

$$\mathrm{UT}^\circ = 360 \times \mathrm{UT}_{\mathrm{day}} \qquad (2.19)$$

2.9.3 Julian dates[1]

Calendar times are expressed in UT, and although the time interval between any two events may be measured as the difference in their calendar times, the calendar time notation is not suited to computations where the timing of many events has to be computed. What is required is a reference time to which all events can be related in decimal days. Such a reference time is provided by the Julian zero time reference, which is 12 noon (12:00 UT) on January 1, in the year 4713 B.C.! Of course this date would not have existed as such at the time; it is a hypothetical starting point which can be established by counting backward according to a certain formula. For details of this intriguing time reference see Wertz, 1984. The important point is that ordinary calendar times are easily converted to Julian dates, measured on a continuous time scale of Julian days. To do this, first determine the day of the year, keeping in mind that day zero, denoted as Jan 0, is December 31. For example, noon on December 31 is denoted as Jan 0.5, and noon on January 1 is denoted as Jan 1.5. It may seem strange that the last day of December should be denoted as "day zero in January," but it will be seen that this makes the day count correspond to the actual calendar day.

The next step is to convert the UT of interest to a fractional day as given by Eq. (2.18). Then, determine from tables the Julian day for Jan 0.0 of the year in question. A partial listing of these Julian days (JD) is given in Table 2.2. Denoting the Julian day for Jan 0.0 as $\mathrm{JD}_{0.0}$ gives the time in Julian days as

---

[1]It should be noted that the Julian date is not associated with the Julian calendar introduced by Julius Caesar.

TABLE 2.2   Julian Dates at the Beginning of Each
Year (Jan 0.0 UT) for the Years 1986–2000

Year	Julian date (days)
	2,400,000+
1986	46,430.5
1987	46,795.5
1988	47,160.5
1989	47,526.5
1990	47,891.5
1991	48,256.5
1992	48,621.5
1993	48,987.5
1994	49,352.5
1995	49,717.5
1996	50,082.5
1997	50,448.5
1998	50,813.5
1999	51,178.5
2000	51,543.5

$$JD = JD_{0.0} + \text{day number} + UT_{\text{day}} \qquad (2.20)$$

To assist in finding the Julian date for any given day, Table 2.3 lists the day number for noon on the last day of each month.

Example 2.10   Find the Julian date corresponding to 3 h UT on October 11, 1986.

solution   From Table 2.2, noon, September 30, is day number 273.5, since 1986 is not a leap year. Hence noon on October 11 is day number 273.5 + 11 = 284.5. The start of October 11 is therefore day number 284. UT 0300 h is 3/24 = 0.125 day, therefore the day number is 284.125. Add to this the Julian date for Jan 0.0, 1986, obtained from Table 2.1 to get 2,400,000 + 46,430.5 + 284.125 = **2,446,714.625.**

There are a number of methods available for calculating the Julian day (see for example, Wertz, 1984, p. 10, and Duffett-Smith, 1986, p. 9). In the latter reference a general method is given for calculating the Julian day for any date, for use with hand-held calculators. The Mathcad routine given in the following example is based on this but is limited to dates later than October 15, 1582.

TABLE 2.3 Day Number for Noon on
the Last Day of the Month

Date	Day number
January 31	31.5
February 28	59.5 (60.5)
March 31	90.5 (91.5)
April 30	120.5 (121.5)
May 31	151.5 (152.5)
June 30	181.5 (182.5)
July 31	212.5 (213.5)
August 31	243.5 (244.5)
September 30	273.5 (274.5)
October 31	304.5 (305.5)
November 30	334.5 (335.5)
December 31	365.5 (366.5)

Numbers in parentheses are for leap years.

Example 2.11

Find the Julian Day for 13h UT on 18 December 1993.

solution

Enter the year y, the month m, the day d, and the hour in Universal Time (UT)
expressed as a fraction of a day.

$$y := 1993 \qquad m := 12 \qquad d := 18 +$$

$$y := \text{if}(m \le 2, y - 1, y)$$
$$m := \text{if}(m \le 2, m + 12, m)$$

$$A := \text{floor}\left(\right)$$

$$B := 2 - A + \text{floor}\left(\right)$$

$$C := \text{floor}(365.25 \cdot y)$$
$$D := \text{floor}(30.6001 \cdot (M = 1))$$
$$JD := B + C + D + d + 1720994.5 \qquad JD = 2449340.0417$$

$$==========$$

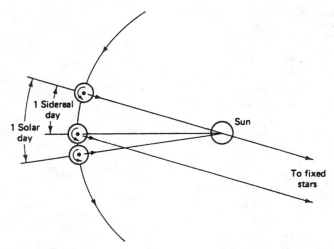

Figure 2.7 A sidereal day, or one rotation of the earth relative to fixed stars, is shorter than the solar day.

2.9.4 Sidereal time

Sidereal time is time measured relative to the fixed stars (Fig. 2.7). It will be seen that one complete rotation of the earth relative to the fixed stars is not a complete rotation relative to the sun. This is because the earth moves in its orbit around the sun.

The *sidereal day* is defined as one complete rotation of the earth relative to the fixed stars. One sidereal day has 24 sidereal hours, one sidereal hour has 60 sidereal minutes, and one sidereal minute has 60 sidereal seconds. Care must be taken to distinguish between sidereal times and mean solar times which use the same basic subdivisions. The relationships between the two systems, given in Bate et al., 1971, are

$$1 \text{ mean solar day} = 1.0027379093 \text{ mean sidereal days}$$

$$= 24^h \ 03^m \ 56^s .55536 \text{ sidereal time} \qquad (2.21)$$

$$= 86,636.55536 \text{ mean sidereal seconds}$$

$$1 \text{ mean sidereal day} = 0.9972695664 \text{ mean solar days}$$

$$= 23^h \ 56^m \ 04^s .09054 \text{ mean solar time} \qquad (2.22)$$

$$= 86,164.09054 \text{ mean solar seconds}$$

Measurements of longitude on the earth's surface require the use of sidereal time (discussed further in Sec. 2.9.7). The use of 23 h, 56 min as an approximation for the mean sidereal day will be used later in determining the height of the geostationary orbit.

2.9.5 The orbital plane

In the orbital plane, the position vector **r** and the velocity vector **v** specify the motion of the satellite, as shown in Fig. 2.8. For present purposes, only the magnitude of the position vector is required. From the geometry of the ellipse (see App. B) this is found to be

$$r = \tag{2.23}$$

The true anomaly v is a function of time, and determining it is one of the more difficult steps in the calculations.

The usual approach to determining v proceeds in two stages. First, the mean anomaly M at time t is found. This is a simple calculation:

$$M = n(t - T) \tag{2.24}$$

Here, n is the mean motion as previously defined in Eq. (2.9) and T is the time of perigee passage.

The time of perigee passage T can be eliminated from Eq. (2.24) if one is working from the elements specified by NASA. For the NASA elements

$$M_0 = n(t_0 - T)$$

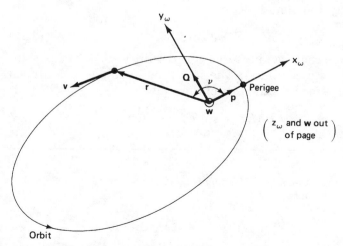

Figure 2.8 Perifocal coordinate system (**PQW** frame).

Therefore

$$T = t_0 - \qquad (2.25)$$

Hence, substituting this in Eq. (2.24) gives

$$M = M_0 + n(t - t_0) \qquad (2.26)$$

Consistent units must be used throughout. For example, with n in degrees/day, time $(t - t_0)$ must be in days and M_0 in degrees, and M will then be in degrees.

Example 2.12   Calculate the time of perigee passage for the NASA elements given in Table 2.1.

solution   The specified values at epoch are mean motion $n = 14.22296917$ rev/day, mean anomaly $M_0 = 297.7604°$, and $t_0 = 221.07019100$ days. In this instance it is only necessary to convert the mean motion to degrees/day, which is $360n$. Applying Eq. (2.25) gives

$$T = 221.07019100 - \frac{297.7604}{14.22296917 \times 360}$$

$$= 221.012 \text{ days}$$

Once the mean anomaly M is known, the next step is to solve an equation known as *Kepler's equation*. Kepler's equation is formulated in terms of an intermediate variable E, known as the *eccentric anomaly*, and is usually stated as

$$M = E - e \sin E \qquad (2.27)$$

This rather innocent looking equation requires an iterative solution, preferably by a computer. The following example in Mathcad shows how to solve for E as the root of the equation

$$M - (E - e \sin E) = 0 \qquad (2.28)$$

Example 2.13

Given that the mean anomaly is 205 degrees and the eccentricity .0025, calculate the eccentric anomaly.

solution

$$M := 205 \cdot \deg \qquad\qquad e := .0025$$

E : = π   ...this is the initial guess value for E.

E : = root(M − E + e·sin(E), E)   ...this is the root equation which Mathcad solves for E.

E = 204.938·deg

Once E is found, v can be found from an equation known as *Gauss' equation,* which is

$$\tan \quad = \sqrt{\frac{1+e}{1-e}} \tan \qquad (2.29)$$

It may be noted that r, the magnitude of the radius vector, can also be obtained as a function of E, and is

$$r = a(1 - e \cos E) \qquad (2.30)$$

For near-circular orbits where the eccentricity is small, an approximation for v directly in terms of M is

$$v \cong M + 2e \sin M + \quad e^2 \sin 2M \qquad (2.31)$$

Example 2.14   For satellite no. 14452 the eccentricity is given in the NASA prediction bulletin as 9.5981×10^{-3} and the mean anomaly at epoch as $204.9779°$. The mean motion is 14.2171404 rev/day. Calculate the true anomaly and the magnitude of the radius vector 5 s after epoch. The semimajor axis is known to be 7194.9 km.

solution

$$n = \frac{14.2171404 \times 2\pi}{86400} \cong 0.001 \text{ rad/s}$$

$$M = 204.9779 + 0.001 \times \frac{180}{\pi} \times 5$$

$$= 205.27° \quad \text{or} \quad 3.583 \text{ rad}$$

Since the orbit is near-circular (small eccentricity), Eq. (2.26) may be used to calculate the true anomaly v as

$$v \cong 3.583 + 2 \times 9.5981 \times 10^{-3} \times \sin 205.27 + \tfrac{5}{4} \times 9.5981^2 \times 10^{-6} \sin 2 \times 205.27$$

$$= 3.575 \text{ rad}$$

$$= 204.81°$$

Applying Eq. (2.19) gives r as

$$r = \frac{7194.9 \times (1 - 9.5981^2 \times 10^{-6})}{1 + (9.5981 \times 10^{-3}) \times \cos 204.81} \cong 7257 \text{ km}$$

The magnitude r of the position vector r may be calculated by either Eq. (2.23) or Eq. (2.30). It may be expressed in vector form in the *perifocal coordinate system.* Here, the orbital plane is the fundamental plane, and the origin is at the center of the earth (only earth-orbiting satellites are being considered). The positive x axis lies in the orbital plane and passes through the perigee. Unit vector **P** points along the positive x axis as shown in Fig. 2.7. The positive y axis is rotated 90° from the x axis in the orbital plane, in the direction of satellite motion, and the unit vector is shown as **Q**. The positive z axis is normal to the orbital plane such that coordinates x-y-z form a right-hand set, and the unit vector is shown as **W**. The subscript ω is used to distinguish the xyz coordinates in this system, as shown in Fig. 2.8. The position vector in this coordinate system, which will be referred to as the **PQW** *frame,* is given by

$$\mathbf{r} = (r \cos v)\mathbf{P} + (r \sin v)\mathbf{Q} \qquad (2.32)$$

The perifocal system is very convenient for describing the motion of the satellite. If the earth were uniformly spherical, the perifocal coordinates would be fixed in space, that is, inertial. However, the equatorial bulge causes rotations of the perifocal coordinate system, as described in Sec. 2.8.1. These rotations are taken into account when the satellite position is transferred from perifocal coordinates to *geocentric-equatorial coordinates,* described in the next section.

Example 2.15 Using the values $r = 7257$ km and $v = 204.81°$ obtained in the previous example, express r in vector form in the perifocal coordinate system.

solution

$$r_P = 7257 \times \cos 204.81 = -6587.6 \text{ km}$$
$$r_Q = 7257 \times \sin 204.81 = -3045.3 \text{ km}$$

Hence

$$\mathbf{r} = -6587.6\mathbf{P} - 3045.3\mathbf{Q} \quad \text{km}$$

2.9.6 The geocentric-equatorial coordinate system

The *geocentric-equatorial coordinate system* is an inertial system of axes, the reference line being fixed by the fixed stars. The reference line is the line of Aries described in Sec. 2.5. (The phenomenon known as the precession of the equinoxes is ignored here. This is a very slow rotation of this reference frame, amounting to approximately 1.396971° per Julian century, where a Julian century consists of 36,525 mean solar days). The fundamental plane is the earth's equatorial plane. Figure 2.9 shows the part of the ellipse above the equatorial plane and the orbital angles Ω, ω, and i. It should be kept in mind that Ω and ω may be slowly varying with time as shown by Eqs. (2.12) and (2.13).

The unit vectors in this system are labeled **I, J, K,** and the coordinate system is referred to as the **IJK** *frame,* with positive **I** pointing along the line of Aries. The transformation of vector **r** from the **PQW** frame to the **IJK** frame is most easily expressed in matrix form, the components being indicated by the appropriate subscripts

$$\begin{bmatrix} r_I \\ r_J \\ r_K \end{bmatrix} = \tilde{\mathbf{R}} \begin{bmatrix} r_P \\ r_Q \end{bmatrix} \qquad (2.33a)$$

where the transformation matrix $\tilde{\mathbf{R}}$ is given by

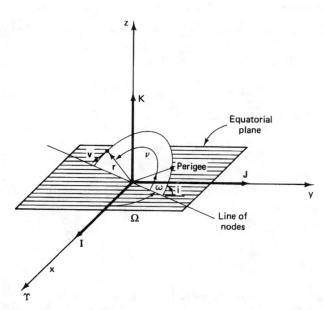

Figure 2.9 Geocentric-equatorial coordinate system (**IJK** frame).

$$\tilde{\mathbf{R}} = \begin{bmatrix} & & \\ & & \\ & & \end{bmatrix}$$

(2.33*b*)

This gives the components of the position vector **r** for the satellite, in the **IJK,** or inertial, frame. It should be noted that the angles Ω and ω take into account the rotations resulting from the earth's equatorial bulge, as described in Sec. 2.8.1. Because matrix multiplication is most easily carried out by computer, the following example is completed in Mathcad.

Example 2.16

Calculate the magnitude of the position vector in the **PQW** frame for the orbit specified below. Calculate also the position vector in the **IJK** frame, and its magnitude. Confirm that this remains unchanged from the value obtained in the **PQW** frame.

solution

The given orbital elements are:

$\Omega := 300 \cdot \deg \quad \omega := 60 \cdot \deg \quad i := 65 \cdot \deg \quad r_P := -6500 \cdot km \quad r_Q := 4000 \cdot km$

$r := \sqrt{r_P{}^2 + r_Q{}^2} \quad$ ...from eq. (2.32)

$r = 7632.2 \cdot km$

$========$

Equation (2.33) is

$$\begin{bmatrix} \\ \end{bmatrix} := \left(\right) \cdot ()$$

$r_I = -4685.3 \cdot km$

$r_J = 5047.7 \cdot km \qquad$ ...these are the values obtained by Mathcad.

$$r_K = -3289.1 \cdot km$$

The magnitude is $$|r| = 7632.2 \cdot km$$
                                      ==========

This is seen to be the same as that obtained from the **P** and **Q** components.

___

2.9.7 Earth station referred to the IJK frame

The earth station's position is given by the geographic coordinates of latitude λ_E and longitude ϕ_E. (Unfortunately there does not seem to be any standardization of the symbols used for latitude and longitude. In some texts, as here, the Greek lambda is used for latitude and the Greek phi for longitude. In other texts the reverse of this happens. One minor advantage of the former is that latitude and lambda both begin with the same *la* which makes the relationship easy to remember.)

Care must also be taken regarding the sign conventions used for latitude and longitude because different systems are sometimes used, depending on the application. In this book, north latitudes will be taken as positive numbers and south latitudes as negative numbers, zero latitude of course being the equator. Longitudes east of the Greenwich meridian will be taken as positive numbers, and longitudes west as negative numbers.

The position vector of the earth station relative to the **IJK** frame is **R** as shown in Fig. 2.10. The angle between **R** and the equatorial plane, denoted by ψ_E in Fig. 2.19, is closely related, but not quite equal to, the earth station latitude. More will be said about this angle shortly. **R** is obviously a function of the rotation of the earth, and so first it is necessary to find the position of the Greenwich meridian relative to the **I** axis as a function of time. The angular distance from the **I** axis to the Greenwich meridian is measured directly as *Greenwich sidereal time* (GST), also known as the *Greenwich hour angle,* or GHA. Sidereal time is described in Sec. 2.9.4.

GST may be found using values tabulated in some almanacs (see Bate et al., 1971), or it may be calculated using formulas given in Wertz, 1984. In general, sidereal time may be measured in time units of sidereal days, hours, and so on, or it may be measured in degrees, minutes, and seconds. The formula for GST in degrees is

$$GST = 99.6910 + 36,000.7689 \times T + 0.0004 \times T^2 + UT° \quad (2.34)$$

Here, UT° is universal time expressed in degrees as given by Eq. (2.19). The symbol T also stands for time, but it is the time reckoned in Julian centuries, from a known reference.

A Julian century consists of 36,525 mean solar days, and the reference time for T is taken as Jan 0.5, 1900, the corresponding Julian

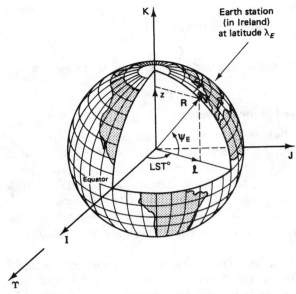

Figure 2.10 Position vector **R** of the earth relative to the **IJK** frame.

date being 2,415,020.0 days. Thus, if JD represents the universal time in Julian days, then T is calculated as

$$T = \frac{JD - 2{,}415{,}020.0}{36{,}525} \qquad (2.35)$$

The reference time is easily changed. For example, suppose it is wished to use as a reference time Jan 0.0, 1986. Let JD_{1986} represent the Julian day reckoned from Jan 0.0, 1986 (see Table 2.2). Then Eq. (2.35) becomes

$$T = \frac{JD_{1986} + 2{,}446{,}430.5 - 2{,}415{,}020.0}{36{,}525}$$

$$= \qquad (2.36)$$

Once GST is known, the *local sidereal time* (LST) is found by adding east longitude of the station in degrees. East longitude for the earth station will be denoted as EL. Recall that previously longitude was expressed in positive degrees east and negative degrees west. For east longitudes, $EL = \phi_E$, while for west longitudes, $EL = 360° + \phi_E$. For example, for an earth station at east longitude 40°, EL

= 40°. For an earth station at west longitude 40°, EL = 360 + (−40) = 320°. Thus, the local sidereal time in degrees is given by

$$LST = GST + EL \qquad (2.37)$$

The procedure is illustrated in the following examples.

Example 2.17 Calculate the GST for 15 h UT on October 23, 1986.

solution From Table 2.2, noting that 1986 is not a leap year, noon on September 30 is day number 273.5. Hence noon on October 23 is day number 273.5 + 23 = 296.5. The start of October 23 is therefore day number 296, and 15 h UT is 0.625 day. Therefore the Julian day number is 296.625.

$$T = \frac{296.625 + 31{,}410.5}{36{,}525} = 0.8680938$$

The individual terms in Eq. (2.34) are

Term	Degrees
Constant term	99.69
36,000.7689×0.8680938	31,252.044
0.0004×(0.8680938)²	0.0003
15 h UT	225
GST (sum of terms)	31,576.735

This, expressed as modulo 360, is 256.735°.

Example 2.18 Find the LST for Thunder Bay, longitude 89.26° W for 15 h UT October 23, 1986.

solution

$$EL = 360 - 89.26 = 270.74°$$

This is added onto the GST calculated in the previous example to give

$$LST = 270.74 + 256.735 = 527.475°$$

This, expressed as modulo 360, is 167.475°.

Knowing the LST enables the position vector **R** of the earth station to be located with reference to the **IJK** frame as shown in Fig. 2.10. However, when **R** is resolved into its rectangular components, account must be taken of the oblateness of the earth. The earth may be modeled as an *oblate spheroid,* in which the equatorial plane is circular, and any meridional plane (i.e., any plane containing the earth's polar axis) is elliptical, as illustrated in Fig. 2.11. For one particular model, known as a *reference ellipsoid,* the semimajor axis of the ellipse is equal to the equatorial radius, the semiminor axis is equal to the polar radius, and the surface of the ellipsoid represents the *mean sea level.* Denoting the semimajor axis by a_E and the semiminor axis by b_E and using the known values for the earth's radii gives

$$a_E = 6378.1414 \text{ km} \tag{2.38}$$

$$b_E = 6356.755 \text{ km} \tag{2.39}$$

From these values, the eccentricity of the earth is seen to be

$$e_E = \frac{\sqrt{a_E^2 - b_E^2}}{a_E} = 0.08182 \tag{2.40}$$

In Figs. 2.10 and 2.11, what is known as the *geocentric latitude* is shown as ψ_E. This differs from what is normally referred to as *latitude.* An imaginary plumb line dropped from the earth station makes an angle λ_E with the equatorial plane, as shown in Fig. 2.11. This is known as the *geodetic latitude* and for all practical purposes here this can be taken as the geographic latitude of the earth station.

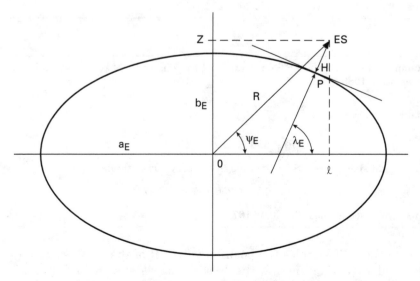

Figure 2.11 Reference ellipsoid for the earth showing the geocentric latitude ψ_E and the geodetic latitude λ_E.

With the height of the earth station above mean sea level denoted by *H,* the geocentric coordinates of the earth station position are given in terms of the geodetic coordinates by (Thompson 1966)

$$N = \frac{a_E}{\sqrt{1 - e_E^2 \sin^2\lambda_E}} \qquad (2.41)$$

$$R_I = (N + H) \cos \lambda_E \cos \text{LST} = l \cos \text{LST} \qquad (2.42)$$

$$R_J = (N + H) \cos \lambda_E \sin \text{LST} = l \sin \text{LST} \qquad (2.43)$$

$$R_K = [N(1 - e_E^2) + H] \sin \lambda_E = z \qquad (2.44)$$

Example 2.19

Using the LST value obtained in Example 2.14, find the components of the radius vector to the earth station at Thunder Bay, given that the latitude is 48.42 degrees, and the height above sea level is 200 m.

solution

The given data are: LST : = 167.475·deg λ_E : = 48.42·deg   H : = 200·m

The required earth constants are: a_E : = 6378.1414·km e_E : = .08182

$$l := \left(\frac{a_E}{\sqrt{1 - e_E^2 \cdot \sin(\lambda_E)^2}} + H \right) \cdot \cos(\lambda_E) \qquad \text{...Eq. (2.42)}$$

$$z := \left[\frac{a_E \cdot (1 - e_E^2)}{\sqrt{1 - e_E^2 \cdot \sin(\lambda_E)^2}} + H \right] \cdot \sin(\lambda_E) \qquad \text{...Eq. (2.43)}$$

For check purposes, the values are:

$$l = 4241 \cdot \text{km} \qquad z = 4748.2 \cdot \text{km}$$

$$R := \begin{pmatrix} l \cdot \cos(\text{LST}) \\ l \cdot \sin(\text{LST}) \\ z \end{pmatrix} \qquad \text{...this gives the R components in matrix form.}$$

The values are:

$$R = \begin{pmatrix} \quad \\ \quad \\ \quad \end{pmatrix} \cdot \text{km}$$

The magnitude of the R vector is:

$$|R| = 6366.4 \cdot \text{km}$$

At this point, both the satellite radius vector **r** and the earth station radius vector **R** are known in the **IJK** frame, for any position of satellite and earth. From the vector diagram shown in Fig. 2.12a the range vector **ρ** is obtained as

$$\rho = r - R \tag{2.45}$$

This gives **ρ** in the **IJK** frame. It then remains to transform **ρ** to the observer's frame, known as the topocentric-horizon frame, shown in Fig. 2.12b.

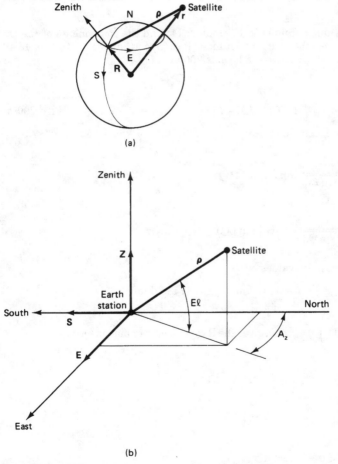

(a)

(b)

Figure 2.12 Topocentric-horizon coordinate system (**SEZ** frame): (a) overall view; (b) detailed view.

2.9.8   The topocentric-horizon coordinate system

The position of the satellite, as measured from the earth station, is usually given in terms of the azimuth and elevation angles, and the range ρ. These are measured in the *topocentric-horizon coordinate system* illustrated in Fig. 2.12*b*. In this coordinate system, the fundamental plane is the observer's horizon plane. In the notation given in Bate et al., 1971, the positive x axis is taken as south, the unit vector being denoted by **S**. The positive y axis points east, the unit vector being **E**. The positive z axis is "up," pointing to the observer's zenith, the unit vector being **Z**. (*Note:* This is not the same z as that used in Sec. 2.9.7). The frame is referred to as the **SEZ** *frame*, which of course rotates with the earth.

As shown in the previous section, the range vector ρ is known in the **IJK** frame, and it is now necessary to transform this to the **SEZ** frame. Again, this is a standard transformation procedure given by (see Bate et al., 1971)

$$\begin{bmatrix} \rho_S \\ \rho_E \\ \rho_Z \end{bmatrix} = \begin{bmatrix} \sin \psi_E \cos \text{LST} & \sin \psi_E \sin \text{LST} & -\cos \psi_E \\ -\sin \text{LST} & \cos \text{LST} & 0 \\ \cos \psi_E \cos \text{LST} & \cos \psi_E \sin \text{LST} & \sin \psi_E \end{bmatrix} \begin{bmatrix} \rho_I \\ \rho_J \\ \rho_K \end{bmatrix} \quad (2.46)$$

From Fig. 2.11, the geocentric angle ψ_E is seen to be given by

$$\psi_E = \arctan \frac{z}{l} \quad (2.47)$$

The coordinates l and z given in Eqs. (2.42) and (2.44) are known in terms of the earth station height and latitude, and hence the range vector is known in terms of these quantities and the LST. As a point of interest, for zero height, the angle ψ_E is known as the *geocentric latitude,* and is given by

$$\tan \psi_{E(H=0)} = (1 - e_E^2) \tan \lambda_E \quad (2.48)$$

Here, e_E is the earth's eccentricity, equal to 0.08182. The difference between the geodetic and geocentric latitudes reaches a maximum at a geocentric latitude of 45°, when the geodetic latitude is 45.192°.

Finally, the magnitude of the range and the antenna look angles are obtained from

$$\rho = \sqrt{\rho_S^2 + \rho_E^2 + \rho_Z^2} \quad (2.49)$$

$$\text{EL} = \arcsin \left(\frac{\rho_Z}{\rho} \right) \quad (2.50)$$

TABLE 2.4

ρ_S	ρ_E	Azimuth degrees
−	+	α
+	+	$180 - \alpha$
+	−	$180 + \alpha$
−	−	$360 - \alpha$

We define an angle α as

$$\alpha = \arctan \frac{|\rho_E|}{|\rho_S|} \qquad (2.51)$$

Then the azimuth depends on which quadrant α is in and is given by Table 2.4.

Example 2.20

The **IJK** range vector components for a certain satellite, at GST = 240 degrees, are as given below. Calculate the corresponding range and the look angles for an earth station the coordinates for which are, latitude 48.42 degrees N, longitude 89.26 degrees **W**, height above mean sea level 200m.
Given data:

$$\rho_I := -1280 \cdot km \qquad \rho_J := -1278 \cdot km \qquad \rho_K := 66 \cdot km$$

$$GST := 240 \cdot deg \qquad \lambda_E := 48.42 \cdot deg \qquad \phi_E := -89.26 \cdot deg \qquad H := 200 \cdot m$$

The required earth constants are:

$$a_E := 6378.1414 \cdot km \qquad e_E := .08182$$

$$l := \left(\frac{a_E}{\sqrt{1 - e_E{}^2 \cdot \sin(\lambda_E)^2}} + H \right) \cdot \cos(\lambda_E) \qquad \dots eq. (2.42)$$

$$z := \left[\frac{a_E \cdot (1 - e_E{}^2)}{\sqrt{1 - e_E{}^2 \cdot \sin(\lambda_E)^2}} + H \right] \cdot \sin(\lambda_E) \qquad \dots eq. (2.43)$$

(The values for check purposes are:

$$l = 4241 \cdot km \qquad z = 4748.2 \cdot km)$$

$$\psi_E := atan\left(\frac{z}{l}\right) \qquad eq. (2.47) \qquad \psi_E = 48.2 \cdot deg$$

$$LST := 240 \cdot deg + \phi_E \qquad \dots eq. (2.37)$$

$$D := \begin{bmatrix} (\sin(\psi_E) \cdot \cos(LST)) & (\sin(\psi_E) \cdot \sin(LST)) & -\cos(\psi_E) \\ (-\sin(LST)) & (\cos(LST)) & 0 \\ (\cos(\psi_E) \cdot \cos(LST)) & (\cos(\psi_E) \cdot \sin(LST)) & (\sin(\psi_E)) \end{bmatrix}$$

$$D = \begin{pmatrix} & & \\ & & \\ & & \end{pmatrix} \qquad \text{...the D-values are given for check purposes.}$$

$$\begin{bmatrix} \\ \\ \end{bmatrix} := D \cdot \begin{bmatrix} \\ \\ \end{bmatrix} \qquad \text{...eq. (2.46)}$$

$$\begin{bmatrix} \\ \\ \end{bmatrix} = \begin{pmatrix} \\ \\ \end{pmatrix} \cdot \text{km} \qquad \text{...the values are given for check purposes.}$$

In Mathcad the magnitude is given simply by $\qquad |\rho| = 1810 \cdot \text{km}$

$$========$$

$$El := \text{asin}\left(\frac{\rho_Z}{\rho}\right) \qquad \text{...eq. (2.50)} \qquad El = 12 \cdot \text{deg}$$

$$========$$

$$\alpha := \text{atan}\left(\left|\frac{\rho_E}{\rho_S}\right|\right) \qquad \text{...eq. (2.51)}$$

The azimuth is determined by setting the quadrant conditions (see Table 2.4) as:

$$Az_a := \text{if}[(\rho_S < 0 \cdot \text{m}) \cdot (\rho_E > 0 \cdot \text{m}), \alpha, 0]$$

$$Az_b := \text{if}[(\rho_S > 0 \cdot \text{m}) \cdot (\rho_E > 0 \cdot \text{m}), 180 \cdot \text{deg} - \alpha, 0]$$

$$Az_c := \text{if}[(\rho_S > 0 \cdot \text{m}) \cdot (\rho_E < 0 \cdot \text{m}), 180 \cdot \text{deg} + \alpha, 0]$$

$$Az_d := \text{if}[(\rho_S < 0 \cdot \text{m}) \cdot (\rho_E < 0 \cdot \text{m}), 360 \cdot \text{deg} - \alpha, 0]$$

Since all but one of these are zero, the azimuth is given by

$$Az := Az_a + Az_b + Az_c + Az_d \qquad\qquad Az = 100.5 \cdot \text{deg}$$

$$=========$$

Note that the range could also have been obtained from

$$\sqrt{\rho_I{}^2 + \rho_J{}^2 + \rho_K^2} = 1810 \cdot \text{km}$$

2.9.9 The subsatellite point

The point on the earth vertically under the satellite is referred to as the *subsatellite point*. The latitude and longitude of the subsatellite point and the height of the satellite above the subsatellite point can be determined from a knowledge of the radius vector **r.** Figure 2.13 shows

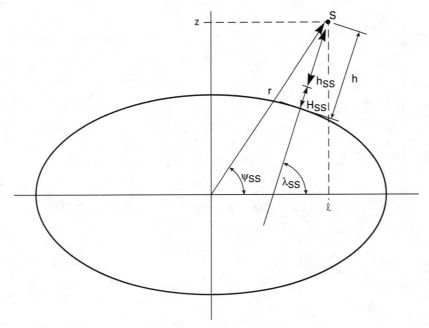

Figure 2.13   Geometry for determining the subsatellite point.

the meridian plane which cuts the subsatellite point. The height of the terrain above the reference ellipsoid at the subsatellite point is denoted by H_{SS} and the height of the satellite above this by h_{SS}. Thus the total height of the satellite above the reference ellipsoid is

$$h = H_{SS} + h_{SS} \tag{2.52}$$

Now, the components of the radius vector **r** in the **IJK** frame are given by Eq. (2.33). Figure 2.13 is seen to be similar to Fig. 2.11, with the difference that r replaces R, the height to the point of interest is h rather than H, and the subsatellite latitude λ_{SS} is used. Thus Eqs. (2.41) through (2.43) may be written for this situation as

$$N = \frac{a_E}{\sqrt{1 - e_E^2 \sin^2 \lambda_{SS}}} \tag{2.53}$$

$$r_I = (N + h) \cos \lambda_{SS} \cos \text{LST} \tag{2.54}$$

$$r_J = (N + h) \cos \lambda_{SS} \sin \text{LST} \tag{2.55}$$

$$r_K = [N(1 - e_E^2) + h] \sin \lambda_{SS} \tag{2.56}$$

We now have three equations in three unknowns, LST, λ_E, and h, and these can be solved as shown in the following Mathcad example. In addition, by analogy with the situation shown in Fig. 2.10, the east longitude is obtained from Eq. (2.37) as

$$EL = LST - GST \qquad (2.57)$$

where GST is the Greenwich sidereal time.

Example 2.21

Determine the sub-satellite height, latitude and LST for the satellite in Example 2.16.

solution

From Example 2.16, the components of the radius vector are:

$$\begin{bmatrix} \\ \\ \end{bmatrix} := \begin{pmatrix} \\ \\ \end{pmatrix}$$

The required earth constants are:

$$a_E = 6378.1414 \cdot km \qquad e_E := .08182$$

In order to solve Eqs. (2.53) through (2.56) by means of the Mathcad solve-block, guess values must be provided for the unknowns. Also, rather than using Eq. (2.53) it is easier to write N directly into Eqs. (2.54) through (2.56). The magnitude of the radius vector is:

$$r := \sqrt{r_I{}^2 + r_J{}^2 + r_K{}^2}$$

Guess value for LST:

$$LST := \pi$$

Guess value for latitude:

$$\lambda_E := \text{atan}\left(\frac{r_K}{r_I}\right)$$

Guess value for height:

$$h := r - a_E$$

The Mathcad solve block can now be used:
Given

$$r_I = \left(\frac{a_E}{\sqrt{1 - e_E^2 \cdot \sin(\lambda_E)^2}} + h\right) \cdot \cos(\lambda_E) \cdot \cos(LST)$$

$$r_J = \left(\frac{a_E}{\sqrt{1 - e_E^2 \cdot \sin(\lambda_E)^2}} + h\right) \cdot \cos(\lambda_E) \cdot \sin(LST)$$

$$r_K = \left[\frac{a_E \cdot (1 - e_E^2)}{\sqrt{1 - e_E^2 \cdot \sin (\lambda_E)^2}} + h \right] \cdot \sin (\lambda_E)$$

$$\left[\qquad \right] : = \text{Find} (\lambda_E, h, \text{LST})$$

$$\lambda_E = -25.65 \cdot \text{deg} \qquad h = 1258 \cdot \text{km} \qquad \text{LST} = 132.9 \cdot \text{deg}$$

==========   ========   =========

2.9.10 Predicting satellite position

The basic factors affecting satellite position are outlined in the previous sections. The NASA two-line elements are generated by prediction models contained in Spacetrack report no. 3 (ADC USAF, 1980), which also contains Fortran IV programs for the models. Readers desiring highly accurate prediction methods are referred to this report. Spacetrack report no. 4 (ADC USAF, 1983) gives details of the models used for atmospheric density analysis.

2.10 Sun-Synchronous Orbit

Some details of the Tiros-N/NOAA satellites used for search and rescue (Sarsat) operations are given in Sec. 1.4. These satellites operate in sun-synchronous orbits. The orientation of a sun-synchronous orbit remains fixed relative to the sun as illustrated in Fig. 2.14, the angle Φ remaining constant.

Figure 2.15 shows an alternative view, from above the earth's north pole. The angle Φ is equal to $\Omega - \alpha$ and to the local solar time expressed in degrees as will be explained shortly. From this view, the earth rotates daily around a fixed axis in space, and the sun appears to move in space, relative to the fixed stars, because of the earth's yearly orbit around the sun. The mean yearly orbit of 360° takes 365.24 mean solar days, and hence the daily shift is 360/365.24 = 0.9856°. The angle α, shown in Fig. 2.15 and known as the *right ascension of the mean sun*, moves eastward by this amount each day.

For the satellite orbit to be sun-synchronous, the right ascension of the ascending node Ω must also increase eastward by this amount. Use is made of the regression of the nodes to achieve sun synchronicity. As shown in Sec. 2.8.1 by Eqs. (2.12) and (2.14), the rate of regression of the nodes, and the direction are determined by the orbital elements a, e, and i. These can be selected to give the required regression of 0.9856° east per day.

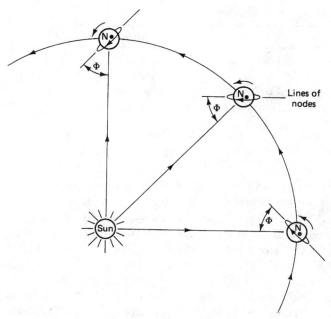

Figure 2.14   Sun-synchronous orbit.

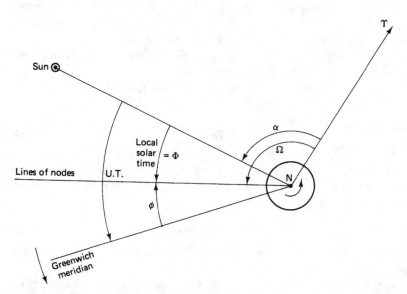

Figure 2.15   The condition for sun synchronicity is that the local solar time should be constant. Local solar time = $\Omega - \alpha$, which is also equal to the angle Φ shown in Fig. 2.14. This is the local solar time at the ascending node, but a similar situation applies at other latitude crossings.

TABLE 2.5   Tiros-N Series Orbital Parameters

	833-km orbit	870-km orbit
Inclination	98.739°	98.899°
Nodal period	101.58 min	102.37 min
Nodal regression	25.40°/orbit W	25.59°/orbit W
Nodal precession	0.986°/day E	0.986°/day E
Orbits per day	14.18	14.07

SOURCE: Schwalb, 1982a.

The orbital parameters for the Tiros-N satellites are listed in Table 2.5. These satellites follow near-circular, near-polar orbits.

From Fig. 2.15 it will be seen that with sun synchronicity the angle $\Omega - \alpha$ remains constant. This is the angle Φ shown in Fig. 2.14. Solar time is measured by the angle between the sun line and the meridian line as shown in Fig. 2.15, known as the *hour angle*. For example, universal time discussed in Sec. 2.9.2 is the hour angle between the sun and the Greenwich meridian as shown in Fig. 2.15. Likewise, local solar time is the hour angle between the sun and the local meridian. The local solar time for the line of nodes is seen to be $\Omega - \alpha$ and, as shown, for a sun synchronous orbit this is constant. In this case the latitude is zero (equator), but a similar argument can be applied for the local solar time at any latitude. What this means in practical terms is that a satellite in sun-synchronous orbit crosses a given latitude at the same local solar time and hence under approximately the same solar lighting conditions each day. This is a desirable feature for weather and surveillance satellites.

Local solar time is not the same as standard time. Letting ϕ represent the longitude of the ascending node in Fig. 2.15 gives

$$\text{Local solar time} = UT + \phi \tag{2.58}$$

As before, ϕ is negative if west and positive if east. UT is related to standard time by a fixed amount. For example universal time is equal to eastern standard time plus 5 hours. If the correction between standard time and UT is S hours, then

$$UT = \text{standard time} + S \tag{2.59}$$

This can be substituted in Eq. (2.58) to find the relationship between local standard time and local solar time.

If a satellite in sun-synchronous orbit completes an integral number of orbits per day, it will also be earth-synchronous. This means that equatorial crossings separated in time by a 1-day period will occur at the same longitude, and hence at the same standard time. As it is, the Tiros-N satellites do not have an integral number of orbits per day, and although the local solar time of crossings remains unchanged, the standard time will vary, as will the longitude. The same arguments as used here for equatorial crossings can be applied to any latitude.

The nodal regression given in Table 2.5 is the number of degrees rotated by the earth during one orbit of the satellite, and is approximately equal to 360° divided by the number of orbits per day. It clearly is different from the rate of nodal precession.

2.11   The Geostationary Orbit

The geostationary orbit is the orbit in which a satellite appears stationary relative to the earth. This is the most widely used of all orbits, for the very practical reason that an earth station antenna pointed at a geostationary satellite automatically follows it, and elaborate tracking systems are not required.

The geostationary orbit lies in the equatorial plane, meaning that the inclination is zero. This follows, since any finite inclination implies that there are ascending and decending nodes, which in turn means that the satellite would have to cross lines of latitude and by definition would not be geostationary. Another requirement is that the satellite must orbit the earth in the same direction as the earth spins, and at the same speed. Since the earth rotates at constant speed, the orbital speed must also be constant, and Kepler's second law therefore requires the orbit to be circular.

Kepler's third law may be used to find the required altitude of the geostationary orbit above the equator. As already deduced, the geostationary orbit lies in the earth's equatorial plane and is circular. Figure 2.16 shows the plan view where a_E is the equatorial radius of

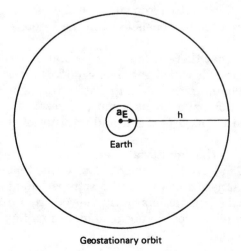

Geostationary orbit

Figure 2.16 The geostationary orbit is circular and lies in the earth's equatorial plane at an altitude $h = 35,786$ km above the equator.

the earth and h is the altitude of the orbit above the equator. Because the orbit is circular, $a = b = a_E + h$.

The periodic time P for the geostationary orbit, to the nearest minute, is 23 h, 56 min in mean solar time (ordinary clock time). This is the time taken for the earth to complete one revolution about its N-S axis, measured relative to the fixed stars (sidereal time as described in Sec. 2.9). It is 4 min shorter than the 24-h day or solar day, the difference being accounted for by the earth's movement around the sun. This is discussed further in Sec. 2.9.4, and illustrated in Fig. 2.7. In Sec. 2.9.4 the mean sidereal day is shown to be equal to 0.9972695664 mean solar days and the mean motion is $n = 2\pi/P$. From Sec. 2.4, Kepler's third law is given as $a^3 = \mu/n^2$ where $\mu = 3.986005 \times 10^{14}$ m³/s². Hence

$$a_{GSO} = \left(\frac{\mu P^2}{4\pi^2}\right)^{1/3}$$

$$= 42{,}164 \text{ km}$$

The *geostationary height* is therefore

$$h_{GSO} = a_{GSO} - a_E$$

$$= 42{,}164 - 6378$$

$$= 35{,}786 \text{ km}$$

The value of h_{GSO} is often rounded up to 36,000 km for approximate calculations. Recall that Kepler's third law assumes a perfectly spherical earth, and a more precise value for a_{GSO} can be obtained by allowing for the effects of a nonspherical earth as described in Sec. 2.8.1. In practice the exact value for h_{GSO} need not be known, as station-keeping maneuvers are carried out on a regular basis to keep the satellite on its assigned orbital position.

An important point to grasp is that there is only one geostationary orbit since there is only one value of a which satisfies Eq. (2.3) for a periodic time of 23 h, 56 min. Communications authorities throughout the world regard the geostationary orbit as a natural resource, and its use is carefully regulated through national and international agreements.

The geostationary orbit, being in the equatorial plane, has zero inclination. Although it is possible to have a geosynchronous orbit, that is, one which has the same orbital period as the earth's spin period, at some inclination i, this will not be geostationary. As viewed from a fixed location on earth, such a satellite would appear to move in a figure 8 pattern.

A precise geostationary orbit cannot be attained in practice because of disturbance forces in space and the effects of the earth's equatorial bulge. The gravitational fields of the sun and moon produce a shift of about 0.85°/year in inclination. Also, the earth's *equatorial ellipticity* causes the satellite to drift eastward along the orbit. In practice, station-keeping maneuvers have to be performed quite frequently to correct for these shifts; thus certain command and control earth stations must have tracking facilities. Typically, the satellite is maintained within ± 0.1° in both latitude and longitude. Station keeping is discussed in more detail in Sec. 6.4.

2.11.1   Antenna look angles

The user must be able to determine the azimuth and elevation angles (termed the *look angles*) of the ground station antenna. For large commercially operated ground stations, the look angle settings will be controlled by computer, but the owner of a home satellite receiving system will probably have to make such adjustments manually.

Figure 2.17 shows the geometry involved in determining the look angles. Here, a_E is the earth's equatorial radius, and R is the radius at the earth station. It will be recalled from Eqs. (2.42) and (2.43) that R takes into account the oblateness of the earth and the height H of the earth station above mean sea level. These equations are repeated here for convenience where λ_E is the geodetic latitude of the earth station:

$$l = \left(\frac{a_E}{\sqrt{1 - e_E^2 \sin^2 \lambda_E}} + H \right) \cos \lambda_E$$

$\sigma = 90° + E\,l$

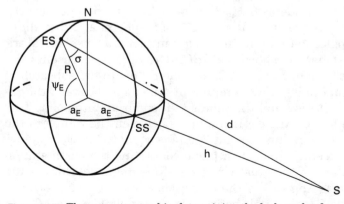

Figure 2.17 The geometry used in determining the look angles for a geostationary satellite.

$$z = \left(\frac{a_E(1 - e_E^2)}{\sqrt{1 - e_E^2 \sin^2 \lambda_E}} + H \right) \sin \lambda_E$$

The geocentric angle is given by Eq. (2.47) as

$$\psi_E = \arctan \frac{z}{l}$$

The magnitude of the earth station radius is

$$R = \sqrt{l^2 + z^2} \tag{2.60}$$

The earth station is denoted by ES. Point S denotes the satellite in geostationary orbit, and point SS the subsatellite point. Height h_{GSO} is the geostationary height, equal to 35,786 km, and distance d is the distance between earth station and satellite, referred to as the *range*.

As already mentioned in Sec. 2.9.7, care must be taken regarding the sign conventions used for latitude and longitude. As was done in Sec. 2.9.7, north latitudes will be taken as positive numbers and south latitudes as negative numbers, zero latitude of course being the equator. Longitudes east of the Greenwich meridian will be taken as positive numbers, and longitudes west as negative numbers.

The problem then is this. Given the earth station coordinates and the satellite longitude, what are the required azimuth and elevation angles for the earth station antenna? This requires the solution of the spherical triangle bounded by the points N, ES, and SS in Fig. 2.17. More correctly, since the earth is not spherical, a sphere of radius R must be considered as shown in Fig. 2.18. The equatorial great circle is denoted as 1-1', the great circle containing the earth station and sub-satellite points is shown as 2-2', the meridian great circle at the earth station longitude is shown as 3-ϕ_{ES}-3', and the meridian great circle at the subsatellite longitude as 3-ϕ_{SS}-3'. Now, spherical triangles are triangles which can be drawn on the surface of a sphere, and they differ from plane triangles in that only angles are involved. The spherical triangle of interest here is shown as ABC, abc. Side a represents the angle between the radius to point 3, and the radius to the subsatellite point, and is equal to 90°. Side b represents the angle between the radius to the earth station and the radius to the subsatellite point. This angle has to be found. Angle c is the angle between the radius to the earth station and the radius to point 3, and is equal to $90 - \psi_E$. Angle A is the angle between the meridian plane containing the earth station point and the plane of the great circle 2-2'. This angle determines the antenna azimuth. Angle B is the angle between the two meridian planes, and angle C is the angle between the great circle plane 2-2' and the meridian plane containing the subsatellite point.

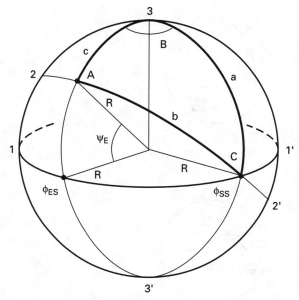

Figure 2.18 The spherical geometry related to Fig. 2.17.

The information which is known about the spherical triangle is

$$a = 90° \tag{2.61}$$

$$c = 90 - \psi_E \tag{2.62}$$

$$B = \phi_E - \phi_{SS} \tag{2.63}$$

It will be shown shortly that the maximum value of B is 81.3°, set by the earth station horizon, or what is termed the *limits of visibility*. The information which has to be determined from the spherical triangle is angles A and b. The particular spherical triangle shown is known as a *quadrantal triangle* because one of its sides is equal to 90° of arc. Special rules, known as *Napier's rules,* may be used to solve the triangle (see for example Wertz, 1984), and only the results will be summarized here. Angle A is given by

$$A = \arctan\left(\frac{-\tan|B|}{\sin \psi_E}\right) \tag{2.64}$$

Once angle A is determined, the azimuth angle Az can be found. Four situations must be considered, as shown in Fig. 2.19. The results are summarized in Table 2.6.

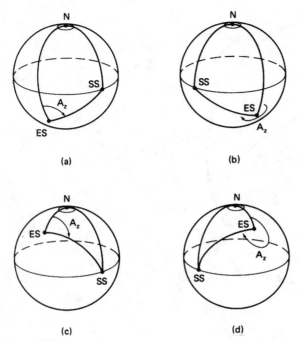

Figure 2.19 Azimuth angles A: (a) earth station in southern hemisphere, west of subsatellite point; (b) earth station in southern hemisphere, east of subsatellite point; (c) earth station in northern hemisphere, west of subsatellite point; (d) earth station in northern hemisphere, east of subsatellite point.

TABLE 2.6   Azimuth Angles from Fig. 2.19

Fig. 2.19	ψ_E	B	A_z degrees
a	< 0	< 0	A
b	< 0	> 0	360 − A
c	> 0	< 0	180 + A
d	> 0	> 0	180 − A

These results do not take into account the case when the earth station is on the equator. Obviously when the earth station is directly under the satellite, the elevation is 90° and azimuth is irrelevant. When the subsatellite point is east of the equatorial earth station (B < 0) the azimuth is 90°, and when west (B > 0), the azimuth is 270°.

In order to find the range and elevation it is necessary first to find angle b of the quadrantal spherical triangle, and then use this in the

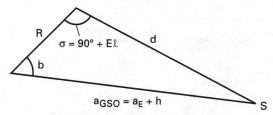

Figure 2.20 The plane triangle obtained from Fig. 2.17.

plane triangle shown in Fig. 2.17 and again in Fig. 2.20. Angle b is obtained by using the rule:

$$\cos b = \cos B \cos \psi_E \tag{2.65}$$

Once the antenna is rotated through the azimuth angle it will be pointing along the horizontal toward the subsatellite point. It must then be elevated by some angle El, in order to point it at the satellite. This accounts for the angle $90° + \text{El}$ shown in Fig. 2.17, and again in the plane triangle in Fig. 2.20. Applying the cosine rule (for plane triangles) gives the range d as

$$d = \sqrt{R^2 + a_{\text{GSO}}^2 - 2Ra_{\text{GSO}} \cos b} \tag{2.66}$$

where it will be recalled that $a_{\text{GSO}} = 42{,}164$ km.

The elevation angle El is obtained by applying the sine rule to the plane triangle, which gives, after some slight manipulation:

$$\text{El} = \arccos \left(\frac{a_{\text{GSO}}}{d} \sin b \right) \tag{2.67}$$

Figure 2.21 shows the look angles for Ku-band satellites as seen from Thunder Bay, Ontario, Canada. The calculations are illustrated in Example 2.22.

Example 2.22

Determine the look angles and the range for a geostationary satellite at 30 degrees, for an earth station at latitude −20 degrees, longitude −30 degrees. The earth station is situated 1000m above mean sea level.

solution

Given data

$$\lambda_E := -20 \cdot \deg \qquad \phi_E := -30 \cdot \deg \qquad \phi_{SS} := 30 \cdot \deg \qquad H := 1000 \cdot m$$

$$a_E := 6378.1414 \cdot km \qquad e_E := .08182 \qquad a_{\text{GSO}} := 42164.14 \cdot km$$

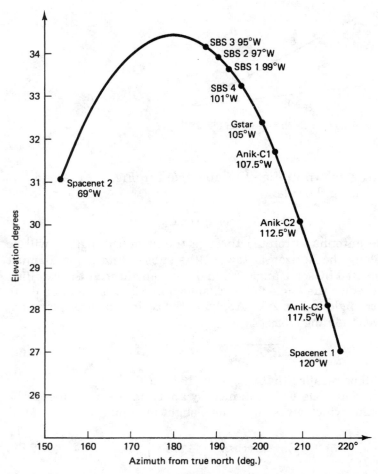

Figure 2.21 Azimuth-elevation angles for an earth-station location 48.42 deg N, 89.26 deg W (Thunder Bay, Ont.). Ku-band satellites are shown.

$$l := \left(\frac{a_E}{\sqrt{1 - e_E^2 \cdot \sin(\lambda_E)^2}} + H \right) \cdot \cos(\lambda_E) \qquad \text{...eq. (2.43)}$$

$$z := \left[\frac{a_E \cdot (1 - e_E^2)}{\sqrt{1 - e_E^2 \cdot \sin(\lambda_E)^2}} + H \right] \cdot \sin(\lambda_E) \qquad \text{...eq. (2.44)}$$

$$R := \sqrt{l^2 + z^2} \qquad \text{...eq. (2.60)}$$

$$\psi_E := \text{atan}\left(\frac{z}{l} \right) \qquad \text{...eq. (2.47)}$$

$$B := \phi_E - \phi_{SS} \qquad \qquad \text{...eq. (2.63)}$$

$$A := \text{atan}\left(\frac{-\tan(|B|)}{\sin(\psi_E)}\right) \qquad \qquad \text{...eq. (2.64)}$$

The conditions set out in Table 2.5 can be formulated as follows:

$$\text{Aza} := \text{if}[(\psi_E < 0){\cdot}(B < 0), \, A, \, 0]$$

$$\text{Azb} := \text{if}[(\psi_E < 0){\cdot}(B > 0), \, 2{\cdot}\pi - A, \, 0]$$

$$\text{Azc} := \text{if}[(\psi_E > 0){\cdot}(B < 0), \, \pi + A, \, 0]$$

$$\text{Azd} := \text{if}[(\psi_E > 0){\cdot}(B > 0), \, \pi - A, \, 0]$$

Since only one of these will be other than zero, the azimuth can be found by setting

$$Az := \text{Aza} + \text{Azb} + \text{Azc} + \text{Azd} \qquad Az = 78.9{\cdot}\text{deg}$$
$$\phantom{Az := \text{Aza} + \text{Azb} + \text{Azc} + \text{Azd} \qquad } \underline{\overline{========}}$$

$$b := \text{acos}(\cos(\psi_E){\cdot}\cos(B)) \qquad \qquad \text{...eq. (2.65)}$$

$$d := \sqrt{R^2 + a_{GSO}{}^2 - 2{\cdot}R{\cdot}a_{GSO}{\cdot}\cos(b)} \qquad \text{...eq. (2.66)} \qquad d = 39568{\cdot}\text{km}$$
$$\phantom{d := \sqrt{R^2 + a_{GSO}{}^2 - 2{\cdot}R{\cdot}a_{GSO}{\cdot}\cos(b)} \qquad \text{...eq. (2.66)} \qquad } \underline{\overline{========}}$$

$$El := \text{acos}\left(\frac{a_{GSO}}{d}{\cdot}\sin(b)\right) \qquad \qquad \text{...eq. (2.67)} \qquad El = 19.9{\cdot}\text{deg}$$
$$\phantom{El := \text{acos}\left(\frac{a_{GSO}}{d}{\cdot}\sin(b)\right) \qquad \qquad \text{...eq. (2.67)} \qquad } \underline{\overline{========}}$$

For a typical home installation, practical adjustments will be made to align the antenna to a known satellite for maximum signal. Thus the look angles need not be determined with great precision, but are calculated to give the expected values for a satellite whose longitude is close to the earth station longitude. In some cases, especially with direct broadcast satellites (DBS), the home antenna is aligned to one particular satellite and no further adjustments are necessary. In these situations it is permissible to assume a spherical earth of mean radius 6371 km and to ignore the earth station elevation. Also, the assumption that $\psi_E = \lambda_E$ is valid. Thus the steps leading to Eq. (2.60) can be ignored, and $R = 6371$ km substituted in Eqs. (2.66) and (2.67). Where ψ_E appears in Eqs. (2.64) and (2.65) (and in Table 2.5), it can be replaced by λ_E. These approximations simplify the calculations considerably without introducing any significant change in the look angles, as shown in Example 2.23.

Example 2.23

Repeat Example 2.22 using a spherical earth approximation of radius 6371 km, and ignore earth station altitude.

solution

Given data

$$\lambda_E := -20 \cdot deg \qquad \phi_E := -30 \cdot deg \qquad \phi_{SS} := 30 \cdot deg \qquad R := 6371 \cdot km$$

$$a_{GSO} := 42164 \cdot km$$

$$B := \phi_E - \phi_{SS} \qquad \qquad ...eq.\ (2.63)$$

$$A := atan\left(\frac{-tan(|B|)}{sin(\lambda_E)}\right) \qquad ...eq.\ (2.64)$$

From Example 2.22 it is known that Az = A and

$$A = 78.8 \cdot deg$$
$$========$$

$$b := acos\left(cos(\lambda_E) \cdot cos(B)\right) \qquad ...eq.\ (2.65)$$

$$d := \sqrt{R^2 + a_{GSO}^2 - 2 \cdot R \cdot a_{GSO} \cdot cos(b)} \qquad ...eq.\ (2.66) \qquad d = 39572.3 \cdot km$$
$$========$$

$$El := acos\left(\frac{a_{GSO}}{d} \cdot sin(b)\right) \qquad ...eq.\ (2.67) \qquad El = 19.9 \cdot deg$$
$$========$$

Thus it is seen that only the range d shows any appreciable change.

2.11.2   The polar mount antenna

Where the home antenna has to be steerable, expense usually precludes the use of separate azimuth and elevation actuators. Instead, a single actuator is used which moves the antenna in a circular arc. This is known as a *polar mount antenna*. The antenna pointing can only be accurate for one satellite, and some pointing error must be accepted for satellites on either side of this. With the polar mount antenna, the dish is mounted on an axis termed the *polar axis,* such that the antenna boresight is normal to this axis, as shown in Fig. 2.22a. The polar mount is aligned along a true north line as shown in Fig. 2.22 with the boresight pointing due south. The angle between the polar mount and the local horizontal plane is set equal to the earth station latitude λ_E; simple geometry shows that this makes the boresight lie parallel to the equatorial plane. Next the dish is tilted at an angle δ relative to the polar mount until the boresight is pointing at a satellite position due south of the earth station. Note that there does not need to be an actual satellite at this position. (The angle of tilt is often referred to as the *declination* which must not be confused with the magnetic declination used in correcting compass readings. The term *angle of tilt* will be used for δ in this text.)

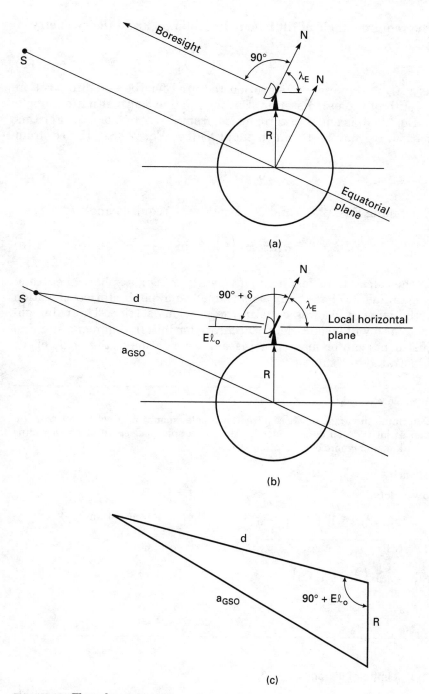

Figure 2.22 The polar mount antenna.

The required angle of tilt is found as follows. From the geometry of Fig. 22.2b,

$$\delta = 90° - El_0 - \lambda_E \tag{2.68}$$

where El_0 is the angle of elevation required for the satellite position due south of the earth station. But for the due south situation, angle B in Eq. (2.63) is equal to zero, hence, from Eq. (2.65), $b = \psi_E$. For this situation, ψ_E may be taken to equal λ_E [see Eq. (2.48)]. Hence, from Eq. (2.67)

$$\cos El_0 = \frac{a_{GSO}}{d} \sin \lambda_E \tag{2.69}$$

Combining Eqs. (2.68) and (2.69) gives the required angle of tilt as

$$\delta = 90° - \arccos\left(\frac{a_{GSO}}{d} \sin \lambda_E\right) - \lambda_E \tag{2.70}$$

In the calculations leading to d (see Fig. 2.22c), a spherical earth of mean radius 6371 km may be assumed and earth station elevation may be ignored, as was done in the previous section. The value obtained for δ will be sufficiently accurate for initial alignment and fine adjustments can be made if necessary. Calculation of the angle of tilt is illustrated in Example 2.24.

Example 2.24

Determine the angle of tilt required for a polar mount used with an earth station at latitude 49 degrees north. Assume a spherical earth of mean radius 6371 km, and ignore earth station altitude.

solution

Given data

$$\lambda_E := 49 \cdot deg \qquad a_{GSO} := 42164 \cdot km \qquad R := 6371 \cdot km$$

$$d := \sqrt{R^2 + a_{GSO}^2 - 2 \cdot R \cdot a_{GSO} \cdot \cos(\lambda_E)} \qquad \text{...eq. (2.66)}$$

$$El_0 := a\cos\left(\frac{a_{GSO}}{d} \cdot \sin(\lambda_E)\right) \qquad \text{...eq. (2.67)}$$

$$\delta := 90 \cdot deg - El_0 - \lambda_E \qquad\qquad \delta = 7 \cdot deg$$

$$========$$

2.11.3 Limits of visibility

There will be east and west limits on the geostationary arc visible from any given earth station. The limits will be set by the geographic coordinates of the earth station and the antenna elevation. The low-

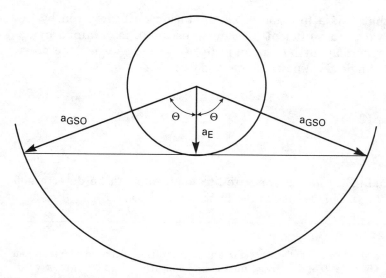

Figure 2.23 Illustrating the limits of visibility.

est elevation in theory is zero, when the antenna is pointing along the horizontal. A quick estimate of the longitudinal limits can be made by considering an earth station at the equator, with the antenna pointing either west or east along the horizontal as shown in Fig. 2.23. The limiting angle is given by:

$$\theta = \arccos \frac{a_E}{a_{GSO}}$$

$$= \arccos \frac{6378}{42{,}160} \tag{2.71}$$

$$= 81.3°$$

Thus, for this situation an earth station could see satellites over a geostationary arc bounded by $\pm 81.3°$ about the earth station longitude.

In practice, to avoid reception of excessive noise from the earth, some finite minimum value of elevation is used, which will be denoted here by El_{min}. A typical value is 5°. The limits of visibility will also depend on the earth station latitude. As in Fig. 2.20, let S represent the angle subtended at the satellite when the angle $\sigma_{min} = 90° + El_{min}$. Applying the sine rule gives

$$S = \arcsin\left(\frac{R}{a_{GSO}} \sin \sigma_{min}\right) \tag{2.72}$$

Recall that R is a function of earth station latitude given by Eq. (2.60). However, a sufficiently accurate estimate is obtained by assuming a spherical earth of mean radius 6371 km as was done previously. Once angle S is known, angle b is found from

$$b = 180 - \sigma_{min} - S \qquad (2.73)$$

From Eq. (2.65)

$$B = \arccos\left(\frac{\cos b}{\cos \lambda_E}\right) \qquad (2.74)$$

Once angle B is found, the satellite longitude can be determined from Eq. (2.63). This is illustrated in Example 2.25.

Example 2.25

Determine the limits of visibility for an earth station situated at mean sea level, at latitude 48.42 degrees north, and longitude 89.26 degrees west. Assume a minimum angle of elevation of 5 degrees.

solution

Given data

$$\lambda_E := 48.42 \cdot deg \qquad \phi_E := -89.26 \cdot deg \qquad El_{min} := 5 \cdot deg \qquad a_{GSO} := 42164 \cdot km$$
$$R := 6371 \cdot km$$

$$\sigma_{min} := 90 \cdot deg + El_{min}$$

$$S := asin\left(\frac{R}{a_{GSO}} \cdot \sin(\sigma_{min})\right) \qquad \text{...eq. (2.72)}$$

$$b := 180 \cdot deg - \sigma_{min} - S \qquad \text{...eq. (2.73)}$$

$$B := acos\left(\frac{\cos(b)}{\cos(\lambda_E)}\right) \qquad \text{...eq. (2.74)}$$

The satellite limit east of the earth station is at:

$$\phi_E + B = -20 \cdot deg$$

and west of the earth station at:

$$\phi_E - B = -158 \cdot deg$$
$$= = = = = = = = =$$

2.11.4 Earth eclipse of satellite

If the earth's equatorial plane coincided with the plane of the earth's orbit around the sun (the ecliptic plane), geostationary satellites

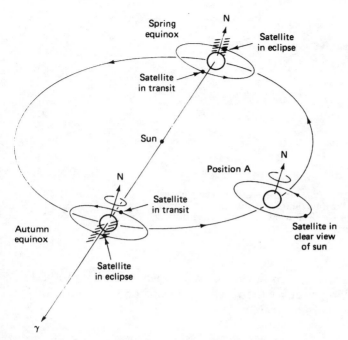

Figure 2.24 Showing satellite eclipse and satellite sun-transit around spring and autumn equinoxes.

would be eclipsed by the earth once each day. As it is, the equatorial plane is tilted at an angle of 23.4° to the ecliptic plane, and this keeps the satellite in full view of the sun for most days of the year, as illustrated by position A, Fig. 2.24. Around the spring and autumnal equinoxes, when the sun is crossing the equator, the satellite does pass into the earth's shadow at certain periods, these being periods of eclipse as illustrated in Fig. 2.24. The spring equinox is the first day of spring, and the autumnal equinox the first day of autumn.

Eclipses begin 23 days before equinox and end 23 days after equinox. The eclipse lasts about 10 min at the beginning and end of the eclipse period, and increases to a maximum duration of about 72 min at full eclipse (Spilker, 1977). During an eclipse the solar cells do not function and operating power must be supplied from batteries. This is discussed further in Sec. 6.2, and Fig. 6.3 shows eclipse time as a function of days of the year.

Where the satellite longitude is east of the earth station, the satellite enters eclipse during daylight (and early evening) hours for the earth station, as illustrated in Fig. 2.25. This can be undesirable if the satellite has to operate on reduced battery power. Where the satellite longitude is west of the earth station, eclipse does not occur

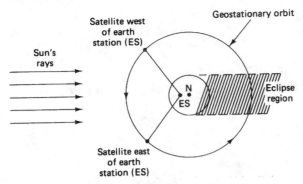

Figure 2.25   A satellite east of the earth station enters eclipse during daylight (busy) hours at the earth station. A satellite west of the earth station enters eclipse during night and early morning (nonbusy) hours.

until the earth station is in darkness, when usage is likely to be low. Thus satellite longitudes which are west, rather than east, of the earth station are more desirable.

2.11.5   Sun transit outage

Another event which must be allowed for during the equinoxes is the transit of the satellite between earth and sun (see Fig. 2.24), such that the sun comes within the beamwidth of the earth station antenna. When this happens, the sun appears as an extremely noisy source which completely blanks out the signal from the satellite. This effect is termed *sun transit outage*, and it lasts for short periods each day for about 6 days around the equinoxes. The occurrence and duration of the sun transit outage depends on the latitude of the earth station, a maximum outage time of 10 min being typical.

2.12   Launching Orbits

Satellites may be *directly injected* into low-altitude orbits, up to about 200 km altitude, from a launch vehicle. Launch vehicles may be classified as *expendable* or *reusable*. Typical of the expendable launchers are the U.S. Atlas-Centaur and Delta rockets and the European Space Agency Ariane rocket. Japan, China, and Russia all have their own expendable launch vehicles, and one may expect to see competition for commercial launches among the countries which have these facilities.

Until the tragic mishap with the Space Shuttle in 1986, this was to be the primary transportation system for the United States. As a reusable launch vehicle, the shuttle, also referred to as the Space

Transportation System (STS), was planned to eventually replace expendable launch vehicles for the United States (Mahon and Wild, 1984).

Where an orbital altitude greater than about 200 km is required, it is not economical in terms of launch vehicle power to perform direct injection, and the satellite must be placed into transfer orbit between the initial low earth orbit and the final high-altitude orbit. In most cases, the transfer orbit is selected to minimize the energy required for transfer, and such an orbit is known as a *Hohmann transfer* orbit. The time required for transfer is longer for this orbit than all other possible transfer orbits.

Assume for the moment that all orbits are in the same plane and that transfer is required between two circular orbits as illustrated in Fig. 2.26. The Hohmann elliptical orbit is seen to be tangent to the low-altitude orbit at perigee and to the high-altitude orbit at apogee. At the perigee, in the case of rocket launch, the rocket injects the satellite with the required thrust into the transfer orbit. With the STS, the satellite must carry a perigee kick motor which imparts the required thrust at perigee. Details of the expendable vehicle launch are shown in Fig. 2.27 and of the STS launch, in Fig. 2.28. At apogee, the apogee kick motor (AKM) changes the velocity of the satellite to place it into a circular orbit in the same plane. As shown in Fig. 2.27, it takes 1 to 2 months for the satellite to be fully operational (although not shown in Fig. 2.28, the same conditions apply). Throughout the launch and acquisition phases a network of ground stations, spread across the earth, is required to perform the tracking, telemetry, and command (TT&C) functions.

Velocity changes in the same plane change the geometry of the orbit but not its inclination. In order to change the inclination, a velocity

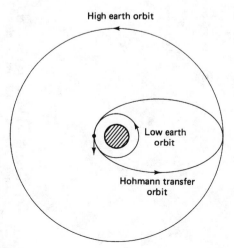

Figure 2.26 Hohmann transfer orbit.

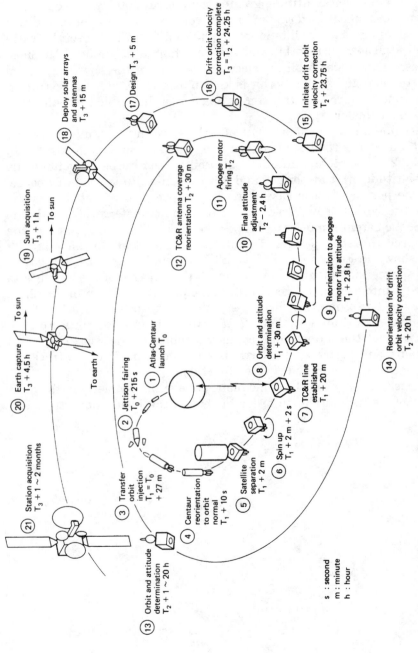

Figure 2.27 From launch to station of Intelsat V (by Atlas/Centaur). (© *KDD Engineering & Consulting Inc., Tokyo. From "Satellite Communications Technology," edited by K. Miya, 1981.*)

① Atlas-Centaur launch T_0

② Jettison fairing $T_0 + 215$ s

③ Transfer orbit injection $T_1 = T_0 + 27$ m

④ Centaur reorientation to orbit normal $T_1 + 10$ s

⑤ Satellite separation $T_1 + 2$ m

⑥ Spin up $T_1 + 2$ m + 2 s

⑦ TC&R line established $T_1 + 20$ m

⑧ Orbit and attitude determination $T_1 + 30$ m

⑨ Reorientation to apogee motor fire attitude $T_1 + 2.8$ h

⑩ Final attitude adjustment $T_2 - 2.4$ h

⑪ Apogee motor firing T_2

⑫ TC&R antenna coverage reorientation $T_2 + 30$ m

⑬ Orbit and attitude determination $T_2 + 1 \sim 20$ h

⑭ Reorientation for drift orbit velocity correction $T_2 + 20$ h

⑮ Initiate drift orbit velocity correction $T_2 + 23.75$ h

⑯ Drift orbit velocity correction complete $T_3 = T_2 + 24.25$ h

⑰ Design $T_3 + 5$ m

⑱ Deploy solar arrays and antennas $T_3 + 15$ m

⑲ Sun acquisition $T_3 + 1$ h

⑳ Earth capture $T_3 + 4.5$ h

㉑ Station acquisition $T_3 + 1 \sim 2$ months

To sun

To earth

s : second
m : minute
h : hour

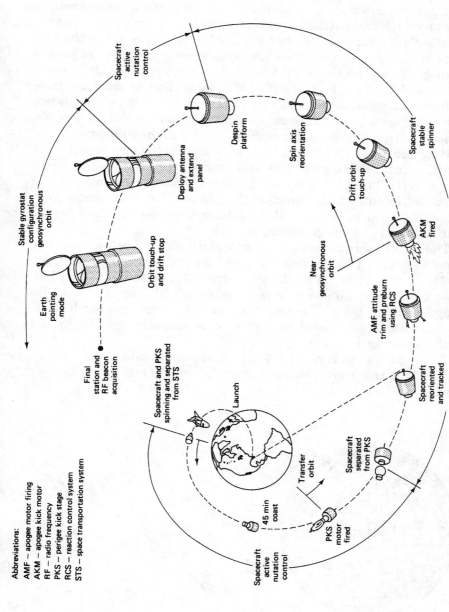

Abbreviations:

AMF — apogee motor firing
AKM — apogee kick motor
RF — radio frequency
PKS — perigee kick stage
RCS — reaction control system
STS — space transportation system

Spacecraft active nutation control

Stable gyrostat configuration geosynchronous orbit

Earth pointing mode

Deploy antenna and extend panel

Despin platform

Spin axis reorientation

Spacecraft stable spinner

Orbit touch-up and drift stop

Drift orbit touch-up

AKM fired

Near geosynchronous orbit

AMF attitude trim and preburn using RCS

Final station and RF beacon acquisition

Spacecraft and PKS spinning and separated from STS

Launch

Spacecraft reoriented and tracked

Transfer orbit

Spacecraft separated from PKS

45 min coast

PKS motor fired

Spacecraft active nutation control

Figure 2.28 STS-7/Anik C2 mission scenario. (*From "Anik C2 Launch Handbook," courtesy Telesat Canada.*)

change is required normal to the orbital plane. Changes in inclination can be made at either one of the nodes, without affecting the other orbital parameters. Since energy must be expended to make any orbital changes, a geostationary satellite should be initially launched with as low an orbital inclination as possible. It will be shown shortly that the smallest inclination obtainable at initial launch is equal to the latitude of the launch site. Thus the farther away from the equator a launch site is, the less useful it is, since the satellite has to carry extra fuel to effect a change in inclination. Russia does not have launch sites south of 45° N, which makes the launching of geostationary satellites a much more expensive operation for Russia than for other countries which have launch sites closer to the equator.

Prograde (direct) orbits, Fig. 2.4, have an easterly component of velocity, and these launches gain from the earth's rotational velocity. For a given launcher size, a significantly larger payload can be launched in an easterly direction than is possible with a retrograde (westerly) launch. In particular, easterly launches are used for the initial launch into the geostationary orbit.

The relationship between inclination, latitude, and azimuth may be seen as follows (this analysis is based on that given in Bate et al., 1971). Figure 2.29a shows the geometry at the launch site A at latitude λ (the slight difference between geodetic and geocentric latitudes may be ignored here). The dotted line shows the satellite earth track, the satellite having been launched at some azimuth angle Az. Angle i is the resulting inclination.

The spherical triangle of interest is shown in more detail in Fig. 2.29b. This is a right spherical triangle, and Napier's rule for this gives

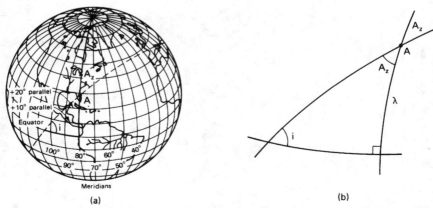

Meridians

(a)   (b)

Figure 2.29 (a) Launch site A, showing launch azimuth (Az); (b) enlarged version of the spherical triangle shown in (a). λ is the latitude of the launch site.

$$\cos i = \cos \lambda \sin Az \qquad (2.75)$$

For a prograde orbit (see Fig. 2.4 and Sec. 2.5) $0 \le i < 90°$ and hence $\cos i$ is positive. Also $-90° \le \lambda \le 90°$, and hence $\cos \lambda$ is also positive. It follows therefore from Eq. (2.75) that $0 \le Az \le 180°$, or the launch azimuth must be easterly in order to obtain a prograde orbit, confirming what was already known.

For a fixed λ, Eq. (2.75) also shows that to minimize the inclination i, $\cos i$ should be a maximum, which requires $\sin Az$ to be maximum, or $Az = 90°$. Equation (2.75) shows that under these conditions

$$\cos i_{min} = \cos \lambda \qquad (2.76)$$

or

$$i_{min} = \lambda \qquad (2.77)$$

Thus, the *lowest* inclination possible on initial launch is equal to the latitude of the launch site. This result confirms the converse statement made in Sec. 2.5 under *inclination* that the greatest latitude north or south is equal to the inclination. From Cape Kennedy the smallest initial inclination which can be achieved for easterly launches is approximately 28°.

2.13 Problems

2.1. State Kepler's three laws of planetary motion. Illustrate in each case their relevance to artificial satellites orbiting the earth.

2.2. Using the results of App. B show that for any point P, the sum of the focal distances to S and S' is equal to $2a$.

2.3. Show that for the ellipse the differential element of area $dA = r^2 \, dv/2$ where dv is the differential of the true anomaly. Using Kepler's second law, show that the ratio of the speeds at apoapsis and periapsis (or apogee and perigee for an earth-orbiting satellite) is equal to $(1 - e)/(1 + e)$.

2.4. A satellite orbit has an eccentricity of 0.2 and a semimajor axis of 10,000 km. Find the values of (a) the latus rectum; (b) the minor axis; (c) the distance between foci.

2.5. For the satellite in Prob. 2.4, find the length of the position vector when the true anomaly is 130°.

2.6. The orbit for an earth-orbiting satellite orbit has an eccentricity of 0.15 and a semimajor axis of 9000 km. Determine (a) its periodic time; (b) the apogee height; (c) the perigee height. Assume a mean value of 6371 km for the earth's radius.

2.7. For the satellite in Prob. 2.6, at a given observation time during a south to north transit the height above ground is measured as 2000 km. Find the corresponding true anomaly.

2.8. The semimajor axis for the orbit of an earth-orbiting satellite is found to be 9500 km. Determine the mean anomaly 10 min after passage of perigee.

2.9. The following conversion factors are exact: one foot = 0.3048 meters; one statute mile = 1609.344 meters; one nautical mile = 1852 meters. A satellite travels in an unperturbed circular orbit of semimajor axis $a = 27,000$ km. Determine its tangential speed in (a) km/sec, (b) mi/hr, and (c) knots.

2.10. Explain what is meant by *apogee height* and *perigee height*. The Cosmos 1675 satellite has an apogee height of 39,342 km and a perigee height of 613 km. Determine the semimajor axis and the eccentricity of its orbit. Assume a mean earth radius of 6371 km.

2.11. The Aussat 1 satellite in geostationary orbit has an apogee height of 35,795 km and a perigee height of 35,779 km. Assuming a value of 6378 km for the earth's equatorial radius, determine the semimajor axis and the eccentricity of the satellite's orbit.

2.12. Explain what is meant by the ascending and descending nodes. In what units would these be measured, and, in general, would you expect them to change with time?

2.13. Explain what is meant by (a) line of apsides and (b) line of nodes. Is it possible for these two lines to be coincident?

2.14. With the aid of a neat sketch, explain what is meant by each of the angles: *inclination*; *argument of perigee*; *right ascension of the ascending node*. Which of these angles would you expect, in general, to change with time?

2.15. The inclination of an orbit is 67°. What is the greatest latitude, north and south, reached by the subsatellite point? Is this orbit retrograde or prograde?

2.16. Describe briefly the main effects of the earth's equatorial bulge on a satellite orbit. Given that a satellite is in a circular equatorial orbit for which the semimajor axis is equal to 42,165 km, calculate (a) the mean motion, (b) the rate of regression of the nodes, and (c) the rate of rotation of argument of perigee.

2.17. A satellite in polar orbit has a perigee height of 600 km and an apogee height of 1200 km. Calculate (a) the mean motion, (b) the rate of regression of the nodes, and (c) the rate of rotation of the line of apsides. The polar radius of the earth may be assumed equal to 6357 km.

2.18. What is the fundamental unit of universal coordinated time? Express the following universal times in (a) days and (b) degrees: 0 h, 5 min, 24 s; 6 h, 35 min, 20 s; your present time.

2.19. Determine the Julian days for the following dates and times: 0300 h, January 3, 1986; midnight March 10, 1987; noon, February 23, 1988; 1630 h, March 1,1988.

2.20. Find, for the times and dates given in Prob. 2.19, (a) T in Julian centuries and (b) the corresponding Greenwich sidereal time.

2.21. Find the month, day, and UT for the following Julian dates: (a) day

3.00, year 1976; (b) day 186.125, year 1976; (c) day 300.12157650, year 1986; (d) day 3.29441845, year 1987; (e) day 31.1015, year 2000.

2.22. Find the Greenwich sidereal time (GST) corresponding to the Julian dates given in Prob. 2.21.

2.23. The Molnya 3-(25) satellite has the following parameters specified: perigee height 462 km; apogee height 40,850 km; period 736 min; inclination 62.8°. Using an average value of 6371 km for the earth's radius, calculate (a) the semimajor axis and (b) the eccentricity. (c) Calculate the nominal mean motion n_0. (d) Calculate the mean motion. (e) Using the calculated value for a, calculate the anomalistic period and compare with the specified value, (f) the rate of regression of the nodes, and (g) the rate of rotation of the line of apsides.

2.24. Repeat the calculations in Prob. 2.23 for an inclination of 63.435°.

2.25. Determine the orbital condition necessary for the argument of perigee to remain stationary in the orbital plane. The orbit for a satellite under this condition has an eccentricity of 0.001 and a semimajor axis of 27,000 km. At a given epoch the perigee is exactly on the line of Aries. Determine the satellite position relative to this line after a period of 30 days from epoch.

2.26. For a given orbit, K as defined by Eq. (2.11) is equal to 0.112 rev/day. Determine the value of inclination required to make the orbit sun-synchronous.

2.27. A satellite has an inclination of 90° and an eccentricity of 0.1. At epoch, which corresponds to time of perigee passage, the perigee height is 2643.24 km directly over the north pole. Determine (a) the satellite mean motion. For 1 day after epoch determine (b) the true anomaly, (c) the magnitude of the radius vector to the satellite, and (d) the latitude of the subsatellite point.

2.28. The following elements apply to a satellite in inclined orbit: $\Omega_0 = 0°$; $\omega_0 = 90°$; $M_0 = 309°$; $i = 63°$; $e = 0.01$; $a = 7130$ km. An earth station is situated at 45° N, 80° W, and at zero height above sea level. Assuming a perfectly spherical earth of uniform mass and radius 6371 km, and given that epoch corresponds to a GST of 116°, determine at epoch the orbital radius vector in the (a) **PQW** frame; (b) **IJK** frame; (c) the position vector of the earth station in the **IJK** frame; (d) the range vector in the **IJK** frame; (e) the range vector in the **SEZ** frame; and (f) the earth station look angles.

2.29. A satellite moves in an inclined elliptical orbit, the inclination being 63.45°. State with explanation the maximum northern and southern latitudes reached by the subsatellite point. The nominal mean motion of the satellite is 14 rev/day, and at epoch the subsatellite point is on the ascending node at 100° W. Calculate the longitude of the subsatellite point 1 day after epoch. The eccentricity is 0.01.

2.30. A "no-name" satellite has the following parameters specified: perigee height 197 km; apogee height 340 km; period 88.2 min; inclination 64.6°. Repeat the calculations in Prob. 2.23 for this satellite.

2.31. Given that $\Omega_0 = 250°$, $\omega_0 = 85°$, and $M_0 = 30°$ for the satellite in Prob.

2.30, calculate, for 65 min after epoch ($t_0 = 0$) the new values of Ω, ω, and M. Find also the true anomaly and radius.

2.32. From the NASA bulletin given in App. C, determine the date and the semimajor axis.

2.33. Determine, for the satellite listed in the NASA bulletin of App. C, the rate of regression of the nodes, the rate of change of the argument of perigee, and the nominal mean motion n_0.

2.34. From the NASA bulletin in App. C, verify that the orbital elements specified in Part I are for a nominal S-N equator crossing, and verify from this the times of the succeeding S-N equator crossings given in Part II.

2.35. A satellite in exactly polar orbit has a slight eccentricity (just sufficient to establish the idea of a perigee). The anomalistic period is 110 min. Assuming that the mean motion is $n = n_0$ calculate the semimajor axis. Given that at epoch the perigee is exactly over the north pole, determine the position of the perigee relative to the north pole after one anomalistic period and the time taken for the satellite to make one complete revolution relative to the north pole.

2.36. A satellite is in an exactly polar orbit with apogee height 7000 km and perigee height 600 km. Assuming a spherical earth of uniform mass and radius 6371 km, calculate (*a*) the semimajor axis, (*b*) the eccentricity, and (*c*) the orbital period. (*d*) At a certain time the satellite is observed ascending directly overhead from an earth station on latitude 49° N. Given that the argument of perigee is 295° calculate the true anomaly at the time of observation.

2.37. For the satellite elements shown in Fig. 2.6, determine approximate values for the latitude and longitude of the subsatellite point at epoch.

2.38. Explain what is meant by the geostationary orbit. How do the geostationary orbit and a geosynchronous orbit differ?

2.39. (*a*) Explain why there is only one geostationary orbit. (*b*) Show that the range d from an earth station to a geostationary satellite is given by $d = \sqrt{(R \sin \mathrm{El})^2 + h(2R + h)} - R \sin \mathrm{El}$ where R is the earth's radius (assumed spherical), h is the height of the geostationary orbit above the equator, and El is the elevation angle of the earth station antenna.

2.40. Determine the latitude and longitude of the farthest north earth station which can link with any given geostationary satellite. The longitude should be given relative to the satellite longitude, and a minimum elevation angle of 5° should be assumed for the earth station antenna. A spherical earth of mean radius 6371 km may be assumed.

2.41. An earth station at latitude 30° S is in communication with an earth station on the same longitude at 30° N, through a geostationary satellite. The satellite longitude is 20° east of the earth stations. Calculate the antenna look angles for each earth station and the round-trip time, assuming this consists of propagation delay only.

2.42. Determine the maximum possible longitudinal separation which can exist between a geostationary satellite and an earth station while maintaining line-of-sight communications, assuming the minimum angle of elevation of the earth station antenna is 5°. State also the latitude of the earth station.

2.43. An earth station is located at latitude 35° N and longitude 100° W. Calculate the antenna look angles for a satellite at 67° W.

2.44. An earth station is located at latitude 12° S and longitude 52° W. Calculate the antenna look angles for a satellite at 70° W.

2.45. An earth station is located at latitude 35° N and longitude 65° E. Calculate the antenna look angles for a satellite at 19° E.

2.46. An earth station is located at latitude 30° S and longitude 130° E. Calculate the antenna look angles for a satellite at 156° E.

2.47. Calculate for your home location the look angles required to receive from the satellite (a) immediately east and (b) immediately west of your longitude.

2.48. CONUS is the acronym used for the 48 contiguous states. Allowing for a 5° elevation angle at earth stations, verify that the geostationary arc required to cover CONUS is 55–136° W.

2.49. Referring to Prob. 2.48, verify that the geostationary arc required for CONUS plus Hawaii is 85–136° W and for CONUS plus Alaska is 115–136° W.

2.50. By taking the Mississippi River as the dividing line between east and west, verify that the western region of the United States would be covered by satellites in the geostationary arc from 136–163° W and the eastern region by 25–55° W. Assume a 5° angle of elevation.

2.51. (a) An earth station is located at latitude 35° N. Assuming a polar mount antenna is used, calculate the angle of tilt. (b) Would the result of Prob. 2.51 apply to polar mounts used at the earth stations specified in Probs. 2.43 and 2.45?

2.52. Repeat Prob. 2.51 (a) for an earth station located at latitude 12° S. Would the result of Prob. 2.52 apply to a polar mount used at the earth station specified in Prob. 2.44?

2.53. Repeat Prob. 2.51 (a) for an earth station located at latitude 30° S. Would the result of Prob. 2.53 apply to a polar mount used at the earth station specified in Prob. 2.46?

2.54. Calculate the angle of tilt required for a polar mount antenna used at your home location.

2.55. The borders of a certain country can be roughly represented by a triangle with coordinates 39° E, 33.5° N; 43.5° E, 37.5° N; 48.5° E, 30° N. If a geostationary satellite has to be visible from *any point* in the country determine the limits of visibility (i.e., the limiting longitudinal positions for a satellite on the geostationary arc). Assume a minimum angle of elevation for the earth station antenna of 5°, and show which geographic location fixes which limit.

2.56. Explain what is meant by the *earth eclipse* of an earth-orbiting satellite. Why is it preferable to operate with a satellite positioned west, rather than east, of earth station longitude?

2.57. Explain briefly what is meant by *sun transit outage*.

Radio Wave Propagation

3.1 Introduction

A signal traveling between an earth station and a satellite must pass through the earth's atmosphere, including the ionosphere, as shown in Fig. 3.1, and this can introduce certain impairments, which are summarized in Table 3.1.

Some of the more important of these impairments will be described in this chapter.

3.2 Atmospheric Losses

Losses occur in the earth's atmosphere as a result of energy absorption by the atmospheric gases. These losses are treated quite separately from those which result from adverse weather conditions, which of course are also atmospheric losses. To distinguish between these, the weather related losses are referred to as *atmospheric attenuation* and the absorption losses simply as *atmospheric absorption*.

The atmospheric absorption loss varies with frequency as shown in Fig. 3.2. The figure is based on statistical data (CCIR Report 719-1, 1982). Two absorption peaks will be observed, the first one at a frequency of 22.3 GHz, resulting from resonance absorption in water vapor (H_2O), and the second one at 60 GHz, resulting from resonance absorption in oxygen (O_2). However, at frequencies well clear of these peaks the absorption is quite low. The graph in Fig. 3.2 is for vertical incidence, that is, for an elevation angle of 90° at the earth station antenna. Denoting this value of absorption loss as $[AA]_{90}$ decibels, then for elevation angles down to 10°, an approximate formula for the absorption loss is (CCIR Report 719-1, 1982)

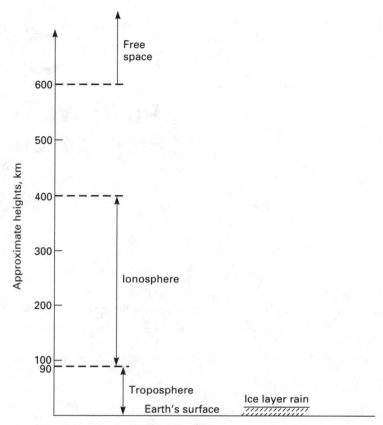

Figure 3.1 Layers in the earth's atmosphere.

$$[AA] = [AA]_{90} \, \text{cosec} \, \theta \qquad (3.1)$$

where θ is the angle of elevation. An effect known as *atmospheric scintillation* can also occur. This is a fading phenomenon, the fading period being several tens of seconds (Miya, 1981). It is caused by differences in the atmospheric refractive index, which in turn results in focusing and defocusing of the radio waves, which follow different ray paths through the atmosphere. It may be necessary to make an allowance for atmospheric scintillation, through the introduction of a fade margin in the link power-budget calculations.

3.3 Ionospheric Effects

Radio waves traveling between satellites and earth stations must pass through the ionosphere. The ionosphere is the upper region of the

TABLE 3.1   Propagation Concerns for Satellite Communications Systems

Propagation impairment	Physical cause	Prime importance
Attenuation and sky noise increases	Atmospheric gases, cloud, rain	Frequencies above about 10 GHz
Signal depolarization	Rain, ice crystals	Dual-polarization systems at C and Ku bands (depends on system configuration)
Refraction, atmospheric multipath	Atmospheric gases	Communication and tracking at low elevation angles
Signal scintillations	Tropospheric and ionospheric refractivity fluctuations	Tropospheric at frequencies above 10 GHz and low elevation angles; ionospheric at frequencies below 10 GHz
Reflection multipath, blockage	Earth's surface, objects on surface	Mobile satellite services
Propagation delays, variations	Troposphere, ionosphere	Precise timing and location systems; time-division multiple access (TDMA) systems
Intersystem interference	Ducting, scatter, diffraction	Mainly C band at present; rain scatter may be significant at higher frequencies

SOURCE: Brussard and Rogers, 1990.

earth's atmosphere, which has been ionized, mainly by solar radiation. The free electrons in the ionosphere are not uniformly distributed but form in layers. Furthermore, clouds of electrons (known as *traveling ionospheric disturbances*) may travel through the ionosphere and give rise to fluctuations in the signal that can only be determined on a statistical basis. The effects include *scintillation, absorption, variation in the direction of arrival, propagation delay, dispersion, frequency change,* and *polarization rotation* (CCIR Report 263-5, 1982). All of these effects decrease as frequency increases, most in inverse proportion to the frequency squared, and only the polarization rotation and the scintillation effects are of major concern for satellite communications. Polarization rotation is described in Sec. 4.5.

Ionospheric scintillations are variations in the amplitude, phase, polarization, or angle of arrival of radio waves. They are caused by irregularities in the ionosphere which change with time. The main effect of scintillations is fading of the signal. The fades can be quite severe, and they may last up to several minutes. As with fading caused by atmospheric scintillations, it may be necessary to include a fade margin in the link power-budget calculations to allow for ionospheric scintillation.

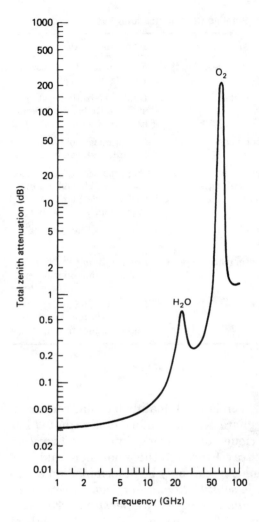

Figure 3.2 Total zenith attenuation at ground level: pressure = 1 atm, temperature = 20°C, and water vapor = 7.5 g/m^3 (*Adapted from CCIR Report 719-2. With permission from International Telecommunication Union.*)

3.4   Rain Attenuation

Rain attenuation is a function of *rain rate*. By rain rate is meant the rate at which rainwater would accumulate in a rain gauge situated at the ground in the region of interest (for example at an earth station). In calculations relating to radio wave attenuation, the rain rate is measured in millimeters per hour. Of interest is the percentage of time that specified values are exceeded. The time percentage is usually that of a year; for example, a rain rate of 0.001 percent means that the rain rate would be exceeded for 0.001 percent of a year, or about 5.4 min during any one year. In this case the rain rate would be de-

TABLE 3.2   Specific Attenuation Coefficients

Frequency, GHz	a_h	a_v	b_h	b_v
1	0.0000387	0.0000352	0.912	0.88
2	0.000154	0.000138	0.963	0.923
4	0.00065	0.000591	1.121	1.075
6	0.00175	0.00155	1.308	1.265
7	0.00301	0.00265	1.332	1.312
8	0.00454	0.00395	1.327	1.31
10	0.0101	0.00887	1.276	1.264
12	0.0188	0.0168	1.217	1.2
15	0.0367	0.0335	1.154	1.128
20	0.0751	0.0691	1.099	1.065
25	0.124	0.113	1.061	1.03
30	0.187	0.167	1.021	1

SOURCE: Ippolito, 1986, p. 46.

noted by $R_{.001}$. In general, the percentage time is denoted by p and the rain rate by R_p. The *specific attenuation* α is

$$\alpha = aR_p^b \quad \text{dB/km} \tag{3.2}$$

where a and b depend on frequency and polarization. Values for a and b are available in tabular form in a number of publications. The values in Table 3.2 have been abstracted from Table 4-3 of Ippolito, 1986. The subscripts h and v refer to horizontal and vertical polarizations respectively.

Once the specific attenuation is found, the total attenuation is determined as

$$A = \alpha L \quad \text{dB} \tag{3.3}$$

where L is the *effective path length* of the signal through the rain. Because the rain density is unlikely to be uniform over the actual path length, an effective path length must be used rather than the actual (geometric) length. Figure 3.3 shows the geometry of the situation. The geometric, or slant, path length is shown as L_S. This depends on the antenna angle of elevation θ and the *rain height* h_R, which is the height at which freezing occurs. Figure 3.4 shows curves for h_R for different climatic zones. In this figure, three methods are labeled: Method 1—*maritime climates,* Method 2—*tropical climates,* Method 3—*continental climates.* For the last, curves are shown for p values of 0.001, 0.01, 0.1, and 1 percent.

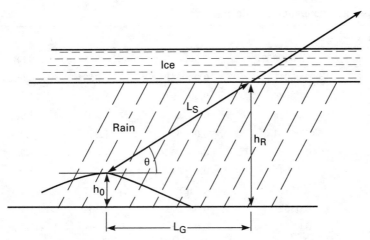

Figure 3.3 Path length through rain.

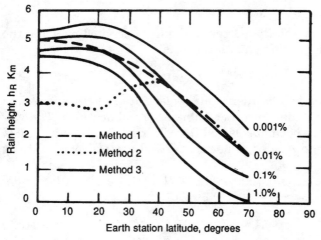

Figure 3.4 Rain height as a function of earth station latitude for different climatic zones.

For small angles of elevation ($\theta < 10°$) the determination of L_S is complicated by earth curvature (see CCIR Report 564-2). However, for $\theta \geq 10°$ a flat earth approximation may be used, and from Fig. 3.3 it is seen that

$$L_S = \frac{h_R - h_O}{\sin \theta} \qquad (3.4)$$

The effective path length is given in terms of the slant length by

$$L = L_S r_p \tag{3.5}$$

where r_p is a *reduction factor* which is a function of the percentage time p and L_G, the horizontal projection of L_S. From Fig. 3.3 the horizontal projection is seen to be

$$L_G = L_S \cos \theta \tag{3.6}$$

The reduction factors are given in Table 3.3. With all these factors together into one equation, the rain attenuation in dB is given by

$$A_p = a R_p^b L_S r_p \quad \text{dB} \tag{3.7}$$

TABLE 3.3 Reduction Factors

For p = 0.001%	$r_{.001} = \dfrac{10}{10 + L_G}$
For p = 0.01%	$r_{0.01} = \dfrac{90}{90 + 4L_G}$
For p = 0.1%	$r_{0.1} = \dfrac{180}{180 + L_G}$
For p = 1%	$r_1 = 1$

SOURCE: Ippolito, 1986.

Interpolation formulas which depend on the climatic zone being considered are available for values of p other than those quoted above (see, for example, Ippolito, 1986). Polarization shifts resulting from rain are described in Sec. 4.6.

Example 3.1

Calculate, for a frequency of 12 GHz and for horizontal and vertical polarizations, the rain attenuation which is exceeded for 0.01% of the time in any year, for a point rain rate of 10 mm/h. The earth station altitude is 600 m, and the antenna elevation angle is 50 degrees. The rain height is 3 km.

solution

The given data are shown below. Because the CCIR formulas contain hidden conversion factors, units will not be attached to the data, and it is understood that all lengths and heights are in km, and rain rate is in mm/h. The elevation angle must however be stated in degrees in order for the sine and cosine functions to be properly evaluated.

$$\theta : = 50 \cdot \deg \qquad h_O : = .6 \qquad h_r : = 3 \qquad R : = 10$$

From Table 3.2 at f = 12 GHz:

$$a_h : = .0188 \qquad a_v : = .0168 \qquad b_h : = 1.217 \qquad b_v : = 1.2$$

To speed up the calculations these coefficients will be defined using an indexed subscript i:

i: = 1.2

$$a_i := \qquad b_i :=$$

a_h
a_v

b_h
b_v

The specific attenuation in dB/km is:

$$\alpha := \overrightarrow{(a \cdot R^b)}$$

The required lengths are:

$$L_S := \frac{h_r - h_O}{\sin(\theta)} \qquad L_G := L_S \cdot \cos(\theta)$$

From Table 3.3 the rate reduction factor is:

$$r_{01} := \frac{90}{90 + 4 \cdot L_G}$$

The attenuation is:

$$AdB := \alpha \cdot L_S \cdot r_{01}$$

The values, rounded off to one decimal place are:

$$AdB_i$$

Horizontal	0.9
Vertical	0.8

The corresponding equations for circular polarization are

$$a_c = \frac{a_h + a_v}{2} \tag{3.8a}$$

$$b_c = \frac{a_h b_h + a_v b_v}{2a_c} \tag{3.8b}$$

The attenuation for circular polarization is compared to that for linear polarization in the following example.

Example 3.2

Repeat Example 3.1 for circular polarization.

solution

The given data are (see Example 3.1 for note on units):

$$\theta: = 48 \cdot \deg \qquad h_O: = .6 \qquad h_r: = 3 \qquad R: = 10$$
$$a_h: = .0188 \qquad a_v: = .0168 \qquad b_h: = 1.217 \qquad b_v: = 1.2$$

The corresponding values for circular polarization are:

$$a_c: = \frac{a_h + a_v}{2} \qquad\qquad b_c: = \frac{a_h \cdot b_h + a_v \cdot b_v}{2 \cdot a_c}$$

The attenuation coefficient is:

$$\alpha: = a_c \cdot R_p^{b_c}$$

The required lengths are:

$$L_S: = \frac{h_r - h_O}{\sin(\theta)} \qquad\qquad L_G: = L_S \cdot \cos(\theta)$$

The reduction factor is:

$$r_{01}: = \frac{90}{90 + 4 \cdot L_G}$$

The attenuation is:

$$AdB: = \alpha \cdot L_S \cdot r_{01} \qquad\qquad AdB: = 0.8$$

$$=======$$

3.5 Other Propagation Impairments

Hail, ice, and snow have little effect on attenuation because of the low water content. Ice can cause depolarization, described briefly in Chap. 4. The attenuation resulting from clouds can be calculated like that for rain (Ippolito, 1986, p. 56), although the attenuation is generally much less. For example, at a frequency of 10 GHz and a water content of 0.25 g/m³, the specific attenuation is about 0.05 dB/km, and about 0.2 dB/km for a water content of 2.5 g/m³.

3.6 Problems

3.1 With reference to Table 3.1, identify the propagation impairments which most affect transmission in the C band.

3.2 Repeat Prob. 3.1 for Ku-band transmissions.

3.3 Calculate the approximate value of atmospheric attenuation for a satellite transmission at 14 GHz, for which the angle of elevation of the earth station antenna is 15°.

3.4 Calculate the approximate value of atmospheric attenuation for a satellite transmission at 6 GHz, for which the angle of elevation of the earth station antenna is 30°.

3.5 Describe the major effects the ionosphere has on the transmission of satellite signals at frequencies of (*a*) 4 GHz and (*b*) 12 GHz.

3.6 Explain what is meant by *rain rate* and how this is related to specific attenuation.

3.7 Compare the specific attenuations for vertical and horizontal polarization at a frequency of 4 GHz and a point rain rate of 8 mm/h which is exceeded for 0.01 percent of the year.

3.8 Repeat Prob. 3.7 for a frequency of 12 GHz.

3.9 Explain what is meant by *effective path length* in connection with rain attenuation.

3.10 For a satellite transmission path, the angle of elevation of the earth station antenna is 35° and the earth station is situated at mean sea level. The signal is vertically polarized at a frequency of 18 GHz. The rain height is 1 km, and a rain rate of 10 mm/h is exceeded for 0.001 percent of the year. Calculate the rain attenuation under these conditions.

3.11 Repeat Prob. 3.10 for exceedance values of (*a*) 0.01 percent and (*b*) 0.1 percent.

3.12 Given that for a satellite transmission $\theta = 22°$, $R_{.01} = 15$ mm/h, $h_0 = 600$ m, $h_R = 1500$ m, and horizontal polarization is used, calculate the rain attenuation for a signal frequency of 14 GHz.

3.13 Determine the specific attenuation for a circularly polarized satellite signal at a frequency of 4 GHz where a point rain rate of 8 mm/h is exceeded for 0.01 percent of the year.

3.14 A circularly polarized wave at a frequency of 12 GHz is transmitted from a satellite. The point rain rate for the region is $R_{.01} = 13$ mm/h. Calculate the specific attenuation.

3.15 Given that for Prob. 3.13 the earth station is situated at altitude 500 m and the rain height is 2 km, calculate the rain attenuation. The angle of elevation of the path is 35°.

3.16 Given that for Prob. 3.14 the earth station is situated at altitude 200 m and the rain height is 2 km, calculate the rain attenuation. The angle of elevation of the path is 25°.

4

Polarization

4.1 Definition

In the *far field zone* of a transmitting antenna, the radiated wave takes on the characteristics of a *transverse electromagnetic* (TEM) wave. By far field zone is meant at distances greater than $2D^2/\lambda$ from the antenna, where D is the largest linear dimension of the antenna and λ is the wavelength. For a parabolic antenna of 3 m diameter transmitting a 6-GHz wave ($\lambda = 5$ cm), the far field zone begins at approximately 360 m. The TEM designation is illustrated in Fig. 4.1, where it can be seen that both the magnetic field **H** and the electric field **E** are transverse to the direction of propagation, denoted by the propagation vector **k.**

E, H, and **k** represent vector quantities and it is important to note their relative directions. When one looks along the direction of propa-

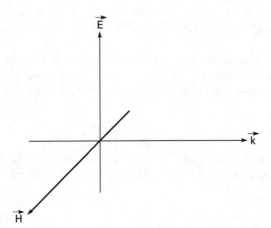

Figure 4.1 Vector diagram for a transverse electromagnetic (TEM) wave.

gation, the rotation from **E** to **H** is in the direction of rotation of a right-hand-threaded screw, and the vectors are said to form a *right-hand set*. The wave always retains the directional properties of the right-hand set, even when reflected, for example. One way of remembering how the right-hand set appears is to note that the letter E comes before H in the alphabet, and rotation is from **E** to **H** when looking along the direction of propagation.

At great distances from the transmitting antenna, such as are normally encountered in radio systems, the TEM wave can be considered to be plane. This means that the **E** and **H** vectors lie in a plane which is at right angles to the vector **k**. The vector **k** is said to be normal to the plane. The magnitudes are related by $E = HZ_0$, where $Z_0 = 120\pi$ ohms.

The direction of the line traced out by the tip of the electric field vector determines the *polarization* of the wave. Keep in mind that the electric and magnetic fields are varying as functions of time. The magnetic field varies exactly in phase with the electric field, and its amplitude is proportional to the electric field amplitude, so it is only necessary to consider the electric field in this discussion. The tip of the **E** vector may trace out a straight line, in which case the polarization is referred to as *linear*. Other forms of polarization, specifically elliptical and circular, will be introduced later.

In the early days of radio, there was little chance of ambiguity in specifying the direction of polarization in relation to the surface of the earth. Most transmissions utilized linear polarization and were along terrestrial paths. Thus *vertical polarization* meant that the electric field was perpendicular to the earth's surface, and *horizontal polarization* meant that it was parallel to the earth's surface. Although the terms *vertical* and *horizontal* are used with satellite transmissions, the situation is not quite so clear. A linear polarized wave transmitted by a geostationary satellite may be designated vertical if its electric field is parallel to the earth's polar axis, but even so the electric field will be parallel to the earth at the equator. This situation will be clarified shortly.

Suppose for the moment that horizontal and vertical are simply taken to mean two directions which are at right angles to one another, as shown in Fig. 4.2a. A vertically polarized electric field can be described as

$$\mathbf{E}_v = \hat{a}_v E_v \sin \omega t \tag{4.1}$$

where $\hat{a}_v$ is the unit vector in the vertical direction and E_v is the peak value or magnitude of the electric field. Likewise, a horizontally polarized wave could be described by

$$\mathbf{E}_h = \hat{a}_h E_h \sin \omega t \tag{4.2}$$

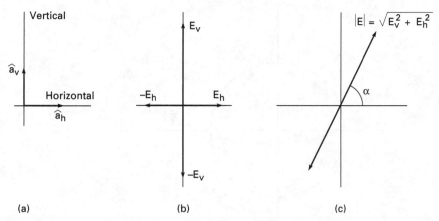

Figure 4.2   Horizontal and vertical components of linear polarization.

These two fields would trace out the straight lines shown in Fig. 4.2b. Now consider the situation where both fields are present simultaneously. These would add vectorially, and the resultant would be a vector **E,** Fig. 4.2c, at an angle to the horizontal given by

$$\alpha = \arctan \frac{E_v}{E_h} \tag{4.3}$$

Note that **E** is still linearly polarized, but cannot be classified as simply horizontal or vertical. Arguing back from this, it is evident that **E** can be resolved into vertical and horizontal components, a fact which is of great importance in practical transmission systems.

More formally, **E**$_v$ and **E**$_h$ are said to be *orthogonal*. The dictionary definition of orthogonal is *at right angles* but a wider meaning will be attached to the word later.

Consider now the situation where the two fields are equal in magnitude (denoted by E) but one leads the other by 90° in phase. The equations describing these are

$$\mathbf{E}_v = \hat{a}_v E \sin \omega t \tag{4.4a}$$

$$\mathbf{E}_h = \hat{a}_h E \cos \omega t \tag{4.4b}$$

Applying Eq. (4.3) in this case yields $\alpha = \omega t$. The magnitude of the resultant vector is $\sqrt{2}E$.

The tip of the resultant electric field vector traces out a circle as shown in Fig. 4.3a, and the resultant wave is said to be *circularly polarized*. The direction of circular polarization is defined by the sense of rotation of the electric vector, but this also requires that the way

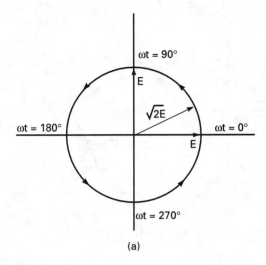

(a)

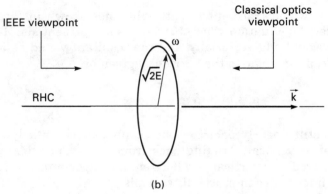

(b)

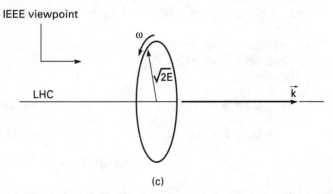

(c)

Figure 4.3   Circular polarization.

the vector is viewed must be specified. The Institute of Electrical and Electronics Engineers (IEEE) defines *right-hand circular* (RHC) *polarization* as a rotation in the clockwise direction when the wave is viewed along the direction of propagation, that is, when viewed from "behind," as shown in Fig. 4.3*b*. *Left-hand circular* (LHC) *polarization* is when the rotation is in the counterclockwise direction when viewed along the direction of propagation, as shown in Fig. 4.3*c*. LHC and RHC polarizations are orthogonal.

As a caution it should be noted that the classical optics definition of circular polarization is just the opposite of the IEEE definition. The IEEE definition will be used throughout this text.

Exercise

Given that Eq. (4.4) represents RHC polarization, show that the following equations represent LHC polarization:

$$\mathbf{E}_v = \hat{a}_v E \sin \omega t \tag{4.5a}$$

$$\mathbf{E}_h = -\hat{a}_h E \cos \omega t \tag{4.5b}$$

In the more general case, a wave may be *elliptically polarized*. This occurs when the two linear components are

$$\mathbf{E}_v = \hat{a}_v E_v \sin \omega t \tag{4.6a}$$

$$\mathbf{E}_h = \hat{a}_h E_h \sin (\omega t + \delta) \tag{4.6b}$$

Here, E_v and E_h are not equal in general and δ is a fixed phase angle. It is left as an exercise for the student to show that when $E_v = 1$, $E_h = 1/3$, and $\delta = 30°$, the polarization ellipse is as shown in Fig. 4.4.

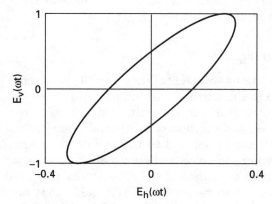

Figure 4.4   Elliptical polarization.

4.2   Antenna Polarization

An antenna which transmits a wave with a given sense of polarization is said to be polarized in that sense, and a receiving antenna which is matched to the polarization of the incoming wave to deliver maximum power is also polarized in that sense. The reciprocity theorem for antennas ensures that an antenna which is designed to transmit a given sense of polarization will also be matched to that polarization for reception. For example, in a terrestrial system, a horizontal half-wave dipole antenna will produce horizontally polarized waves when transmitting, and the same antenna will receive maximum power from a horizontally polarized wave. A combination of two half-wave dipoles, one horizontal and one vertical and fed with signals 90° apart in phase, will produce a circularly polarized wave. The combined output from these two dipoles when receiving will be maximum when the received signal is circularly polarized.

A wave with a given polarization will transfer no energy to a receiving antenna which is orthogonally polarized to the wave. Thus, a vertically polarized receiving antenna will receive nothing from a horizontally polarized wave, and an LHC polarized receiving antenna will receive nothing from an RHC polarized wave.

Exercise

Show that a linearly polarized antenna will always receive a signal from a circularly polarized wave, but with a 3-dB loss.

The use of orthogonal polarization allows signals to overlap in frequency without interference, a technique known as *frequency reuse,* and one which is widely used in satellite communications. Unfortunately, certain *depolarization effects* can occur in the transmission signals, so the technique is less than perfect. These depolarization effects are discussed later in the section on cross-polarization.

4.3   Polarization of Satellite Signals

As mentioned above, the directions "horizontal" and "vertical" are easily visualized with reference to the earth. Consider, however, the situation where a geostationary satellite is transmitting a linear polarized wave. In this situation the usual definition of horizontal polarization is where the electric field vector is parallel to the equatorial plane, and vertical polarization is where the electric field vector is parallel to the earth's polar axis. It will be seen that at the subsatellite point on the equator, both polarizations will result in electric fields that are parallel to the local horizontal plane, and care must be taken therefore not to use "horizontal" as defined for terrestrial systems. For

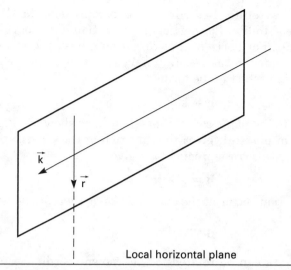

Figure 4.5 The reference plane for the direction of propagation and the local gravity direction.

other points on the earth's surface within the footprint of the satellite beam, the polarization vector (the unit vector in the direction of the electric field) will be at some angle relative to a reference plane. Following the work of Hogg and Chu, 1975, the reference plane will be taken to be that which contains the direction of propagation and the local gravity direction (a "plumb line"). This is shown in Fig. 4.5.

With the propagation direction denoted by **k** and the local gravity direction at the ground station by **r,** the direction of the normal to the reference plane is given by the vector cross-product:

$$\mathbf{f} = \mathbf{k} \times \mathbf{r} \tag{4.7}$$

With the unit polarization vector at the earth station denoted by **p,** the angle between it and **f** is obtained from the vector dot product as

$$\eta = \arccos\left(\frac{\mathbf{p} \cdot \mathbf{f}}{|\mathbf{f}|}\right) \tag{4.8}$$

Since the angle between a normal and its plane is 90°, the angle between **p** and the reference plane is $\xi = |90° - \eta|$ and

$$\xi = \left|\arcsin\left(\frac{\mathbf{p} \cdot \mathbf{f}}{|\mathbf{f}|}\right)\right| \tag{4.9}$$

This is the desired angle. Keep in mind that the polarization vector is always at right angles to the direction of propagation.

The next step is to relate the polarization vector **p** to the defined polarization at the satellite. Let unit vector **e** represent the defined polarization at the satellite. For vertical polarization **e** lies parallel to the earth's N–S axis. For horizontal polarization **e** lies in the equatorial plane at right angles to the geostationary radius a_{GSO} to the satellite. A cross-product vector can be formed,

$$\mathbf{g} = \mathbf{k} \times \mathbf{e} \qquad (4.10)$$

where **g** is normal to the plane containing **e** and **k,** as shown in Fig. 4.6. The cross-product of **g** with **k** gives the direction of the polarization in this plane. Denoting this cross-product by **h** gives

$$\mathbf{h} = \mathbf{g} \times \mathbf{k} \qquad (4.11)$$

The unit polarization vector at the earth station is therefore given by

$$\mathbf{p} = \frac{\mathbf{h}}{|\mathbf{h}|} \qquad (4.12)$$

All of these vectors can be related to the known coordinates of the earth station and satellite shown in Fig. 4.7. With the longitude of the satellite as the reference, the satellite is positioned along the positive x axis at

$$x_s = a_{GSO} \qquad (4.13)$$

The coordinates for the earth station position vector **R** are (ignoring the slight difference between geodetic and geocentric latitudes and assuming the earth station to be at mean sea level)

$$R_x = R \cos \lambda \cos B \qquad (4.14a)$$

$$R_y = R \cos \lambda \sin B \qquad (4.14b)$$

$$R_z = R \sin \lambda \qquad (4.14c)$$

where $B = \phi_E - \phi_{SS}$ as defined in Eq. (2.63).

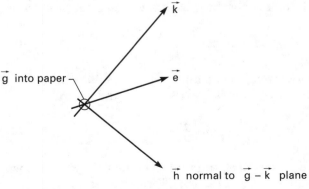

Figure 4.6   Vectors $\mathbf{g} = \mathbf{k} \times \mathbf{e}$ and $\mathbf{h} = \mathbf{g} \times \mathbf{k}$.

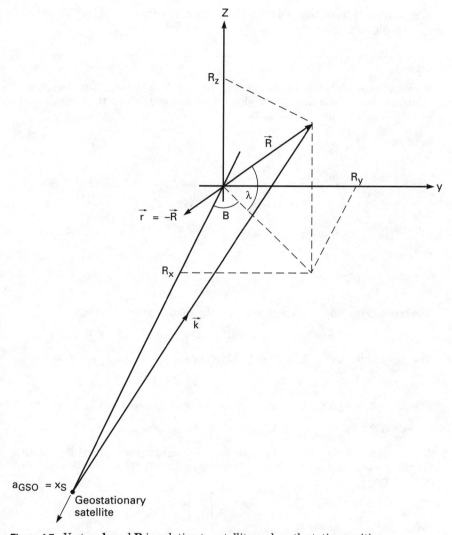

Figure 4.7 Vectors **k** and **R** in relation to satellite and earth station positions.

The local gravity direction is $\mathbf{r} = -\mathbf{R}$. The coordinates for the direction of propagation **k** are

$$k_x = r_x - a_{\text{GSO}} \qquad (4.15a)$$

$$k_y = r_y \qquad (4.15b)$$

$$k_z = r_z \qquad (4.15c)$$

The calculation of the polarization angle is illustrated in the following example.

Example

A geostationary satellite is stationed at 105 deg W and transmits a vertically polarized wave. Determine the angle of polarization at an earth station at latitude 18 deg. N longitude 73 deg. W.

solution

Given data:

$$\lambda: = 18 \cdot \text{deg} \qquad \phi_E: = -73 \cdot \text{deg} \qquad \phi_S: = -105 \cdot \text{deg}$$
$$a_{gso}: = 42164.14 \cdot \text{km} \quad R_E: = 6378.14 \cdot \text{km}$$

Calculations

$$R: = 6371 \cdot \text{km} \qquad\qquad \text{(Spherical earth of mean radius R assumed)}$$

$$B: = \phi_E - \phi_S \qquad\qquad ...\text{eq. (2.63)}$$

The geocentric-equatorial coordinates for the earth station position vector are:

$$R_x: = R \cdot \cos(\lambda) \cdot \cos(B) \qquad R_y: = R \cdot \cos(\lambda) \cdot \sin(B) \qquad R_z: = R \cdot \sin(\lambda)$$

The coordinates for the local gravity direction are

$$r_x: = -R_x \qquad\qquad r_y: = -R_y \qquad\qquad r_z: = -R_z$$

The geocentric-equatorial coordinates for the propagation direction are:

$$k_x: = R_x - a_{gso} \qquad\qquad k_y: = R_y \qquad\qquad k_z: = R_z$$

For vertical polarization at the satellite the geocentric-equatorial coordinates for the polarization vector are $x = 0$, $y = 0$ and $z = 1$. Thus, the three known vectors are:

$$e: = \begin{pmatrix} 0 \\ 0 \\ 1 \end{pmatrix} \qquad k: = \begin{bmatrix} k_x \\ k_y \\ k_z \end{bmatrix} \qquad r: = \begin{bmatrix} r_x \\ r_y \\ r_z \end{bmatrix}$$

The required cross-products are:

$$f: = k \times r \qquad g: = k \times e \qquad h: = g \times k$$

and the polarization vector is

$$p: = \frac{h}{|h|}$$

Hence the required angle is:

$$\xi: = \text{asin}\left(\frac{p \cdot f}{|f|}\right) \qquad\qquad \xi = 58.6 \cdot \text{deg}$$

========

4.4   Cross-Polarization Discrimination

The propagation path between a satellite and earth station passes through the ionosphere, and possibly through layers of ice crystals in the upper atmosphere and rain, all of which are capable of altering the polarization of the wave being transmitted. An orthogonal component may be generated from the transmitted polarization, an effect referred to as *depolarization*. This can cause interference where orthogonal polarization is used to provide isolation between signals, as in the case of frequency reuse.

Two measures are in use to quantify the effects of polarization interference. The most widely used measure is called *cross-polarization discrimination* (XPD). Figure 4.8a shows how this is defined. The transmitted electric field is shown having a magnitude E_1 before it enters the medium which causes depolarization. At the receiving antenna the electric field may have two components, a *copolar* component, having magnitude E_{11}, and a *cross-polar* component, having magnitude E_{12}. The cross-polarization discrimination in decibels is defined as

$$\text{XPD} = 20 \log \frac{E_{11}}{E_{12}} \qquad (4.16)$$

The second situation is shown in Fig. 4.8b. Here, two orthogonally polarized signals, with magnitudes E_1 and E_2, are transmitted. After traversing the depolarizing medium, copolar and cross-polar components exist for both waves. The *polarization isolation* is defined by the ratio of received copolar power to received cross-polar power and thus takes into account any additional depolarization introduced by the receiving system (Ippolito, 1986). Since received power is proportional to the square of the electric field strength, the polarization isolation is defined as

$$I = 20 \log \frac{E_{11}}{E_{21}} \qquad (4.17)$$

When the transmitted signals have the same magnitudes $(E_1 = E_2)$ and where the receiving system introduces negligible depolarization, then I and XPD give identical results.

For clarity, linear polarization is shown in Fig. 4.8, but the same definitions for XPD and I apply for any other system of orthogonal polarization.

4.5   Ionospheric Depolarization

The ionosphere is the upper region of the earth's atmosphere that has been ionized, mainly by solar radiation. The free electrons in the ionosphere are not uniformly distributed but form layers.

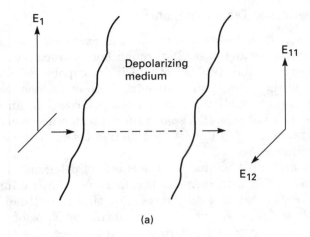

(a)

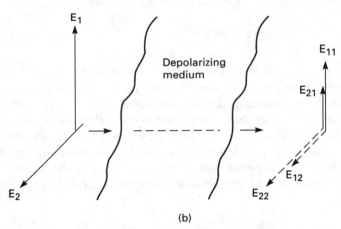

(b)

Figure 4.8 Vectors defining (*a*) cross-polarization discrimination (XPD) and (*b*) polarization isolation (I).

Furthermore, clouds of electrons (known as *traveling ionospheric disturbances*) may travel through the ionosphere and give rise to fluctuations in the signal. One of the effects of the ionosphere is to produce a rotation of the polarization of a signal, an effect known as *Faraday rotation*.

When a linearly polarized wave traverses the ionosphere, it sets in motion the free electrons in the ionized layers. These electrons move in the earth's magnetic field and therefore they experience a force (similar to that which a current-carrying conductor experiences in the magnetic field of a motor). The direction of electron motion is no longer parallel to the electric field of the wave, and as the electrons react back on the wave the net effect is to shift the polarization. The angular shift in polarization (the Faraday rotation) is dependent on the length of the path in the ionosphere, the strength of the earth's magnetic field in the ionized region, and the electron density in the region. Faraday rotation is inversely proportional to frequency squared, and is not considered to be a serious problem for frequencies above about 10 GHz.

Suppose a linearly polarized wave produces an electric field E at the receiver antenna when no Faraday rotation is present. The received power is proportional to E^2. A Faraday rotation of θ_F degrees will result in the copolarized component (the desired component) of the received signal being reduced to $E_{co} = E \cos \theta_F$, the received power in this case being proportional to E_{co}^2. The *polarization loss* PL in decibels is

$$PL = 20 \log \frac{E_{co}}{E} \qquad (4.18)$$

$$= 20 \log (\cos \theta_F)$$

At the same time, a cross-polar component $E_x = E \sin \theta_F$ is created, and hence the XPD is

$$XPD = 20 \log \frac{E_{co}}{E_x} \qquad (4.19)$$

$$= 20 \log (\cot \theta_F)$$

Maximum values quoted by Miya, 1981, for Faraday rotation are $9°$ at 4 GHz, and $4°$ at 6 GHz. In order to counter the depolarizing effects of Faraday rotation, circular polarization may be used. With circular polarization a Faraday shift simply adds to the overall rotation and does not affect the copolar or cross-polar components of electric field. Alternatively, if linear polarization is to be used, polarization tracking equipment may be installed at the antenna.

4.6   Rain Depolarization

The ideal shape of a raindrop is spherical, as this minimizes the energy (the surface tension) required to hold the raindrop together. The shape of small raindrops is close to spherical but the larger drops are better modeled as oblate spheroids with some flattening underneath, as a result of the air resistance. These are sketched in Fig. 4.9*a* and *b*. For vertically falling rain, the axis of symmetry of the raindrops will be

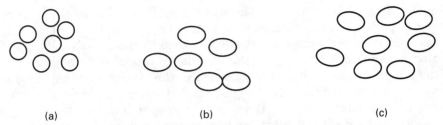

Figure 4.9 Raindrops: (*a*) small spherical, (*b*) flattening resulting from air resistance, and (*c*) angle of tilt randomized through aerodynamic force.

parallel to the local vertical as shown in Fig. 4.9*b*, but more realistically, aerodynamic forces will cause some canting, or tilting, of the drops. Thus there will be a certain randomness in the angle of tilt. The situation for a radio wave passing through rain is sketched in Fig. 4.9*c*.

As shown earlier, a linearly polarized wave can be resolved into two component waves, one vertically polarized and the other horizontally polarized. Consider a wave with its electric vector at some angle τ relative to the major axis of a raindrop, which for clarity is shown horizontal in Fig. 4.10. The vertical component of the electric field lies parallel to the minor axis of the raindrop and therefore encounters less water than the horizontal component. There will be a difference therefore in the attenuation and phase shift experienced by each of the electric field components. These differences are termed the *differential attenuation* and *differential phase shift*, and they result in depolarization of the wave. For the situation shown in Fig. 4.10, the

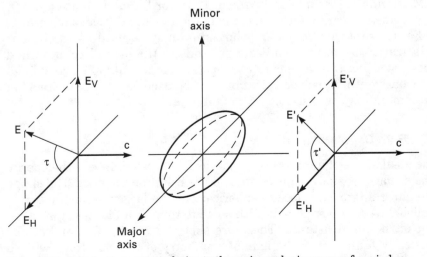

Figure 4.10 Polarization vector relative to the major and minor axes of a raindrop.

angle of polarization of the wave emerging from the rain is altered relative to that of the wave entering the rain. Experience has shown that the depolarization resulting from the differential phase shift is more significant than that resulting from differential attenuation.

The cross-polarization discrimination in decibels associated with rain is given to a good approximation by the empirical relationship (CCIR Report 564-2)

$$\text{XPD} = U - V \log A \qquad (4.20)$$

where U and V are empirically determined coefficients and A is the rain attenuation. U, V, and A must be in decibels in this equation. The attenuation A is as determined in Sec. 3.4. The following formulas are given in the CCIR reference for U and V for the frequency range 8 to 35 GHz:

$$V = \begin{cases} 20 & \text{for } 8 \le f \le 15 \text{ GHz} \\ 23 & \text{for } 15 \le f \le 35 \text{ GHz} \end{cases} \qquad (4.21a)$$

and

$$U = 30 \log f - 10 \log (0.5 - 0.4697 \cos 4\tau) - 40 \log (\cos \theta)) \qquad (4.21b)$$

where f is the frequency in gigahertz, θ is the angle of elevation of the propagation path at the earth station, and τ is the tilt angle of the polarization relative to the horizontal. For circular polariztion $\tau = 45°$. As shown earlier, for a satellite transmission, the angle ξ between the reference plane containing the direction of propagation and the local vertical is a complicated function of position, but the following general points can be observed. When the electric field is parallel to the ground (horizontal), $\tau = 0$, the second term on the right-hand side of the equation for U contributes a $+15$ dB amount to the XPD, whereas with circular polarization the contribution is only about $+0.13$ dB. With the electric field vector in the reference plane containing the direction of propagation and the local vertical, $\tau = 90° - \theta$ (all angles in degrees), and the $\cos 4\tau$ term becomes $\cos 4\theta$.

4.7   Ice Depolarization

As shown in Fig. 3.3 an ice layer is present at the top of a rain region, and as noted in Table 3.1, the ice crystals can result in depolarization. The experimental evidence suggests that the chief mechanism producing depolarization in ice is differential phase shift, with little differential attenuation present. This is because ice is a good dielectric, unlike water which has considerable losses. Ice crystals tend to be needle-shaped or platelike, and if randomly oriented have little effect, but depolarization occurs when they become aligned. Sudden increases in XPD that coincide with lightning flashes are thought to be a result of the lightning producing alignment. An International Radio Consultative Committee (CCIR) recommendation for taking ice depolarization into account is to add a fixed decibel value to the XPD value calculated for rain. Values of 2 dB are suggested for North America and 4 to 5 dB for maritime regions, and it is further suggested that the effects of ice can be ignored for time percentages less than 0.1 percent (Ippolito, 1986).

4.8   Problems

4.1   Explain what is meant by a *plane TEM wave*.

4.2   Two electric fields, in time phase and each of unity magnitude, act at right angles to one another in space. Draw the path traced by the tip of the resultant vector in relation to these two components.

4.3   Repeat Prob. 4.2 for the magnitude of the vectors in the ratio of 3:1.

4.4   Two electric field vectors of equal magnitude are 90° out of time phase with one another. Draw the path traced by the tip of the resultant vector in relation to these two components.

4.5   Repeat Prob. 4.4 for the magnitude of the vectors in the ratio of 3:1.

4.6   With reference to a right-hand set of rectangular coordinates, and given that Eq. (4.4) applies to a plane TEM wave, the horizontal component being directed along the x axis, and the vertical component along the y axis, determine the sense of polarization of the wave.

4.7   Repeat Prob. 4.6 for Eq. (4.5).

4.8   A plane transverse electromagnetic (TEM) wave has a horizontal ($+ x$-directed) component of electric field of magnitude 3 V/m, and a vertical ($+ y$-directed) component of electric field of magnitude 5 V/m. The horizontal component leads the vertical component by a phase angle of 20°. Determine the sense of polarization.

4.9   Repeat Prob. 4.8 for a phase angle lag of 20°.

4.10   Given that the plane (TEM) wave of Prob. 4.8 propagates in free space, determine the magnitude of the magnetic field.

4.11   Explain what is meant by *orthogonal polarization* and the importance of this in satellite communications.

4.12   The TEM wave represented by Eq. (4.4) is received by a linearly polarized antenna. Determine the reduction in emf induced in the antenna compared to what would be obtained with polarization matching.

4.13   A plane TEM wave has a horizontal ($+ x$-directed) component of electric field of magnitude 3 V/m, and a vertical ($+ y$-directed) component electric field of magnitude 5 V/m. The components are in time phase with one another. Determine the angle a linearly polarized antenna must be at with reference to the x axis to receive maximum signal.

4.14   For Prob. 4.13, what would be the reduction in decibels of the received signal if the antenna is placed along the x axis?

4.15   Explain what is meant by *vertical polarization* of a satellite signal. A vertically polarized wave is transmitted from a geostationary satellite and is received at an earth station which is west of the satellite, and in the northern hemisphere. Will the wave received at the earth station be vertically polarized? Give reasons for your answer.

4.16   Explain what is meant by *horizontal polarization* of a satellite signal. A horizontally polarized wave is transmitted from a geostationary satellite and is received at an earth station which is west of the satellite, and in the northern hemisphere. Will the wave received at the earth station be horizontally polarized? Give reasons for your answer.

4.17   A geostationary satellite stationed at 90° W transmits a vertically polarized wave. Determine the polarization of the resulting signal received at an earth station situated at 70° W, 45° N.

4.18   A geostationary satellite stationed at 10° E transmits a vertically polarized wave. Determine the polarization of the resulting signal received at an earth station situated at 5° E, 45 ° N.

4.19   Explain what is meant by *cross-polarization discrimination* and briefly describe the factors which militate against good cross-polarization discrimination.

4.20   Explain the difference between *cross-polarization discrimination* and *polarization isolation.*

4.21   A linearly polarized wave traveling through the ionosphere suffers a Faraday rotation of 9°. Calculate (*a*) the polarization loss and (*b*) the cross-polarization discrimination.

4.22   Why is Faraday rotation of no concern with circularly polarized waves?

4.23   Explain how depolarization is caused by rain.

4.24   A transmission path between an earth station and a satellite has an angle of elevation of 32° with reference to the earth. The transmission is circularly polarized at a frequency of 12 GHz. Given that rain attenuation on the path is 1 dB, calculate the cross-polarization discrimination.

4.25   Repeat Prob. 4.24 for a linearly polarized signal where the electric field vector is parallel to the earth at the earth station.

4.26   Repeat Prob. 4.24 for a linearly polarized signal where the electric field vector lies in the plane containing the direction of propagation and the local vertical at the earth station.

4.27   Repeat Prob. 4.24 for a signal frequency of 18 GHz.

5

Antennas

5.1 Introduction

Antennas can be broadly classified according to function as *transmitting antennas,* and as *receiving antennas.* Although the requirements for each function, or mode of operation, are markedly different, a single antenna may be, and frequently is, used for transmitting and receiving signals simultaneously. Many of the properties of an antenna, such as its directional characteristics, apply equally to both modes of operation, this being a result of the *reciprocity theorem* described in the next section.

Certain forms of interference (see Chap. 11) can present particular problems for satellite systems which are not encountered in other radio systems, and minimizing these requires special attention to those features of the antenna design which control interference.

Another way in which antennas for use in satellite communications can be classified is into *earth station* antennas and *satellite* or *spacecraft* antennas. Although the general principles of antennas may apply to each type, the constraints set by the physical environment lead to quite different designs in each case.

Before looking at antennas specifically for use in satellite systems, some of the general properties and definitions for antennas will be given in this and the next few sections. As already mentioned, antennas form the link between transmitting and receiving equipment and the space propagation path. Figure 5.1a shows the antenna as a radiator. The power amplifier in the transmitter is shown as generating P_T watts. A feeder connects this to the antenna, and the net power reaching the antenna will be P_T minus the losses in the feeder. These losses include ohmic losses and mismatch losses. The power will be further reduced by losses in the antenna, so that the power radiated, shown as P_{rad}, is less than that generated at the transmitter.

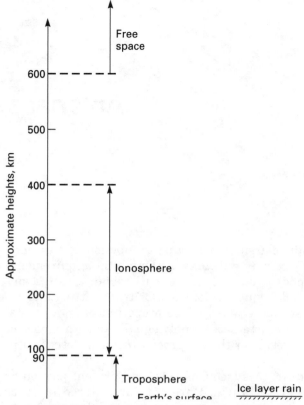

Figure 5.1 (*a*) Transmitting antenna. (*b*) Receiving antenna.

The antenna as a receiver is shown in Fig. 5.1*b*. Power P_{rec} is transferred to the antenna from a passing radio wave. Again, losses in the antenna will reduce the power available for the feeder. Receiver feeder losses will further reduce the power, so that the amount P_R reaching the receiver is less than that received by the antenna.

5.2 Reciprocity Theorem for Antennas

The reciprocity theorem for antennas states that if a current I is induced in an antenna B, operated in the receive mode, by an emf applied at the terminals of antenna A operated in the transmit mode, then the same emf applied to the terminals of B will induce the same current at the terminals of A. This is illustrated in Fig. 5.2. For a proof of the reciprocity theorem see, for example, Glazier and Lamont, 1958.

A number of important consequences result from the reciprocity theorem. All practical antennas have directional patterns, that is

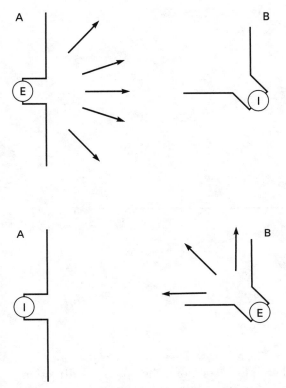

Figure 5.2 Illustrating the reciprocity theorem.

they transmit more energy in some directions than others, and they receive more energy when pointing in some directions than others. The reciprocity theorem requires that *the directional pattern for an antenna operating in the transmit mode is the same as that when operating in the receive mode.*

Another important consequence of the reciprocity theorem is that *the antenna impedance is the same for both modes of operation.*

5.3 Coordinate System

In order to discuss the directional patterns of an antenna it is necessary to set up a coordinate system to which these can be referred. The system in common use is the *spherical* (or *polar*) *coordinate system* illustrated in Fig. 5.3. The antenna is imagined to be at the origin of the coordinates, and a distant point P in space is related to the origin by the coordinates r, θ, and ϕ. Thus, r is the radius vector, the magnitude of which gives the distance between point P and the antenna; ϕ is the angle measured from the x axis to the projection of r in the x-y plane; and θ is the angle measured from the z axis to r.

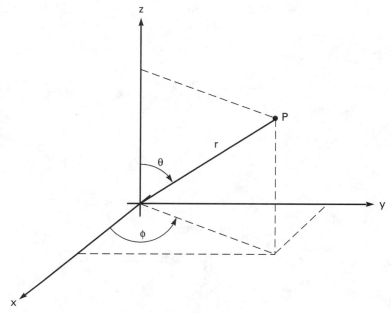

Figure 5.3 The spherical coordinate system.

It is important to note that the x, y, z axes form a *right-hand set*. What this means is that when one looks along the positive z direction, a clockwise rotation is required to move from the positive x axis to the positive y axis. This becomes particularly significant when the polarization of the radio waves associated with antennas is described.

5.4 The Radiated Fields

There are three main components to the radiated electromagnetic fields surrounding an antenna: two near field regions and a far field region. The field strengths of the near field components decrease rapidly with increasing distance from the antenna, one component being inversely related to distance squared, and the other to the distance cubed. At comparatively short distances these components are negligible compared to the radiated component used for radio communications, the field strength of which decreases in proportion to distance. Estimates for the distances at which the fields are significant are shown in Fig. 5.4a. Here, D is the largest dimension of the antenna (for example the diameter of a parabolic dish reflector) and λ is the wavelength. Only the far field region is of interest here, which applies for distances greater than about $2D^2/\lambda$.

In the far field region, the radiated fields form a transverse electromagnetic (TEM) wave in which the electric field is at right angles to the magnetic field, and both are at right angles (transverse) to the di-

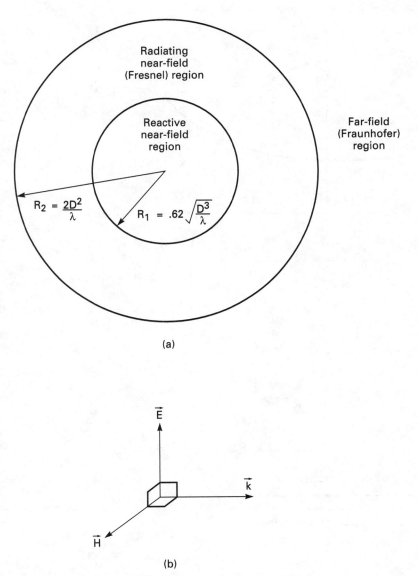

(a)

(b)

Figure 5.4 (*a*) The electromagnetic field regions surrounding an antenna. (*b*) Vector diagrams in the far field region.

rection of propagation. The vector relationship is shown in Fig. 5.4*b*, where **E** represents the electric field, **H** the magnetic field, and **k** the direction of propagation. These vectors form a right-hand set in the sense that, when one looks along the direction of propagation, a clockwise rotation is required to go from **E** to **H**. An important practical point is that the wavefront can be assumed to be plane, that is, **E** and **H** lie in a plane to which **k** is a normal.

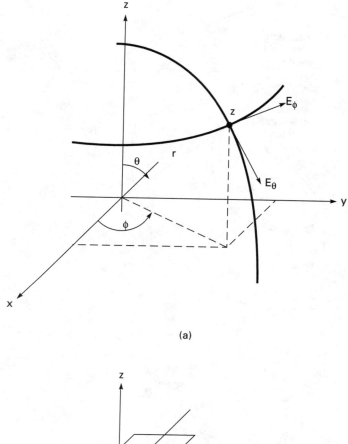

(a)

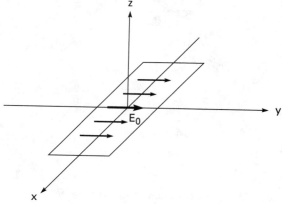

(b)

Figure 5.5 (a) The electric field components E_θ and E_ϕ in the far field region. The reference vector E_O at the origin.

In the far field, the electric field vector can be resolved into two components, which are shown in relation to the coordinate system in Fig. 5.5a. The component labeled E_θ is tangent at point P, to the circular arc of radius r. The component labeled E_ϕ is tangent at point P, to the circle of radius $r \sin \theta$ centered on the z axis (this is similar to a circle of latitude on the earth's surface). Both these components are functions of θ and ϕ and in functional notation would be written as $E_\theta (\theta, \phi)$ and $E_\phi (\theta, \phi)$. The resultant magnitude of the electric field is given by

$$E = \sqrt{E_\theta{}^2 + E_\phi{}^2} \tag{5.1}$$

Peak or rms values may be used in this equation.

The vector E_o shown at the origin of the coordinate system represents the principal electric vector of the antenna itself. For example, for a horn antenna, this would be the electric field vector across the aperture as shown in Fig. 5.5b. For definiteness, the E_o vector is shown aligned with the y axis, as this allows two important planes to be defined:

The H plane is the x-z plane, for which $\phi = 0$.

The E plane is the y-z plane, for which $\phi = 90°$.

Magnetic field vectors are associated with these electric field components. Thus, following the right-hand rule, the magnetic vector associated with the E_θ component will lie parallel with E_ϕ and is normally denoted by H_ϕ, while that associated with E_ϕ will lie parallel (but pointing in the opposite direction) to E_θ and is denoted by H_θ. For clarity the **H** fields are not shown in Fig. 5.5, but the magnitudes of the fields are related through the *wave impedance* Z_W. For radio waves in free space, the value of the wave impedance is (in terms of field magnitudes)

$$Z_W = \frac{E_\phi}{H_\theta} = \frac{E_\theta}{H_\phi} = 120\pi \ \Omega \tag{5.2}$$

The same value can be used with negligible error for radio waves in the earth's atmosphere.

5.5 Power Flux Density

The *power flux density* of a radio wave is a quantity used in calculating the performance of satellite communications links. The concept can be understood by imagining the transmitting antenna to be at the center of a sphere. The power from the antenna radiates outward, normal to the surface of the sphere, and the power flux density is the power flow per unit surface area. Power flux density is a vector quantity, and its magnitude is given by

$$\Psi = \frac{E^2}{Z_W} \tag{5.3}$$

Here, E is the rms value of the field given by Eq. (5.1). The units for Ψ are watts per square meter with E in volts per meter and Z_W in ohms. Because the E field is inversely proportional to distance (in this case the radius of the sphere), the power density is inversely proportional to the square of the distance.

5.6 The Isotropic Radiator and Antenna Gain

The word *isotropic* means, rather loosely, equally in all directions. Thus an *isotropic radiator* is one which radiates equally in all directions. No real antenna can radiate equally in all directions, and the isotropic radiator is therefore hypothetical. It does, however, provide a very useful theoretical standard against which real antennas can be compared. Being hypothetical it can be made 100 percent efficient, meaning that it radiates all the power fed into it. Thus, referring back to Fig. 5.1a, $P_{rad} = P_S$. By imagining the isotropic radiator to be at the center of a sphere of radius r, the power flux density, which is the power flow through unit area, is

$$\Psi_i = \frac{P_S}{4\pi r^2} \tag{5.4}$$

Now the flux density from a real antenna will vary with direction, but with most antennas a well-defined maximum occurs. The *gain* of the antenna is the ratio of this maximum to that for the isotropic radiator at the same radius r:

$$G = \frac{\Psi_M}{\Psi_i} \tag{5.5}$$

A very closely related gain figure is the *directive gain*. This differs from the power gain only in that in determining the isotropic flux density, the actual power P_{rad} radiated by the real antenna is used, rather than the power P_S supplied to the antenna. These two values are related as $P_{rad} = \eta_A P_S$ where η_A is the *antenna efficiency*. Denoting the directive gain by D_G gives

$$G = \eta_A D_G$$

Often, the directive gain is the parameter which can be calculated, and the efficiency is assumed to be equal to unity so that the power gain is also known. Note that η_A does not include feeder mismatch or polarization losses, which are accounted for separately.

The gain as described is called the *isotropic power gain,* sometimes denoted by G_i. The power gain of an antenna may also be referred to some standard other than isotropic. For example, the gain of a reflector-type antenna may be made relative to the antenna illuminating the reflector. Care must be taken therefore to know what reference antenna is being used when gain is stated. The isotropic gain is the most commonly used figure and will be assumed throughout this text (without use of a subscript) unless otherwise noted.

5.7   Radiation Pattern

The *radiation pattern* shows how the gain of an antenna varies with direction. Referring to Fig. 5.3, at a fixed distance r the gain will vary with θ and ϕ and may be written generally as $G(\theta, \phi)$. The radiation pattern is the gain normalized to its maximum value. Denoting the maximum value simply by G as before, the radiation pattern is

$$g(\theta, \phi) = \frac{G(\theta, \phi)}{G} \tag{5.6}$$

The radiation pattern gives the directional properties of the antenna normalized to the maximum value, in this case the maximum gain. The same function gives the power density normalized to the maximum power density. For most satellite antennas, the three-dimensional plot of the radiation pattern shows a well-defined main lobe, as sketched in Fig. 5.6a. In this diagram, the length of a radius line to any point on the surface of the lobe gives the value of the radiation function at that point. It will be seen that the maximum value is normalized to unity, and for convenience this is shown pointing along the positive z axis. Be very careful to observe that axes shown in Fig. 5.6 *do not represent distance.* The distance r is assumed to be fixed at some value in the far field. What is shown is a plot of normalized gain as a function of angles θ and ϕ.

The main lobe represents a *beam* of radiation and the *beamwidth* is specified as the angle subtended by the -3 dB lines. Because in general the beam may not be symmetrical, it is usual practice to give the beamwidth in the H plane ($\phi = 0°$), as shown in Fig. 5.6b, and in the E plane ($\phi = 90°$), as shown in Fig. 5.6c.

Because the radiation pattern is defined in terms of radiated power, the normalized electric field strength pattern will be given by $\sqrt{g(\theta, \phi)}$.

5.8   Beam Solid Angle and Directive Gain

The *beam solid angle,* denoted by Ω_A in Fig. 5.7, is the solid angle subtended by a hypothetical beam having a unity radiation pattern

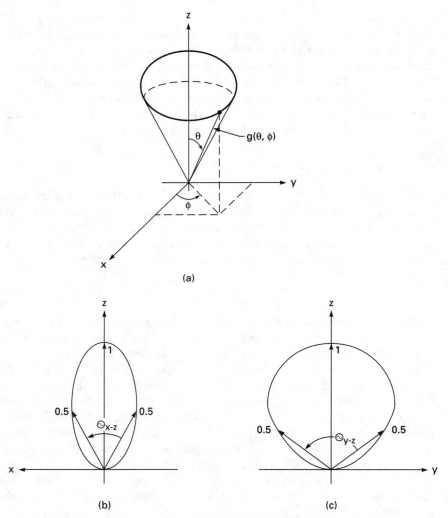

Figure 5.6 (a) A radiation pattern. (b) The beamwidth in the H-plane. (c) The beamwidth in the E-plane.

for all directions enclosed by the beam, and zero radiation outside the beam. The beam solid angle can be calculated for a real antenna from its radiation pattern, and a useful approximate result for narrow-beam antennas (as most satellite antennas are) is

$$\Omega_A \cong \text{HPBW}_E \text{HPBW}_H \tag{5.7}$$

where HPBW_E is the half-power beamwidth in the E plane and HPBW_H is the half-power beamwidth in the H plane, as shown in Fig. 5.6. This equation requires the half-power beamwidths to be expressed in

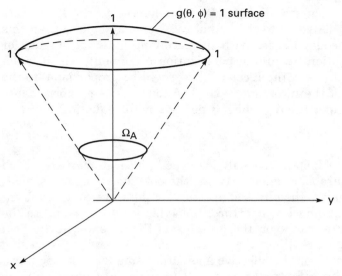

Figure 5.7   The beam solid angle.

radians, and the resulting solid angle is in steradians. The directive gain, introduced in Sec. 5.6, is related to the beam solid angle as

$$D_G = \frac{4\pi}{\Omega_A} \tag{5.8}$$

The usefulness of this relationship is that the half-power beamwidths can be measured, and hence the directive gain can be found. When the half-power beamwidths are expressed in degrees, the equation for the directive gain becomes

$$D_G \cong \frac{41,260}{\text{HPBW}^\circ_E \, \text{HPBW}^\circ_H} \tag{5.9}$$

5.9   Effective Aperture

So far, the properties of antennas have been described in terms of their radiation characteristics. A receiving antenna has directional properties also described by the radiation pattern, but in this case it refers to the ratio of received power normalized to the maximum value.

An important concept used to describe the reception properties of an antenna is that of *effective aperture*. Consider a TEM wave of a given power density Ψ at the receiving antenna. Let the load at the *antenna terminals* be a complex conjugate match, so that maximum power transfer occurs and power P_{rec} is delivered to the load. Note that the power delivered to the actual receiver may be less than this as a result of feeder losses. With the receiving antenna aligned for maximum reception (including polarization alignment, which is described in detail later), the received power will be proportional to the power density of the incoming wave. The constant of proportionality is the effective aperture A_{eff} which is defined by the equation

$$P_{rec} = A_{eff} \Psi \qquad (5.10)$$

For antennas which have easily identified physical apertures, such as horns and parabolic reflector types, the effective aperture is related in a direct way to the physical aperture. If the wave could uniformly illuminate the physical aperture, then this would be equal to the effective aperture. However, the presence of the antenna in the field of the incoming wave alters the field distribution thereby preventing uniform illumination. The effective aperture is smaller than the physical aperture by a factor known as the *illumination efficiency*. Denoting the illumination efficiency by η_I gives

$$A_{eff} = \eta_I A_{physical} \qquad (5.11)$$

The illumination efficiency is usually a specified number, and it can range between about 0.5 and 0.8. Of course it cannot exceed unity, and a conservative value often used in calculations is 0.55.

A fundamental relationship exists between the power gain of an antenna and its effective aperture. This is

$$\frac{A_{eff}}{G} = \frac{\lambda^2}{4\pi} \qquad (5.12)$$

where λ is the wavelength of the TEM wave, assumed sinusoidal (for practical purposes this will be the wavelength of the radio wave carrier). The importance of this equation is that the gain is normally the known (measurable) quantity, but once this is known the effective aperture is also known.

5.10   The Half-Wave Dipole

The half-wave dipole is a basic antenna type which finds limited but essential use in satellite communications. Some radiation occurs in all directions except along the dipole axis itself, and it is this near-omnidirectional property which finds use for telemetry and command signals to and from the satellite, essential during the launch phase when highly directional antennas cannot be employed.

The half-wave dipole is shown in Fig. 5.8*a,* and its radiation pattern in the *x-y* plane and in any one meridian plane in Fig. 5.8*b* and *c.* Because the phase velocity of the radio wave along the wire is some-

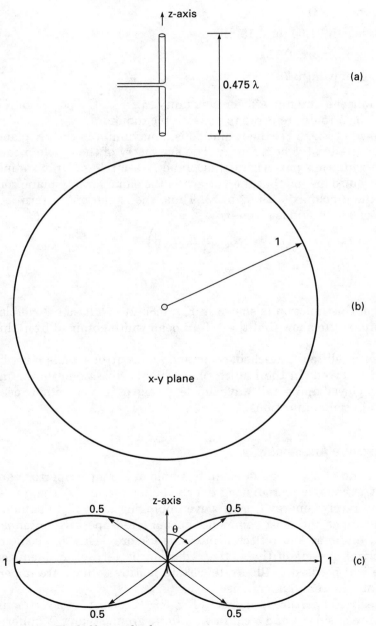

Figure 5.8 The half-wave dipole.

what less than the free-space velocity, the wavelength is also slightly less, and the antenna is cut to about 95 percent of the free-space half-wavelength. This tunes the antenna correctly to resonance. The main properties of the half-wave dipole are

Impedance: 73 Ω

Directive gain: 1.64 (or 2.15 dB)

Effective aperture: $0.13\lambda^2$

-3-dB beamwidth: $78°$

Assuming the antenna efficiency is unity ($\eta_A = 1$), the power gain is also 1.64, or 2.15 dB, referred to an isotropic radiator.

As shown in Fig. 5.8b, the radiation is a maximum in the x-y plane, the normalized value being unity. The symmetry of the dipole means that the radiation pattern in this plane is a circle of unit radius. Symmetry also means that the pattern is the same for any plane containing the dipole axis (the z axis). Thus, the radiation pattern is a function of θ only, and is given by

$$g(\theta) = \frac{\cos^2\left(\dfrac{\pi}{2}\cos\theta\right)}{\sin^2\theta} \tag{5.13}$$

A plot of this function is shown in Fig. 5.8c. It is left as an exercise for the student to show that the -3-dB beamwidth obtained from this pattern is $78°$.

When a satellite is launched, command and control signals must be sent and received. In the launch phase highly directional antennas are not deployed and a half-wave dipole, or one of its variants, is used to maintain communications.

5.11   Aperture Antennas

The open end of a waveguide is an example of a simple aperture antenna. It is capable of radiating energy being carried by the guide, and it can receive energy from a wave impinging upon it. In satellite communications the most commonly encountered aperture antennas are horn antennas and reflector antennas. Before describing some of the practical aspects of these, the radiation pattern of an idealized aperture will be used to illustrate certain features which are important in satellite communications.

The idealized aperture is shown in Fig. 5.9. It consists of a rectangular aperture of sides a and b cut in an infinite ground plane. A uniform electric field exists across the aperture parallel to the sides b, and the

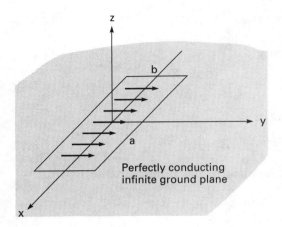

Perfectly conducting
infinite ground plane

Figure 5.9 An idealized aperture radiator.

aperture is centered on the coordinate system shown in Fig. 5.3, with the electric field parallel to the y axis. Radiation from different parts of the aperture adds constructively in some directions and destructively in others, with the result that the radiation pattern exhibits a main lobe and a number of sidelobes. Mathematically, this is shown as follows. At some fixed distance r in the far field region the electric field components described in Sec. 5.4 are given by

$$E_\theta(\theta, \phi) = C \sin \phi \, \frac{\sin X}{X} \frac{\sin Y}{Y} \tag{5.14}$$

$$E_\phi(\theta, \phi) = C \cos \theta \cos \phi \, \frac{\sin X}{X} \frac{\sin Y}{Y} \tag{5.15}$$

Here, C is a constant which depends on the distance r; the lengths a and b, the wavelength λ, and on the electric field strength E_o. For present purposes, it can be set equal to unity. X and Y are variables given by

$$X = \frac{\pi a}{\lambda} \sin \theta \cos \phi \tag{5.16}$$

$$Y = \frac{\pi b}{\lambda} \sin \theta \sin \phi \tag{5.17}$$

It will be seen that even for the idealized and hence simplified aperture situation, the electric field equations are quite complicated. The two principal planes of the coordinate system are defined as the H plane, which is the x-z plane, for which $\phi = 0$, and the E plane, which is the x-y plane, for which $\phi = 90°$. It simplifies matters to examine the radiation pattern in these two planes. Consider first the H plane. With $\phi = 0$ it is readily shown that $Y = 0$, $E_\theta = 0$, and

$$X = \frac{\pi a}{\lambda} \sin \theta \tag{5.18}$$

and, with C set equal to unity,

$$E_\phi(\theta) = \cos \theta \, \frac{\sin X}{X} \tag{5.19}$$

The radiation pattern is given by

$$g(\theta) = |E_\phi(\theta)|^2 \tag{5.20}$$

A similar analysis may be applied to the E plane resulting in $X = 0$, $E_\phi = 0$, and

$$Y = \frac{\pi b}{\lambda} \sin \theta \tag{5.21}$$

$$E_\theta(\theta) = \frac{\sin Y}{Y} \tag{5.22}$$

$$g(\theta) = |E_\theta|^2 \tag{5.23}$$

These radiation patterns are illustrated in Example 5.1.

Example 5.1

Plot the E-plane and H-plane radiation patterns for the uniformly illuminated aperture for which a = 3, b = 2λ.

solution

Since the dimensions are normalized to wavelength this may be set equal to unity: λ: = 1m

a: = 3·λ   b: = 2·λ   dB: = 1

Define a range for θ:

$$\theta := -90 \cdot deg, \; -88 \cdot deg .. 90 \cdot deg$$

For the E-plane, with φ = 90 deg:

$$Y(\theta): = \frac{\pi \cdot b}{\lambda} \cdot \sin(\theta) \qquad E_\theta(\theta): = \frac{\sin(Y(\theta))}{Y(\theta)}$$

$$g_E(\theta): = E(\theta)(\theta)^2 \qquad G_E(\theta): = 10 \cdot \log(g_E(\theta))$$

For the H-plane, with φ = 0 deg:

$$X(\theta): = \frac{\pi \, a}{\lambda} \sin(\theta) \qquad E_\phi(\theta): = \cos(\theta) \, \frac{\sin(X(\theta))}{X(\theta)}$$

$$g_H(\theta): = E_\phi(\theta)^2 \qquad G_H(\theta): = 10 \cdot \log(g_H(\theta))$$

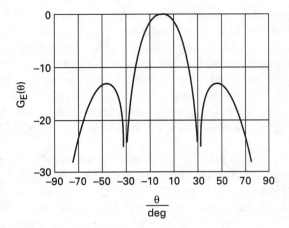

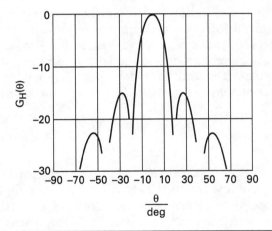

The results of Example 5.1 show the main lobe and the sidelobes. These are a general feature of aperture antennas, and as mentioned above, the pattern is the result of interference phenomena. Mathematically the lobes result from the $(\sin X)/X$ and $(\sin Y)/Y$ terms. One of the main concerns in satellite communications is the reduction of interference caused by sidelobes.

The uniform field distribution assumed above cannot be realized in practice, the actual distribution depending on the manner in which the aperture is energized. In practice, therefore, the radiation pattern will depend on the way the aperture is energized. It is also influenced by the physical construction of the antenna. With reflector-type antennas, for example, the position of the primary feed can change the pattern in important ways.

Another important practical consideration with real antennas is the *cross-polarization* which can occur. This refers to the antenna in the transmit mode radiating, and in the receive mode responding to, an unwanted signal with polarization orthogonal to the desired polarization (see Sec. 4.2). As mentioned in Chap. 4, frequency reuse makes use of orthogonal polarization and any unwanted cross-polarized component will result in interference. The cross-polarization characteristics of some practical antennas will be looked at in the following sections.

The aperture shown in Fig. 5.9 is linearly polarized, the **E** vector being directed along the *y*axis. At some arbitrary point in the far field region, the wave will remain linearly polarized, the magnitude *E* being given by Eq. (5.1). It is only necessary for the receiving antenna to be oriented so that *E* induces maximum signal, with no component orthogonal to *E* so that cross-polarization is absent. Care must be taken, however, in how cross-polarization is defined. The linearly polarized field **E** can be resolved into two vectors, one parallel to the plane containing the aperture vector E_θ, referred to as the *copolar* component, and a second component orthogonal to this, referred to as the *cross-polarized* component. The way in which these components are used in antenna measurements is detailed in Chang, 1989, and Rudge et al., 1982.

5.12 Horn Antennas

The horn antenna is an example of an aperture antenna which provides a smooth transition from a waveguide to a larger aperture that couples more effectively into space. Horn antennas are used directly as radiators aboard satellites to illuminate comparatively large areas of the earth, and they are also widely used as primary feeds for reflector-type antennas both in transmitting and receiving modes. The three most commonly used types of horns are illustrated in Fig. 5.10.

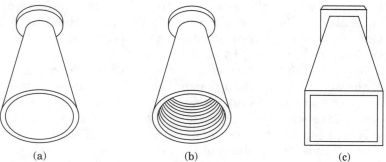

Figure 5.10 Horn antennas: (*a*) smooth-walled conical, (*b*) corrugated, and (*c*) pyramidal.

Conical Horn Antennas. The smooth-walled conical antenna shown in Fig. 5.10 is the simplest horn structure. The term *smooth-walled* refers to the inside wall. The horn may be fed from a rectangular waveguide but this requires a rectangular-to-circular transition at the junction. Feeding from a circular guide is direct and is the preferred method, with the guide operating in the TE_{11} mode. The conical horn antenna may be used with linear or circular polarization, but in order to illustrate some of the important features, linear polarization will be assumed.

The electric field distribution at the horn mouth is sketched in Fig. 5.11 for vertical polarization. The curved field lines can be resolved into vertical and horizontal components as shown. The TEM wave in the far field is linearly polarized, but the horizontal components of the aperture field give rise to cross-polarized waves in the far field region. Because of the symmetry the cross-polarized waves cancel in the principal planes (the E and H planes); however, they produce four peaks, one in each quadrant around the main lobe. Referring to Fig. 5.5, the cross-polarized fields peak in the $\phi = \pm 45°$ planes. The peaks are about -19 dB relative to the peak of the main (copolar) lobe (Olver, 1992).

The smooth-walled horn does not produce a symmetrical main beam, even though the horn itself is symmetrical. The radiation patterns are complicated functions of the horn dimensions. Details will be found in Chang, 1989, where it is shown that the beamwidths in the principal planes can differ widely. This lack of symmetry is a disadvantage where global coverage is required.

By operating a conical horn in what is termed a *hybrid mode,* which is a nonlinear combination of transverse electric (TE) and transverse magnetic (TM), modes, the pattern symmetry is improved, the cross-polarization is reduced, and a more efficient main beam is produced with low sidelobes. It is especially important to reduce the cross-polarization where frequency reuse is employed, as described in Sec. 4.2.

One method of achieving a hybrid mode is to corrugate the inside wall of the horn, thus giving rise to the *corrugated horn* antenna. The cross section of a corrugated horn is shown in Fig. 5.12*a*. The aperture electric field is shown in Fig. 5.12*b*, where it is seen to have a much lower cross-polarized component. This field distribution is

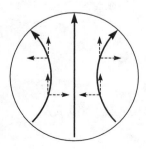

Figure 5.11 Aperture field in a smooth-walled conical horn.

sometimes referred to as a *scalar* field, and the horn as a scalar horn. A development of the scalar horn is the scalar feed, Fig. 5.13, which can be seen on most domestic receiving systems. Here, the flare angle of the horn is 90°, and the corrugations are in the form of a flange surrounding the circular waveguide. The corrugated horn is obviously more difficult to make than the smooth-walled version, and close manufacturing tolerances must be maintained, especially in machining the slots or corrugations, all of which contribute to increased costs. A comprehensive description of the corrugated horn will be found in Olver, 1992, and design details will be found in Chang, 1989.

A hybrid mode can also be created by including a dielectric rod along the axis of the smooth-walled horn, this being referred to as a *dielectric-rod-loaded antenna* (see Miya, 1981).

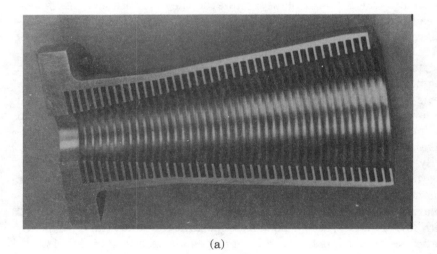

(a)

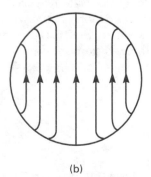

(b)

Figure 5.12 (*a*) Cross-section of a corrugated horn. (*From Alver, 1992. With permission.*) (*b*) Aperture field.

Figure 5.13 A scalar feed.

A *multimode* horn is one which is excited by a linear combination of transverse electric and transverse magnetic fields, the most common type being the *dual-mode* horn which combines the TE_{11} and TM_{11} modes. The advantages of the dual-mode horn are similar to those of the hybrid-mode horn, that is, better main lobe symmetry, lower cross-polarization, and a more efficient main beam with low widelobes. Dual-mode horns have been installed aboard various satellites (see Miya, 1981).

Horns which are required to provide earth coverage from geostationary satellites must maintain low cross-polarization and high gain over a cone angle of $\pm 9°$. This is achieved more simply and economically with dual-mode horns (Hwang, 1992).

Pyramidal Horn Antennas.   The pyramidal horn antenna, illustrated in Fig. 5.14, is primarily designed for linear polarization. In general it has a rectangular cross section $a \times b$ and operates in the TE_{10} waveguide mode, which has the electric field distribution shown in Fig. 5.14.

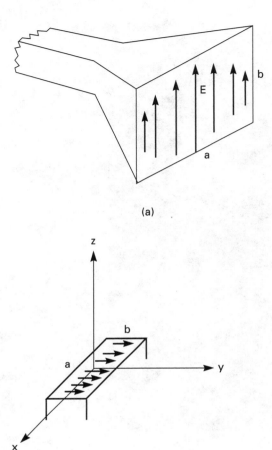

(a)

Figure 5.14 The pyramidal horn.

(b)

In general the beamwidths for the pyramidal horn differ in the E and H planes, but it is possible to choose the aperture dimensions to make these equal. The pyramidal horn can be operated in horizontally and vertically polarized modes simultaneously, giving rise to dual-linear polarization. According to Chang, 1989, the cross-polarization characteristics of the pyramidal horn have not been studied to any great extent, and if required they should be measured.

For any of the aperture antennas discussed, the isotropic gain can be found in terms of the area of the physical aperture by using the relationships given in Eqs. (5.11) and (5.12). For accurate gain determinations the difficulties lie in determining the illumination efficiency η_I which can range from 35 to 80 percent for horns and from 50 to 80 percent for circular reflectors (Balanis, 1982, p. 475). Circular reflectors are discussed in the next section.

5.13   The Parabolic Reflector

Parabolic reflectors are widely used in satellite communications systems to enhance the gain of antennas. The reflector provides a focusing mechanism which concentrates the energy in a given direction. The most commonly used form of parabolic reflector has a circular aperture as shown in Fig. 5.15. This is the type seen in many home installations for the reception of TV signals. The circular aperture configuration is referred to as a *paraboloidal reflector.*

The main property of the paraboloidal reflector is its focusing property, normally associated with light, where parallel rays striking the reflector converge on a single point known as the *focus,* and conversely, rays originating at the focus are reflected as a parallel beam of light. This is illustrated in Fig. 5.16. Light of course is a particular example

Figure 5.15   A parabolic reflector antenna. (*Courtesy Scientific Atlanta Inc.*)

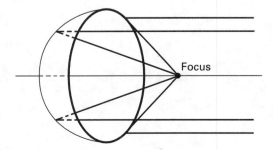

Focus

Figure 5.16   Illustrating the focusing property of a paraboloid reflector.

of an electromagnetic wave, and the same properties apply to electromagnetic waves in general, including the radio waves used in satellite
communications. The ray paths from the focus to the aperture plane
(the plane containing the circular aperture) are all equal in length.

The geometric properties of the paraboloidal reflector of interest
here are most easily demonstrated by means of the *parabola,* which is
the curve traced by the reflector on any plane normal to the aperture
plane and containing the focus. This is shown in Fig. 5.17*a*. The *focal
point* or *focus* is shown as *S,* the *vertex* as *A,* and the axis is the line
passing through *S* and *A. SP* is the *focal distance* for any point *P* and
SA the *focal length,* usually denoted by *f.* (The parabola is examined
in more detail in App. B). A ray path is shown as *SPQ* where *P* is a
point on the curve, and *Q* is a point in the aperture plane. Length *PQ*
lies parallel to the axis. For any point *P* all path lengths *SPQ* are
equal; that is, the distance *SP + PQ* is a constant which applies for all

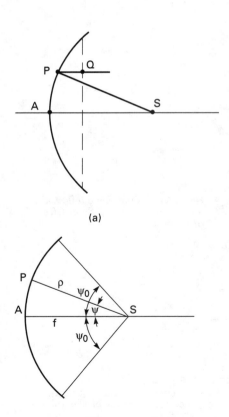

(a)

(b)

Figure 5.17 (*a*) The focal length
f = SA, and a ray path SPQ. (*b*)
The focal distance ρ.

such paths. The path equality means that a wave originating from an isotropic point source has a uniform phase distribution over the aperture plane. This property, along with the parallel beam property, means that the wavefront is plane. Radiation from the paraboloidal reflector appears to originate as a plane wave from the plane normal to the axis and containing the directrix (see App. B). Although the characteristics of the reflector antenna are more readily described in terms of radiation, it should be kept in mind that the reciprocity theorem makes these applicable to the receiving mode as well.

Now although there are near and far field components present in the reflector region (see Fig. 5.4b), the radio link is made through the far field component, and only this need be considered. For this, the reflected wave is a plane wave, while the wave originating from the isotropic source and striking the reflector has a spherical wavefront. The power density in the plane wave is independent of distance. For the spherical wave the power density of the far field component decreases in inverse proportion to the distance squared, and therefore the illumination at the edge of the reflector will be less than that at the vertex. This gives rise to a nonuniform amplitude distribution across the aperture plane, which in effect means that the illumination efficiency is reduced. Denoting the focal distance by ρ and the focal length by f as in Fig. 5.17b, then, as shown in App. B,

$$\frac{\rho}{f} = \sec^2 \frac{\Psi}{2} \tag{5.24}$$

The *space attenuation function* (SAF) is the ratio of the power reaching point P to that reaching point A, and, since the power density is inversely proportional to the square of the distance, the ratio is given by

$$\text{SAF} = \left(\frac{f}{\rho}\right)^2 = \cos^4 \frac{\Psi}{2} \tag{5.25}$$

For satellite applications a high illumination efficiency is desirable. This requires that the radiation pattern of the primary antenna, which is situated at the focus and which illuminates the reflector, should approximate as closely as practical the inverse of the space attenuation factor.

An important ratio is that of aperture diameter to focal length. Denoting the diameter by D (do not confuse with the D used for directivity), then, as shown in App. B,

$$\frac{f}{D} = 0.25 \cot \frac{\Psi_0}{2} \tag{5.26}$$

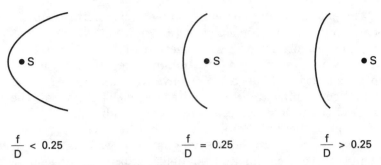

$$\frac{f}{D} < 0.25 \qquad\qquad \frac{f}{D} = 0.25 \qquad\qquad \frac{f}{D} > 0.25$$

Figure 5.18 Position of the focus for various f/D values.

The position of the focus in relation to the reflector for various values of f/D is shown in Fig. 5.18. For $f/D < 0.25$, the primary antenna lies in the space between the reflector and the aperture plane, and the illumination tapers away toward the edge of the reflector. For $f/D > 0.25$, the primary antenna lies outside the aperture plane, which results in more nearly uniform illumination, but *spillover* increases. In the transmitting mode, spillover is the radiation from the primary antenna which is directed toward the reflector but which lies outside the angle $2\Psi_0$. In satellite applications the primary antenna is usually a horn (or an array of horns as will be shown later) pointed toward the reflector. In order to compensate for the space attenuation described above, higher-order modes can be added to the horn feed so that the horn radiation pattern approximates the inverse of the space attenuation function (Chang, 1989).

The radiation from the horn will be a spherical wave and the *phase center* will be the center of curvature of the wavefront. When used as the primary antenna for a parabolic reflector, the horn is positioned so that the phase center lies on the focus.

The focal length can be given in terms of the depth of the reflector and its diameter. It is sometimes useful to know the focal length for setting up a receiving system. The depth l is the perpendicular distance from the aperture plane to the vertex. This relationship is shown in App. B to be

$$f = \frac{D^2}{16l} \tag{5.27}$$

The gain and beamwidths of the paraboloidal antenna are as follows. The physical area of the aperture plane is

$$\text{Area} = \frac{\pi D^2}{4} \tag{5.28}$$

From the relationships given by Eqs. (5.11) and (5.12) the gain is

$$G = \frac{4\pi}{\lambda^2}\, \eta_I\, \text{area}$$

$$= \eta_I \left(\frac{\pi D}{\lambda}\right)^2 \tag{5.29}$$

The radiation pattern for the paraboloidal reflector is similar to that developed in Example 5.1 for the rectangular aperture, in that there is a main lobe and a number of sidelobes, although there will be differences in detail. In practice the sidelobes are accounted for by an envelope function as described in Chap. 11. Useful approximate formulas for the half-power beamwidth and the beamwidth between the first nulls (BWFN) are

$$\text{HPBW} \cong 70\,\frac{\lambda}{D} \tag{5.30}$$

$$\text{BWFN} \cong 2\text{HPBW} \tag{5.31}$$

In these relationships the beamwidths are given in degrees. The paraboloidal antenna described so far is *center-fed* in that the primary horn is pointed toward the center of the reflector. With this arrangement the primary horn and its supports present a partial blockage to the reflected wave. The energy scattered by the blockage is lost from the main lobe, and it can create additional sidelobes. One solution is to use an *offset feed* as described in the next section.

The wave from the primary radiator induces surface currents in the reflector. The curvature of the reflector causes the currents to follow curved paths, so that both horizontal and vertical components are present, even where the incident wave is linearly polarized in one or other of these directions. The situation is sketched for the case of vertical polarization in Fig. 5.19. The resulting radiation consists of copolarized and cross-polarized fields. The symmetry of the arrangement means that the cross-polarized component is zero in the principal

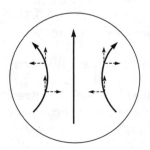

Figure 5.19 Current paths in a paraboloidal reflector for linear polarization.

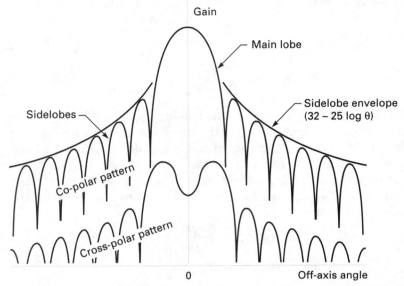

Figure 5.20 Copolar and cross-polar radiation patterns. (*From FCC Report FCC/OST R83-2, 1983.*)

planes (the *E* and *H* planes). Cross-polarization peaks in the $\phi = \pm 45°$ planes, assuming a coordinate system as shown in Fig. 5.5(*a*). Sketches of the copolar and cross-polar radiation patterns for the 45° planes are shown in Fig. 5.20.

5.14   The Offset Feed

Figure 5.21*a* shows a paraboloidal reflector with a horn feed at the focus. In this instance the radiation pattern of the horn is offset so that it illuminates only the upper portion of the reflector. The feed horn and its support can be placed well clear of the main beam so that no blockage occurs. With the center-fed arrangement described in the previous section, the blockage results typically in a 10 percent reduction in efficiency (Brain and Rudge, 1984) and increased radiation in the sidelobes. The offset arrangement avoids this. Figure 5.21*b* shows a development model of an offset antenna intended for use in the European Olympus satellite.

The main disadvantages of the offset feed are that a stronger mechanical support is required to maintain the reflector shape, and, because of the asymmetry, the cross-polarization with a linear polarized feed is worse compared to the center-fed antenna. Polarization compensation can be introduced into the primary feed to correct for the cross-polarization, or a *polarization-purifying grid* can be incorporated into

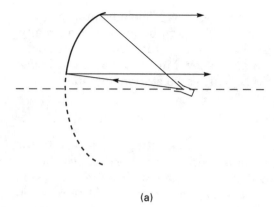

(a)

(b)

Figure 5.21 The offset feed for a paraboloidal reflector. (*From Brain and Rudge 1984. With permission*)

the antenna structure (Brain and Rudge, 1984). The advantages of the offset feed are sufficiently attractive for it to be standard on many satellites (see, for example, Figs. 6.6 and 6.22). It is also used with double-reflector earth station antennas as shown in Fig. 5.24, and is being increasingly used with small receive-only earth station antennas.

5.15 Double-Reflector Antennas

With reflector-type antennas, the feeder connecting the feed horn to the transmit/receive equipment must be kept as short as possible to minimize losses. This is particularly important with large earth stations where the transmit power is large, and where very low receiver noise is required. The single-reflector system described in the previous section does not lend itself very well to achieving this, and more satisfactory, but more costly, arrangements are possible with a double-reflector system. The feed horn is mounted at the rear of the main reflector through an opening at the vertex as illustrated in Fig. 5.22. The rear mount makes for a compact feed, which is an advantage where steerable antennas must be used, and access for servicing is easier. The subreflector, which is mounted at the front of the main reflector, is generally smaller than the feed horn and causes less blockage.

Figure 5.22 A 19-m Cassegrain antenna. (*Courtesy TIW Systems Inc.*)

Two main types are in use, the Cassegrain antenna and the Gregorian antenna, named after the astronomers who first developed them.

Cassegrain Antenna.   The basic Cassegrain form consists of a main paraboloid and a subreflector, which is a hyperboloid (see App. B). The subreflector has two focal points, one of which is made to coincide with that of the main reflector and the other with the phase center of the feed horn, as shown in Fig. 5.23*a*. The

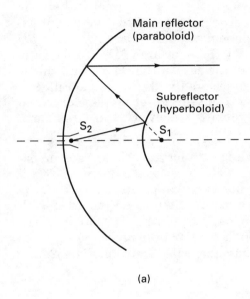

(a)

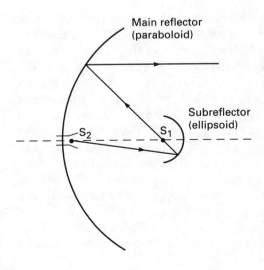

Figure 5.23 Ray paths for (*a*) Cassegrain and (*b*) Gregorian antennas.

(b)

Cassegrain system is equivalent to a single paraboloidal reflector of focal length

$$f_e = \frac{e_h + 1}{e_h - 1} f \tag{5.32}$$

where e_h is the eccentricity of the hyperboloid (see App. B) and f is the focal length of the main reflector. The eccentricity of the hyperboloid is always greater than unity, and typically ranges from about 1.4 to 3. The equivalent focal length therefore is greater than the focal length of the main reflector. The diameter of the equivalent paraboloid is the same as that of the main reflector, and hence the f/D ratio is increased. As shown in Fig. 5.18, a large f/D ratio leads to more uniform illumination, and in the case of the Cassegrain this is achieved without the spillover associated with the single-reflector system. The larger f/D ratio also results in lower cross-polarization (Miya, 1981). The Cassegrain system is widely used in large earth station installations.

Gregorian Antenna. The basic Gregorian form consists of a main paraboloid and a subreflector, which is an ellipsoid (see App. B). As with the hyperboloid, the subreflector has two focal points, one of which is made to coincide with that of the main reflector and the other with the phase center of the feed horn, as shown in Fig. 5.23(*b*). The performance of the Gregorian system is similar in many respects to the Cassegrain. An offset Gregorian antenna is illustrated in Fig. 5.24.

5.16 Shaped Reflector Systems

With the double-reflector systems described, the illumination efficiency of the main reflector can be increased while avoiding the problem of increased spillover, by shaping the surfaces of the subreflector and main reflector. With the Cassegrain system, for example, altering the curvature of the center section of the subreflector to be greater than that of the hyperboloid allows it to reflect more energy toward the edge of the main reflector, which makes the amplitude distribution more uniform. At the same time, the curvature of the center section of the main reflector is made smaller than that required for the paraboloid. This compensates for the reduced path length so that the constant phase condition across the aperture is maintained. The edge of the subreflector surface is shaped in a manner to reduce spillover, and of course the overall design must take into account the radiation pattern of the primary feed. The process, referred to as *reflector shaping,* employs computer-aided design methods. Further details will be found in Miya, 1981, and Rusch, 1992.

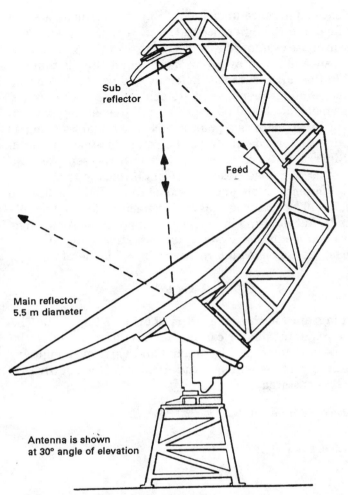

Sub
reflector

Feed

Main reflector
5.5 m diameter

Antenna is shown
at 30° angle of elevation

Figure 5.24   Offset Gregorian antenna. (*From "Radio Electr. Eng.,"*
vol. 54, no. 3, Mar. 1984, p. 112; with permission.)

5.17   Arrays

Beam shaping can be achieved by using an array of basic elements. The
elements are arranged so that their radiation patterns provide mutual
reinforcement in certain directions and cancellation in others. Although
most arrays used in satellite communications are two-dimensional, the
principle is most easily explained with reference to an in-line array
(Fig. 5.25*a* and *b*). As shown previously (Fig. 5.8), the radiation pattern
for a single dipole in the *x-y* plane is circular, and it is this aspect of the
radiation pattern that is altered by the array configuration. Two factors

contribute to this: the difference in distance from each element to some point in the far field and the difference in the current feed to each element. For the coordinate system shown in Fig. 5.25b, the x-y plane, the difference in distance is given by $s \cos \phi$. Although this distance is small compared to the range between the array and point P, it plays a crucial role in determining the phase relationships between the radiation from each element. It should be kept in mind that at any point in the far field the array appears as a point source, the situation being as sketched in Fig. 5.25c. For this analysis, the point P is taken to lie in the x-y plane. Since a distance of one wavelength corresponds to a phase difference of 2π, the phase lead of element n relative to $n - 1$ resulting from the difference in distance is $(2\pi/\lambda)s \cos \alpha$. To illustrate the array principles, it will be assumed that each element is fed by currents of equal magnitude, but differing in phase progressively by some angle α. Positive values of α mean a phase lead, and negative values a phase lag. The total phase lead of element n relative to $n - 1$ is therefore

$$\Psi = \alpha + \frac{2\pi}{\lambda} s \cos \phi \qquad (5.33)$$

The Argand diagram for the phasors is shown in Fig. 5.26. The magnitude of the resultant phasor can be found by first resolving the individual phasors into horizontal (real axis) and vertical (imaginary axis) components, adding these, and finding the resultant. Mathematically this is stated as

$$E_R = E + E \cos \Psi + jE \sin \Psi + E \cos 2\Psi + jE \sin 2\Psi + \cdots$$

$$= \sum_{n=0}^{N-1} E \cos n\Psi + jE \sin n\Psi$$

$$= E \sum_{n=0}^{N-1} e^{jn\Psi} \qquad (5.34)$$

Here N is the total number of elements in the array. A single element would have resulted in a field E, and the array is seen to modify this by the summation factor. The magnitude of summation factor is termed the *array factor* (AF):

$$\mathrm{AF} = \left| \sum_{n=0}^{N-1} e^{jn\Psi} \right| \qquad (5.35)$$

The array factor has a maximum value of N when $\Psi = 0$. Note that this just means that $E_{R\max} = NE$. By recalling that Ψ as given by Eq. (5.33) is a function of the current phase angle α and the angular coordinate ϕ, it is possible to choose the current phase to make array factor show a peak in some desired direction ϕ_0. The required relationship is, from Eq. (5.33),

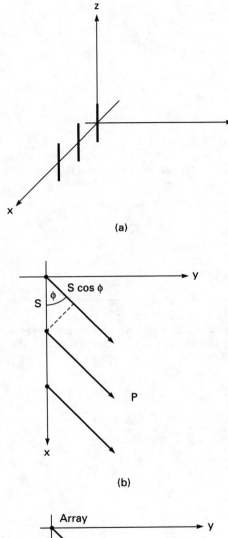

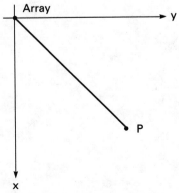

Figure 5.25 An in-line array of dipoles.

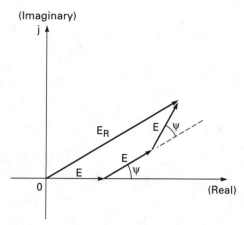

(Imaginary)

Figure 5.26 Phasor diagram for the in-line array of dipoles.

$$\alpha = -\frac{2\pi}{\lambda} s \cos \phi_0 \tag{5.36}$$

This is illustrated in the following example.

Example 5.2

A dipole array has 5 elements equispaced at 0.25 wavelengths. The array factor is required to have a maximum along the positive axis of the array. Plot the magnitude of the array factor as a function of ϕ.

solution

Given data are:

$$N: = 5 \qquad s: = .25 \qquad \phi_0: = 0 \cdot \deg$$

Set the current phase to:

$$\alpha: = -2 \cdot \pi \cdot s \cdot \cos(\phi_0)$$

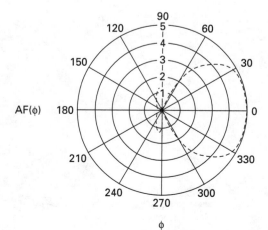

Set up the variable:

$$\phi: = -180 \cdot \deg, -175 \cdot \deg .. 180 \cdot \deg$$

The total phase angle is:

$$\Psi(\phi): = \alpha + 2 \cdot \pi \cdot s \cdot \cos(\phi)$$

The array factor as plotted on the previous page, is:

$$AF(\phi): = \left| \sum_{n=0}^{N-1} e^{j \cdot n \cdot \Psi(\phi)} \right|$$

As the example shows, the array factor has a main peak at $\phi = 0$ and some sidelobes. For this particular example the values were purposely chosen to illustrate what is termed an *end-fire* array where the main beam is directed along the positive axis of the array. Keep in mind that a single dipole would have had a circular pattern.

The current phasing can be altered to make the main lobe appear at $\phi = 90°$, giving rise to a *broadside* array. The symmetry of the dipole array means that two broadside lobes occur, one on each side of the array axis. This is illustrated in the next example.

Example 5.3

Repeat the previous example for $\phi = 90$ degrees.

solution

Given data are:

$$N: = 5 \qquad s: = .25 \qquad \phi_0: = 90 \cdot \deg$$

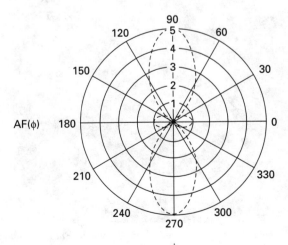

Set the current phase to:

$$\alpha: = -2 \cdot \pi \cdot s \cdot \cos(\phi_0)$$

Set up the variable:

$$\phi: = -180 \cdot \deg, -175 \cdot \deg .. 180 \cdot \deg$$

The total phase angle is:

$$\Psi(\phi): = \alpha + 2 \cdot \pi \cdot s \cdot \cos(\phi)$$

The array factor as plotted on the previous page, is:

$$AF(\phi): = \left| \sum_{n=0}^{N-1} e^{j \cdot n \cdot \Psi(\phi)} \right|$$

Figure 5.27 A multifeed contained beam reflector antenna. (*From Brain and Rudge, 1984. With permission.*)

As these examples show, the current phasing controls the position of the main lobe, and a continuous variation of current can be used to produce a *scanning array*. With the simple dipole array, the shape of the beam changes drastically with changes in the current phasing, and in practical scanning arrays, steps are taken to avoid this. A detailed discussion of arrays will be found in Kummer, 1992.

Arrays may be used directly as antennas, and details of a nine-horn array used to provide an earth coverage beam are given in Hwang, 1992. Arrays are also used as feeders for reflector antennas, and such a horn array is shown in Fig. 5.27.

5.18 Problems

5.1 The power output from a transmitter amplifier is 600 W. The feeder losses amount to 1 dB, and the voltage reflection coefficient at the antenna is 0.01. Calculate the radiated power.

5.2 Explain what is meant by the *reciprocity theorem* as applied to antennas. A voltage of 100 V applied at the terminals of a transmitting dipole antenna results in an induced current of 3 mA in a receiving dipole antenna. Calculate the current induced in the first antenna when a voltage of 350 V is applied to the terminals of the second antenna.

5.3 The position of a point in the coordinate system of Sec. 5.3 is given generally as $r\,(\theta, \phi)$. Determine the $x, y,$ and z coordinates of a point $3(30°, 20°)$.

5.4 What are the main characteristics of a radiated wave in the far field region? The components of a wave in the far field region are $E_\theta = 3$ mV/m, $E_\phi = 4$ mV/m. Calculate the magnitude of the total electric field. Calculate also the magnitude of the magnetic field.

5.5 The **k** vector for the wave specified in Prob. 5.4 is directed along the $+x$ axis. Determine the direction of the resultant electric field in the y-z plane.

5.6 The **k** vector for the wave specified in Prob. 5.4 is directed along the $+z$ axis. Is there sufficient information given to determine the direction of the resultant electric field in the x-y plane? Give reasons for your answer.

5.7 The magnitude of the electric field of a wave in the far field region is 3μV/m. Calculate the power flux density.

5.8 Explain what is meant by the *isotropic power gain* of an antenna. The gain of a reflector antenna relative to a $\frac{1}{2}\lambda$- dipole feed is 49 dB. What is the isotropic gain of the antenna?

5.9 The directive gain of an antenna is 52 dB, and the antenna efficiency is 0.95. What is the power gain of the antenna?

5.10 The radiation pattern of an antenna is given by $g(\theta, \phi) = |\sin\theta \sin\phi|$. Plot the resulting patterns for (*a*) the x-z plane and (*b*) the y-z plane.

5.11 For the antenna in Prob. 5.10, determine the half-power beamwidths, and hence determine the directive gain.

5.12 Explain what is meant by the *effective aperture* of an antenna. A paraboloidal reflector antenna has a diameter of 3 m and an illumination efficiency of 70 percent. Determine (a) its effective aperture and (b) its gain at a frequency of 4 GHz.

5.13 What is the effective aperture of an isotropic antenna operating at a wavelength of 1 cm?

5.14 Determine the half-power beamwidth of a half-wave dipole.

5.15 A uniformly illuminated rectangular aperture has dimensions $a = 4\lambda$, $b = 3\lambda$. Plot the radiation patterns in the principal planes.

5.16 Determine the half-power beamwidths in the principal planes for the uniformly illuminated aperture of Prob. 5.15. Hence determine the gain. State any assumptions made.

5.17 Explain why the smooth-walled conical horn radiates copolar and cross-polar field components. Why is it desirable to reduce the cross-polar field as far as practical, and state what steps can be taken to achieve this.

5.18 When the rectangular aperture shown in Fig. 5.9 is fed from a waveguide operating in the TE_{10} mode, the far field components (normalized to unity) are given by

$$E_\theta(\theta, \phi) = -\frac{\pi}{2} \sin \phi \, \frac{\cos X}{X^2 - \left(\frac{\pi}{2}\right)^2} \, \frac{\sin Y}{Y}$$

$$E_\phi(\theta, \phi) = E_\theta(\theta, \phi) \cos \theta \cot \phi$$

where X and Y are given by Eqs. (5.16) and (5.17). The aperture dimensions are $a = 3\lambda$, $b = 2\lambda$. Plot the radiation patterns in the principal planes.

5.19 Determine the half-power beamwidths in the principal planes for the aperture specified in Prob. 5.18, and hence determine the directive gain.

5.20 A pyramidal horn antenna has dimensions $a = 4\lambda$, $b = 2.5\lambda$ and an illumination efficiency of 70 percent. Determine the gain.

5.21 What are the main characteristics of a parabolic reflector that make it highly suitable for use as an antenna reflector?

5.22 Explain what is meant by the *space attenuation function* in connection with the paraboloidal reflector antenna.

5.23 Figure 5.17b can be referred to x-y rectangular coordinates with A at the origin and the x axis directed from A to S. The equation of the parabola is then $y^2 = 4fx$. Given that $y_{\max} = \pm2.5$ m at $x_{\max} = 0.9$ m, plot the space attenuation function.

5.24   What is the *f/D* ratio for the antenna of Prob. 5.23? Sketch the position of the focal point in relation to the reflector.

5.25   Determine the depth of the reflector specified in Prob. 5.23.

5.26   A 3-m paraboloidal dish has a depth of 1 m. Determine the focal length.

5.27   A 5-m paraboloidal reflector works with an illumination efficiency of 65 percent. Determine its effective aperture and gain at a frequency of 6 GHz.

5.28   Determine the half-power beamwidth for the reflector antenna of Prob. 5.27. What is the beamwidth between the first nulls?

5.29   Describe briefly the *offset feed* used with paraboloidal reflector antennas, stating its main advantages and disadvantages.

5.30   Explain why double-reflector antennas are often used with large earth stations.

5.31   Describe briefly the main advantages to be gained in using an antenna array.

5.32   A basic dipole array consists of five equispaced dipole elements configured as shown in Fig. 5.25. The spacing between elements is 0.3λ. Determine the current phasing needed to produce an end-fire pattern. Provide a polar plot of the array factor.

5.33   What current phasing would be required for the array in Prob. 5.32 to produce a broadside pattern?

5.34   A four-element dipole array, configured as shown in Fig. 5.25, is required to produce maximum radiation in a direction $\phi_0 = 15°$. The elements are spaced by 0.2λ. Determine the current phasing required, and provide a polar plot of the array factor.

The Space Segment

6.1 Introduction

A satellite communications system can be broadly divided into two segments, a ground segment and a space segment. The space segment will obviously include the satellites, but it also includes the ground facilities needed to keep the satellites operational, these being referred to as the *tracking, telemetry, and command* (TT&C) facilities. In many networks it is common practice to employ a ground station solely for the purpose of TT&C.

The equipment carried aboard the satellite can also be classified according to function. The *payload* refers to the equipment used to provide the service for which the satellite has been launched. The *bus* refers not only to the vehicle which carries the payload, but also to the various subsystems which provide the power, attitude control, orbital control, thermal control, and command and telemetry functions required to service the payload.

In a communications satellite, the equipment which provides the connecting link between the satellite's transmit and receive antennas is referred to as the *transponder.* The transponder forms one of the main sections of the payload, the other being the antenna subsystems. In this chapter the main characteristics of certain bus systems and payloads are described.

6.2 The Power Supply

The primary electrical power for operating the electronic equipment is obtained from solar cells. Individual cells can generate only small amounts of power, and therefore arrays of cells in series-parallel connection are required. Figure 6.1 shows the solar cell panels for the HS 376 satellite manufactured by Hughes Space and Communications Company. The spacecraft is 216 cm in diameter and 660 cm long when

Figure 6.1 At a total of 41 satellites, the HS 376 is the world's most purchased commercial communications satellite. The HS 376 series, built by Hughes Space and Communications Company, can provide voice, video, and data telecommunication services. The first HS 376 was launched Nov. 15, 1980, from Cape Canaveral, Fla. In 1982 an HS 376 became the first satellite to be placed into orbit by the Space Shuttle. (*Courtesy Hughes Aircraft Company Space and Communications Group.*)

fully deployed in orbit. During the launch sequence the outer cylinder is telescoped over the inner one, to reduce the overall length. Only the outer panel generates electrical power during this phase. In geostationary orbit the telescoped panel is fully extended, so that both are exposed to sunlight. At the beginning of life the panels produce 940 W dc power, which may drop to 760 W at the end of 10 years. During eclipse, power is provided by two nickel-cadmium long-life batteries, which will deliver 830 W. At the end of life, battery recharge time is less than 16 h.

At the time of writing (1994) the HS 376 spacecraft is the world's most-purchased commercial communications satellite. It is a spin-stabilized spacecraft (the gyroscopic effect of the spin is used for mechanical orientational stability as described in Sec. 6.3). Thus the arrays are only partially in sunshine at any given time, which places a limitation on power.

Higher powers can be achieved with solar panels arranged in the form of rectangular "solar sails." Solar sails must be folded during the launch phase, and extended when in geostationary orbit. Figure 6.2 shows the HS 601 satellite manufactured by Hughes Space and

Figure 6.2 Aussat B1 (renamed Optus B), Hughes' first HS 601 communications satellite, is prepared for environmental testing. Sensors are being attached to the spacecraft to monitor the effects of vibration and thermal-vacuum testing. These tests simulate the vibration of launch as well as the vacuum, heat from the sun, and coldness of space. (*Courtesy Hughes Aircraft Company, Space and Communications Group.*)

Communications Company. As shown, the solar sails are folded up on each side, and when fully extended they stretch to 67 ft (316.5 cm) from tip to tip. The full complement of solar cells is exposed to the sunlight, and the sails are arranged to rotate to track the sun, so they are capable of greater power output than cylindrical arrays having a comparable number of cells. The HS 601 can be designed to provide dc power from 2 to 6 kW. In comparing the power capacity of cylindrical and solar-sail satellites, the crossover point is estimated to be about 2 kW where the solar-sail type is more economical than the cylindrical type (Hyndman, 1991).

As discussed in Sec. 2.9.3, the earth will eclipse a geostationary satellite twice a year, during the spring and autumnal equinoxes. Daily eclipses start approximately 23 days before, and end approximately 23 days after, the equinox for both the spring and autumnal equinoxes, and can last up to 72 min at the actual equinox days. Figure 6.3 shows the graph relating eclipse period to the day of year. In order to maintain service during an eclipse, storage batteries must be provided. Nickel-cadmium (Ni-Cd) batteries continue to be in use as shown in the Hughes HS 376 satellite, but developments in nickel-hydrogen (Ni-H$_2$) batteries offer significant improvement in power/weight ratio. Nickel-hydrogen batteries are used in the Hughes

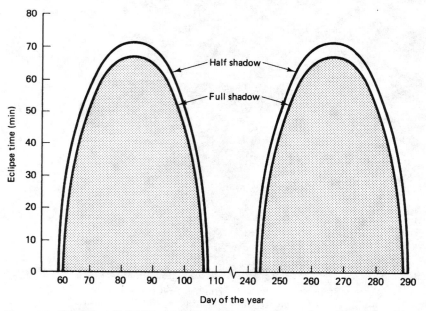

Figure 6.3 Satellite eclipse time as a function of the current day of the year. (*From Spilker, 1977. Reprinted by permission of Prentice-Hall, Engewood Cliffs, New Jersey.*)

HS 601 and in the Intelsat VI (Pilcher, 1982) and Intelsat VII (Lilly, 1990) satellites.

6.3 Attitude Control

The *attitude* of a satellite refers to its orientation in space. Much of the equipment carried aboard a satellite is there for the purpose of controlling its attitude. Attitude control is necessary, for example, to ensure that directional antennas point in the proper directions. In the case of earth environmental satellites, the earth-sensing instruments must cover the required regions of the earth, which also requires attitude control. A number of forces, referred to as *disturbance torques*, can alter the attitude, some examples being the gravitational fields of the earth and the moon, solar radiation, and meteorite impacts. Attitude control must not be confused with *station keeping*, which is the term used for maintaining a satellite in its correct orbital position, although the two are closely related.

To exercise attitude control, there must be available some measure of a satellite's orientation in space and of any tendency for this to shift. In one method, infrared sensors, referred to as *horizon detectors*, are used to detect the rim of the earth against the background of space. With the use of four such sensors, one for each quadrant, the center of the earth can be readily established as a reference point. Any shift in orientation is detected by one or other of the sensors and a corresponding control signal is generated which activates a restoring torque.

Usually, the attitude-control process takes place aboard the satellite, but it is also possible for control signals to be transmitted from earth, based on attitude data obtained from the satellite. Also, where a shift in attitude is desired, an *attitude maneuver* is executed. The control signals needed to achieve this maneuver may be transmitted from an earth station.

Controlling torques may be generated in a number of ways. *Passive attitude control* refers to the use of mechanisms which stabilize the satellite without putting a drain on the satellite's energy supplies; at most, infrequent use is made of these supplies, for example when thruster jets are impulsed to provide corrective torque. Examples of passive attitude control are *spin stabilization* and *gravity gradient stabilization*. The latter depends on the interaction of the satellite with the gravitational field of the central body, and has been used for example with the Radio Astronomy Explorer-2 satellite which was placed in orbit around the moon (Wertz, 1984). For communications satellites, spin stabilization is often used, and this is described in more detail in Sec. 6.3.1.

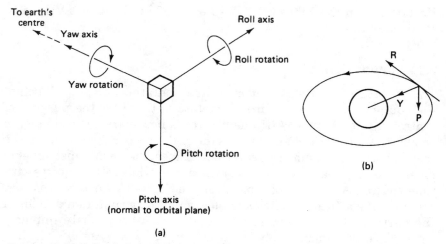

Figure 6.4 (a) Roll, pitch, and yaw axes. The yaw axis is directed toward the earth's center, the pitch axis is normal to the orbital plane, and the roll axis is perpendicular to the other two. (b) *RPY* axes for the geostationary orbit. Here the roll axis is tangential to the orbit, and lies along the satellite velocity vector.

The other form of attitude control is *active control*. With active attitude control there is no overall stabilizing torque present to resist the disturbance torques. Instead, corrective torques are applied as required in response to disturbance torques. Methods used to generate active control torques include momentum wheels, electromagnetic coils, and mass expulsion devices such as gas jets and ion thrusters. The electromagnetic coil works on the principle that the earth's magnetic field exerts a torque on a current-carrying coil, and that this torque can be controlled through control of the current. However, the method is of use only for satellites relatively close to the earth. The use of momentum wheels is described in more detail in Sec. 6.3.2.

The three axes which define a satellite's attitude are its *roll, pitch,* and *yaw* (RPY) axes. These are shown relative to the earth in Fig. 6.4. All three axes pass through the center of gravity of the satellite. For an equatorial orbit, movement of the satellite about the roll axis moves the antenna footprint north and south; movement about the pitch axis moves the footprint east and west; and movement about the yaw axis rotates the antenna footprint.

6.3.1 Spin stabilization

Spin stabilization is used with cylindrical satellites. The satellite is constructed so that it is mechanically balanced about one particular axis and is then set spinning around this axis. For geostationary satellites the spin axis is adjusted to be parallel to the N-S axis of the

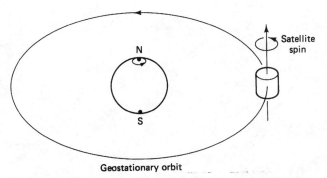

Figure 6.5 Spin stabilization in the geostationary orbit. The spin axis lies along the pitch axis, parallel to the earth's N-S axis.

earth as illustrated in Fig. 6.5. Spin rate is typically in the range of 50 to 100 rev/min.

In the absence of disturbance torques, the spinning satellite would maintain its correct attitude relative to the earth. Disturbance torques are generated in a number of ways, both external and internal to the satellite. Solar radiation, gravitational gradients and meteorite impacts are all examples of external forces which can give rise to disturbance torques. Motor-bearing friction and the movement of satellite elements such as the antennas can also give rise to disturbance torques. The overall effect is that the spin rate will decrease and the direction of the angular spin axis will change. Impulse-type thrusters, or jets, can be used to increase the spin rate again and to shift the axis back to its correct N-S orientation. *Nutation,* which is a form of wobbling, can occur as a result of the disturbance torques and/or from misalignment or unbalance of the control jets. This nutation must be damped out by means of energy absorbers known as *nutation dampers*.

Two forms of spin stabilization are commonly employed. In what is referred to simply as *spin stabilization,* the entire satellite rotates about an axis which for earth-orbiting satellites is the pitch axis. Where an omnidirectional antenna is used (for example as shown for the Intelsat I and II satellites in Fig. 1.1), the antenna, which points along the pitch axis, also rotates with the satellite. Where a directional antenna is used, which is more common for communications satellites, the antenna must be despun, giving rise to a dual-spin construction. An electric motor drive is used for despinning the antenna subsystem.

Figure 6.6 shows the Hughes HS 376 satellite in more detail. The antenna subsystem consists of a parabolic reflector and feed horns

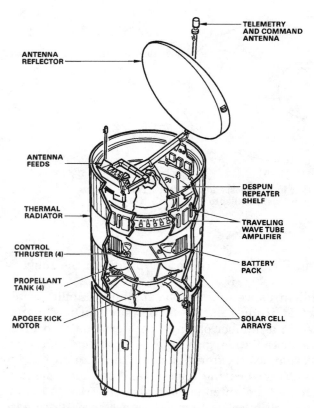

TELEMETRY
AND COMMAND
ANTENNA

ANTENNA
REFLECTOR

ANTENNA
FEEDS

DESPUN
REPEATER
SHELF

THERMAL
RADIATOR

TRAVELING
WAVE TUBE
AMPLIFIER

CONTROL
THRUSTER (4)

BATTERY
PACK

PROPELLANT
TANK (4)

APOGEE KICK
MOTOR

SOLAR CELL
ARRAYS

Figure 6.6 HS 376 Spacecraft configuration. (*Courtesy Hughes Aircraft Company Space and Communications Group.*)

mounted on the despun shelf, which also carries the communications repeaters (transponders). The antenna feeds can therefore be connected directly to the transponders without the need for radio-frequency (RF) rotary joints, while the complete platform is despun. Of course control signals and power must be transferred to the despun section, and a mechanical bearing must be provided. The complete assembly for this is known as the *bearing and power transfer assembly* (BAPTA). Figure 6.7 shows a photograph of the internal structure of the HS 376.

Some dual-spin spacecraft obtain spin stabilization from a spinning flywheel rather than by spinning the satellite itself. These flywheels are termed *momentum wheels,* and their average momentum is referred to as *momentum bias.* Reaction wheels, described in the next section, operate at zero momentum bias.

Figure 6.7 Technicians check the alignment of the Telstar 3 communications satellite, shown without its cylindrical solar panels. The satellite, built for American Telephone and Telegraph Co., carries both traveling wave tube and solid state power amplifiers, as shown on the communications shelf surrounding the center of the spacecraft. The traveling wave tubes are the cylindrical instruments. (*Courtesy Hughes Aircraft Company Space and Communications Group.*)

6.3.2 Three-axis stabilization

In *three-axis stabilization,* as the name suggests, there are stabilizing elements for each of the three axes, roll, pitch, and yaw (Fig. 6.4). Because the body of the satellite remains fixed relative to the earth, three-axis stabilization is also known as *body stabilization.*

Active attitude control is used with three-axis stabilization. This may take the form of control jets (mass-expulsion controllers) fired to correct the attitude of the satellite. Reaction wheels can also be used.

A reaction wheel is a flywheel which is normally stationary but reacts when a disturbance torque tends to shift the spacecraft orientation, by gathering momentum until it absorbs the effect of the disturbance torque. In practice various combinations of wheels and mass-expulsion devices are used (Fig. 6.8). Recall from Sec. 6.3.1 that the flywheels may also be operated at nonzero momentum, this being referred to as momentum bias.

Whether the disturbance torques are random or cyclic, the momentum gathered by a reaction wheel will, on the average, tend to be

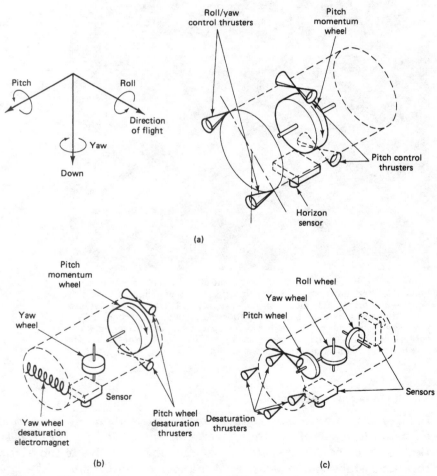

Figure 6.8 Alternative momentum wheel stabilization systems: (*a*) one-wheel; (*b*) two-wheel; (*c*) three-wheel. (*Reprinted with permission from "Spacecraft Attitude Determination and Control," edited by James R. Wertz. Copyright © 1984 by D. Reidel Publishing Company, Dordrecht, Holland.*)

zero. However, there will always be some disturbance torques which cause a cumulative increase in wheel momentum and eventually at some point the wheel *saturates*. In effect it reaches its maximum allowable angular velocity and can no longer take in any more momentum. Mass expulsion devices are then used to unload the wheel, that is, remove momentum from it (in the same way a brake removes energy from a moving vehicle). Of course, operation of the mass-expulsion devices consumes part of the satellite's fuel supply.

6.4 Station Keeping

In addition to having its attitude controlled, it is important that a geostationary satellite be kept in its correct orbital slot. As described in Sec. 2.8.2, the equatorial ellipticity of the earth causes geostationary satellites to drift slowly along the orbit, to one of two stable points, at 75° E and 105° W. To counter this drift, an oppositely directed velocity component is imparted to the satellite by means of jets, which are pulsed once every 2 or 3 weeks. This results in the satellite drifting back through its nominal station position, coming to a stop, and recommencing the drift along the orbit until the jets are pulsed once again. These maneuvers are termed *east-west station-keeping maneuvers*. Satellites in the 6/4-GHz band must be kept within ±0.1° of the designated longitude, and in the 14/12-GHz band, within ±0.05°.

A satellite which is nominally geostationary will also drift in latitude, the main perturbing forces being the gravitational pull of the sun and the moon. These forces cause the inclination to change at a rate of about 0.85°/year. If left uncorrected, the drift would result in a cyclic change in the inclination, going from 0 to 14.67° in 26.6 years (Spilker, 1977), and back to zero, at which the cycle is repeated. To prevent the shift in inclination from exceeding specified limits, jets may be pulsed at the appropriate time to return the inclination to zero. Counteracting jets must be pulsed when the inclination is at zero to halt the change in inclination. These maneuvers are termed *north-south station-keeping maneuvers,* and they are much more expensive in fuel than are east-west station-keeping maneuvers. The north-south station-keeping tolerances are the same as those for east-west station keeping, ±0.1° in the C band and ±0.05° in the Ku band.

Orbital correction is carried out by command from the TT&C earth station, which monitors the satellite position. East-west and north-south station-keeping maneuvers are usually carried out using the same thrusters as are used for attitude control. Figure 6.9 shows typical latitude and longitude variations for the Canadian Anik-C3 satellite which remain after station-keeping corrections are applied.

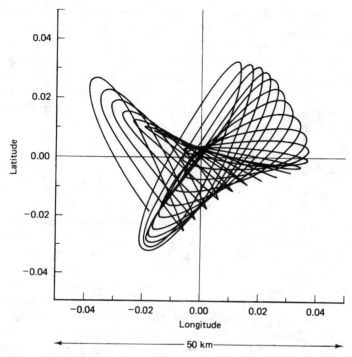

Figure 6.9 Typical satellite motion. (*From Telesat, Canada, 1983; courtesy Telesat, Canada.*)

Satellite altitude will also show variations of about ±0.1 percent of the nominal geostationary height. If, for sake of argument, this is taken as 36,000 km, the total variation in the height is 72 km. A C-band satellite therefore can be anywhere within a box bound by this height and the ±0.1° tolerances on latitude and longitude. Approximating the geostationary radius as 42,000 km (see Sec. 2.11), an angle of 0.2° subtends an arc of approximately 147 km. Thus, both the latitude and longitude sides of the box are 147 km. The situation is sketched in Fig. 6.10, which also shows the relative beamwidths of a 30-m and a 5-m antenna. As shown by Eq. (5.30), the −3-dB beamwidth of a 30-m antenna is about 0.1°, and of a 5-m antenna, about 0.7° at 6 GHz. Assuming 42,000 km for the slant range, the diameter of the 30-m beam at the satellite will be about 85 km. This beam does not encompass the whole of the box and could therefore miss the satellite. Such narrow-beam antennas must therefore track the satellite.

The diameter of the 5-m antenna beam at the satellite will be about 513 km, and this does encompass the box, so that tracking is not re-

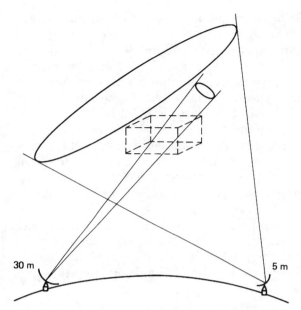

Figure 6.10 The rectangular box shows the positional limits for a satellite in geostationary orbit, in relation to beams from a 30-m antenna and a 5-m antenna.

quired. The positional uncertainty of the satellite also introduces an uncertainty in propagation time, which can be a significant factor in certain types of communications networks.

By placing the satellite in an inclined orbit, the north-south station-keeping maneuvers may be dispensed with. The savings in weight achieved by not having to carry fuel for these maneuvers allows the communications payload to be increased. The satellite is placed in an inclined orbit of about 2.5 to 3°, in the opposite sense to that produced by drift. Over a period of about half the predicted lifetime of the mission, the orbit will change to equatorial and then continue to increase in inclination. However, this arrangement requires the use of tracking antennas at the ground stations.

6.5 Thermal Control

Satellites are subject to large thermal gradients, receiving the sun's radiation on one side while the other side faces into space. In addition, thermal radiation from the earth and the earth's albedo, which is the fraction of the radiation falling on earth which is reflected, can be significant for low-altitude earth-orbiting satellites, although it is negligible for geostationary satellites. Equipment in the satellite also generates heat which has to be removed. The most important consideration is that the satellite's equipment should operate as nearly as possible in a stable temperature environment. Various steps are taken to achieve this. Thermal blankets and shields may be used to provide insulation. Radiation mirrors are often used to remove heat from the communications payload. The mirrored thermal radiator for the Hughes HS 376 satellite can be seen in Fig. 6.1 and in Fig. 6.6. These mirrored drums surround the communications equipment shelves in each case, and provide good radiation paths for the generated heat to escape into the surrounding space. One advantage of spinning satellites compared to body-stabilized is that the spinning body provides an averaging of the temperature extremes experienced from solar flux and the cold background of deep space.

In order to maintain constant temperature conditions, heaters may be switched on (usually on command from ground) to make up for the heat reduction which occurs when transponders are switched off. In Intelsat VI, heaters are used to maintain propulsion thrusters and line temperatures (Pilcher, 1982).

6.6   TT&C Subsystem

The telemetry, tracking, and command subsystem performs several routine functions aboard the spacecraft. The telemetry, or telemetering, function could be interpreted as *measurement at a distance*. Specifically it refers to the overall operation of generating an electrical signal proportional to the quantity being measured, and encoding and transmitting this to a distant station, which for the satellite is one of the earth stations. Data which are transmitted as telemetry signals include attitude information such as that obtained from sun and earth sensors; environmental information such as the magnetic field intensity and direction, the frequency of meteorite impact, and so on; and spacecraft information such as temperatures, power supply voltages, and stored-fuel pressure. Certain frequencies have been designated by international agreement for satellite telemetry transmissions. During the transfer and drift orbital phases of the satellite launch, a special channel is used along with an omnidirectional antenna. Once the satellite is on station, one of the normal communications transponders may be used along with its directional antenna, unless some emergency arises which makes it necessary to switch back to the special channel used during the transfer orbit.

Telemetry and command may be thought of as complementary functions. The telemetry subsystem transmits information about the satellite to the earth station, while the command subsystem receives command signals from the earth station, often in response to telemetered information. The command subsystem demodulates and if necessary decodes the command signals, and routes these to the appropriate equipment needed to execute the necessary action. Thus attitude changes may be made, communication transponders switched in and out of circuits, antennas redirected, and station-keeping maneuvers carried out on command. It is clearly important to prevent unauthorized commands from being received and decoded, and for this reason the command signals are often encrypted. *Encrypt* is derived from a Greek word *kryptein,* meaning *to hide,* and represents the process of concealing the command signals in a secure code. This differs from the normal process of encoding which is one of converting characters in the command signal into a code suitable for transmission.

Tracking of the satellite is accomplished by having the satellite transmit beacon signals which are received at the TT&C earth stations. Tracking is obviously important during the transfer and drift orbital phases of the satellite launch. Once it is on station, the position of a geostationary satellite will tend to be shifted as a result of the various disturbing forces, as described previously. Therefore it is necessary to be able to track the satellite's movement and send correction signals as required. Tracking beacons may be transmitted in the telemetry channel, or by pilot carriers at frequencies in one of the main communications channels, or by special tracking antennas. Satellite range from the ground station is also required from time to time. This can be determined by measurement of the propagation delay of signals especially transmitted for ranging purposes.

It is clear that the telemetry, tracking, and command functions are complex operations which require special ground facilities in addition to the TT&C subsystems aboard the satellite. Figure 6.11 shows in block diagram form the TT&C facilities used by Canadian Telesat for its satellites.

6.7 Transponders

A transponder is the series of interconnected units which forms a single communications channel between the receive and transmit antennas in a communications satellite. Some of the units utilized by a transponder in a given channel may be common to a number of transponders. Thus, although reference may be made to a specific transponder, this must be thought of as an equipment *channel* rather than a single item of equipment.

Before describing in detail the various units of a transponder, the overall frequency arrangement of a typical C-band communications satellite will be examined briefly. The bandwidth allocated for C-band service is 500 MHz, and this is divided into subbands, one for each

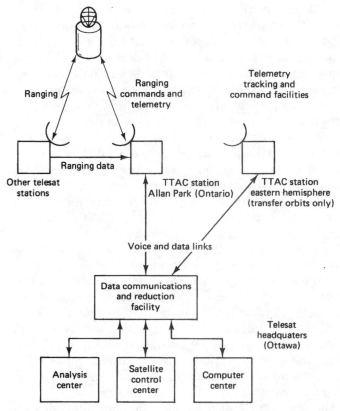

Figure 6.11   Satellite control system. (*From Telesat, Canada, 1983; courtesy Telesat, Canada.*)

transponder. A typical transponder bandwidth is 36 MHz and, allowing for a 4-MHz guardband between transponders, 12 such transponders can be accommodated in the 500-MHz bandwidth. By making use of *polarization isolation,* this number can be doubled. Polarization isolation refers to the fact that carriers, which may be on the same frequency but with opposite senses of polarization, can be isolated from one another by receiving antennas matched to the incoming polarization. With linear polarization, vertically and horizontally polarized carriers can be separated in this way, and with circular polarization, left-hand circular and right-hand circular polarizations can be separated. Because the carriers with opposite senses of polarization may overlap in frequency, this technique is referred to as *frequency reuse.* Figure 6.12 shows part of the frequency and polarization plan for a C-band communications satellite.

Frequency reuse may also be achieved with spot-beam antennas, and these may be combined with polarization reuse to provide an effective

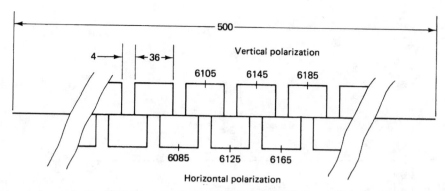

Figure 6.12 Section of an uplink frequency and polarization plan. Numbers refer to frequency in megahertz.

bandwidth of 2000 MHz from the actual bandwidth of 500 MHz.

For one of the polarization groups, Fig. 6.13 shows the channeling scheme for the 12 transponders in more detail. The incoming, or uplink, frequency range is 5.925 to 6.425 GHz. The carriers may be received on one or more antennas, all having the same polarization. The input filter passes the full 500-MHz band to the common receiver while rejecting out-of-band noise and interference such as might be caused by image signals. There will be many modulated carriers within this 500-MHz passband, and all of these are amplified and frequency-converted in the common receiver. The frequency conversion shifts the carriers to the downlink frequency band, which is also 500 MHz wide, extending from 3.7 to 4.2 GHz. At this point the signals are channelized into frequency bands which represent the individual transponder bandwidths. A commonly used value is 36 MHz for each transponder, which along with 4-MHz guard bands between channels allows the 500 MHz available bandwidth to accommodate 12 transponder channels. A transponder may handle one modulated carrier such as a TV signal, or it may handle a number of separate carriers simultaneously, each modulated by its own telephony or other baseband channel.

6.7.1   The wideband receiver

The wideband receiver is shown in more detail in Fig. 6.14. A duplicate receiver is provided so that, if one fails, the other is automatically switched in. The combination is referred to as a *redundant receiver*, meaning that although two are provided, only one is in use at a given time.

The first stage in the receiver is a low-noise amplifier (LNA). This

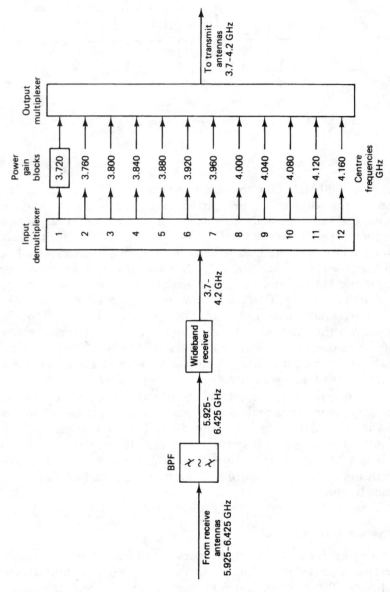

Figure 6.13 Satellite transformer channels. *(Courtesy of CCIR, "CCIR Fixed Satellite Services Handbook," final draft 1984.)*

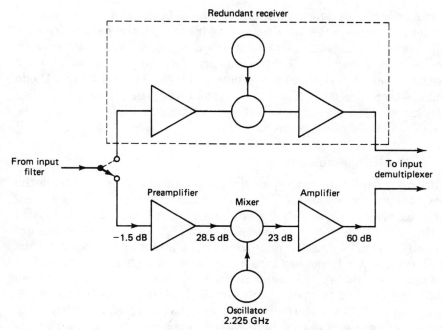

Figure 6.14 Satellite wideband receiver. (*Courtesy of CCIR, "CCIR Fixed Satellite Services Handbook," final draft 1984.*)

amplifier adds little noise to the carrier being amplified, and at the same time it provides sufficient amplification for the carrier to override the higher noise level present in the following mixer stage. In calculations involving noise, it is usually more convenient to refer all noise levels to the LNA input, where the total receiver noise may be expressed in terms of an equivalent noise temperature. In a well-designed receiver the equivalent noise temperature referred to the LNA input is basically that of the LNA alone. The overall noise temperature must take into account the noise added from the antenna, and these calculations are presented in detail in Chap. 10. The equivalent noise temperature of a satellite receiver may be of the order of a few hundred kelvins.

The LNA feeds into a mixer stage, which also requires a local oscillator signal for the frequency-conversion process. The power drive from the local oscillator to the mixer input is about 10 dBm. The oscillator frequency must be highly stable, and have low phase noise. A second amplifier follows the mixer stage to provide an overall receiver gain of about 60 dB. The signal levels in decibels referred to the input are shown in Fig. 6.14 (CCIR, 1984). Splitting the gain between the preamplifier at 6 GHz and the second amplifier at 4 GHz prevents oscillation which might occur if all the gain were to be provided at the same frequency.

The wideband receiver utilizes only solid-state active devices. In

some designs, tunnel-diode amplifiers have been used for the preamplifier at 6 GHz in 6/4-GHz transponders and for the parametric amplifiers at 14 GHz in 14/12-GHz transponders. With advances in
field-effect transistor (FET) technology, FET amplifiers which offer
equal or better performance are now available for both bands. Diode
mixer stages are used. The amplifier following the mixer may utilize
bipolar junction transistors (BJTs) at 4 GHz and FETs at 12 GHz, or
FETs may in fact be used in both bands.

6.7.2 The input demultiplexer

The input demultiplexer separates the broadband input, covering the
frequency range 3.7 to 4.2 GHz, into the transponder frequency channels. Referring to Fig. 6.13, for example, the separate channels labeled
1 through 12 are shown in more detail in Fig. 6.15. The channels are
usually arranged in even-numbered and odd-numbered groups. This
provides greater frequency separation between adjacent channels in a
group, which reduces adjacent channel interference. The output from
the receiver is fed to a power splitter which in turn feeds the two separate chains of circulators. The full broadband signal is transmitted
along each chain, and the channelizing is achieved by means of channel filters connected to each circulator as shown in Fig. 6.15. The
channel numbers correspond to those shown in Fig. 6.13. Each filter
has a bandwidth of 36 MHz and is tuned to the appropriate center frequency as shown in Fig. 6.13. Although there are considerable losses
in the demultiplexer, these are easily made up in the overall gain for
the transponder channels.

6.7.3 The power amplifier

A separate power amplifier provides the output power for each
transponder channel. As shown in Fig. 6.16, each power amplifier is
preceded by an input attenuator. This is necessary to permit the
input drive to each power amplifier to be adjusted to the desired level.
The attenuator has a fixed section and a variable section. The fixed
attenuation is needed to balance out variations in the input attenuation so that each transponder channel has the same nominal attenuation, the necessary adjustments being made during assembly. The
variable attenuation is needed to set the level as required for different types of service (an example being the requirement for input
power back-off discussed below). Because this variable attenuator adjustment is an operational requirement, it must be under the control
of the ground TT&C station.

Traveling wave tube amplifiers (TWTAs) are widely used in
transponders to provide the final output power required to the trans-

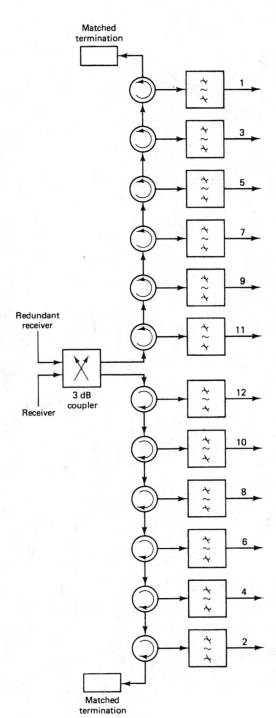

Figure 6.15 Input demulti-plexer. (*Courtesy of CCIR, "CCIR Fixed Satellite Services Handbook," final draft 1984.*)

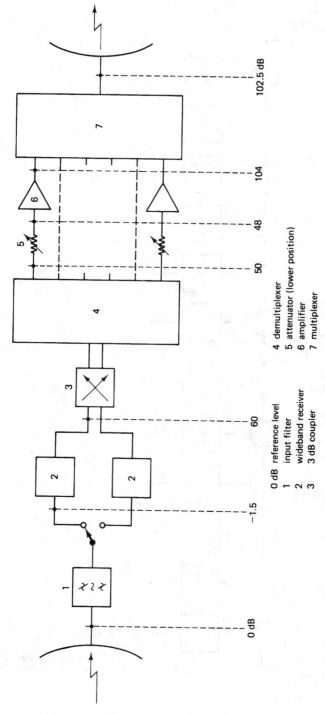

Figure 6.16 Typical diagram of the relative levels in a transponder. (*Courtesy of CCIR, "CCIR Fixed Satellite Services Handbook, final draft 1984."*)

0 dB reference level
1 input filter
2 wideband receiver
3 3 dB coupler

4 demultiplexer
5 attenuator (lower position)
6 amplifier
7 multiplexer

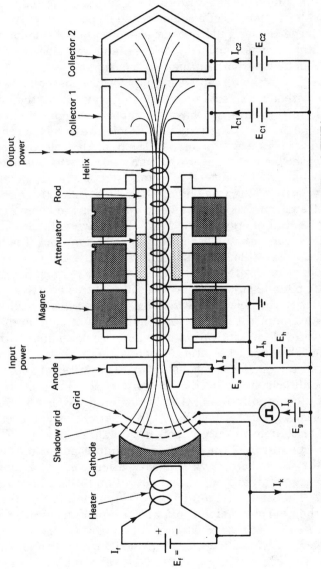

Figure 6.17 Schematic of TWT and power supplies. *(From "Hughes TWT and TWTA Handbook," courtesy Hughes Aircraft Company, Electron Dynamics Division, Torrance, Calif.)*

tube (TWT) and its power supplies. In the TWT, an electron-beam gun assembly consisting of a heater, a cathode, and focusing electrodes is used to form an electron beam. A magnetic field is required to confine the beam to travel along the inside of a wire helix. For high-power tubes such as might be used in ground stations, the magnetic field can be provided by means of a solenoid and dc power supply. The comparatively large size and high power consumption of solenoids makes them unsuitable for use aboard satellites, and lower-power TWTs are used which employ permanent-magnet focusing.

The RF signal to be amplified is coupled into the helix at the end nearest the cathode, and sets up a traveling wave along the helix. The electric field of the wave will have a component along the axis of the helix. In some regions, this field will decelerate the electrons in the beam, and in others it will accelerate them, so that electron bunching occurs along the beam. The average beam velocity, which is determined by the dc potential on the tube collector, is kept slightly greater than the phase velocity of the wave along the helix. Under these conditions an energy transfer takes place, kinetic energy in the beam being converted to potential energy in the wave. The wave will actually travel around the helical path at close to the speed of light, but it is the axial component of wave velocity which interacts with the electron beam. This component is less than the velocity of light approximately in the ratio of helix pitch to circumference. Because of this effective reduction in phase velocity, the helix is referred to as a *slow-wave structure*.

The advantage of the TWT over other types of tube amplifiers is that it can provide amplification over a very wide bandwidth. Input levels to the TWT must be carefully controlled, however, to minimize the effects of certain forms of distortion. The worst of these results from the nonlinear transfer characteristic of the TWT, illustrated in Fig. 6.18. At low input powers, the output-input power relationship is linear; that is, a given decibel change in input power will produce the same decibel change in output power. At higher power inputs, the output power saturates, the point of maximum power output being known as the *saturation point*. The saturation point is a very convenient reference point, and input and output quantities are usually referred to it. The linear region of the TWT is defined as the region bound by the thermal noise limit at the low end, and by what is termed the *1-dB compression point* at the upper end. This is the point where the actual transfer curve drops 1 dB below the extrapolated straight line as shown in Fig. 6.18. The selection of the operating point on the transfer characteristic will be considered in more detail shortly, but first the phase characteristics will be described.

The absolute time delay between input and output signals at a fixed input level is generally not significant. However, at higher input

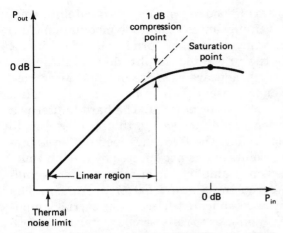

Figure 6.18 Power transfer characteristics of a TWT. The saturation point is used as 0-dB reference for both input and output powers.

levels, where more of the beam energy is converted to output power, the average beam velocity is reduced, and therefore the delay time is increased. Since phase delay is directly proportional to time delay, this results in a phase shift which varies with input level. Denoting the phase shift at saturation by θ_S and in general by θ, the phase difference relative to saturation is $\theta - \theta_S$. This is plotted in Fig. 6.19 as a function of input power. Thus, if the input signal power level changes, phase modulation will result, this being termed *AM/PM conversion*. The slope of the phase shift characteristic gives the phase modulation coefficient, in degrees per decibel. The curve of the slope as a function of input power is also sketched in Fig. 6.19.

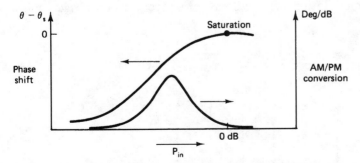

Figure 6.19 Phase characteristics for a TWT. θ is the input-to-output phase shift, and θ_S is the value at saturation. The AM/PM curve is derived from the slope of the phase shift curve.

Frequency modulation (FM) is usually employed in satellite communications circuits. However, unwanted amplitude modulation (AM) can occur from the filtering which takes place prior to the TWT input. The AM process converts the unwanted amplitude modulation to phase modulation (PM), which appears as noise on the FM carrier. Where only a single carrier is present, it may be passed through a *hard limiter* before being amplified in the TWT. The hard limiter is a circuit which clips the carrier amplitude close to the zero baseline to remove any amplitude modulation. The frequency modulation is preserved in the zero crossover points and is not affected by the limiting.

A TWT may also be called on to amplify two or more carriers simultaneously, this being referred to as *multicarrier operation*. The AM/PM conversion is then a complicated function of carrier amplitudes, but in addition, the nonlinear transfer characteristic introduces a more serious form of distortion known as *intermodulation distortion*. The nonlinear transfer characteristic may be expressed as a Taylor series expansion which relates input and output voltages:

$$e_o = ae_i + be_i^2 + ce_i^3 + \cdots \tag{6.1}$$

Here, a, b, c, etc. are coefficients which depend on the transfer characteristic, e_o is the output voltage, and e_i is the input voltage, which consists of the sum of the individual carriers. The *third-order term* is ce_i^3. This and higher-order odd-power terms give rise to intermodulation products, but usually only the third-order contribution is significant. Suppose multiple carriers are present, separated from one another by Δf as shown in Fig. 6.20. Considering specifically the carriers at frequencies f_1 and f_2, these will give rise to frequencies $2f_2 - f_1$ and $2f_1 - f_2$ as a result of the third-order term. (This is demonstrated in App. E.)

Because $f_2 - f_1 = \Delta f$, these two intermodulation products can be written as $f_2 + \Delta f$ and $f_1 - \Delta f$, respectively. Thus the intermodulation products fall on the neighboring carrier frequencies as shown in Fig. 6.20. Similar intermodulation products will arise from other carrier pairs,

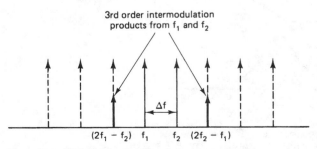

Figure 6.20 Third-order intermodulation products.

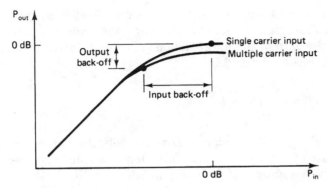

Figure 6.21 Transfer curves for a single carrier, and for one carrier of a multiple-carrier input. Back-off for multiple-carrier operation is relative to saturation for single-carrier input.

and when the carriers are modulated the intermodulation distortion appears as noise across the transponder frequency band. This intermodulation noise is considered further in Sec. 10.10.

In order to reduce the intermodulation distortion, the operating point of the TWT must be shifted closer to the linear portion of the curve, the reduction in input power being referred to as *input back-off*. When multiple carriers are present, the power output around saturation, for any one carrier, is less than that achieved with single-carrier operation. This is illustrated by the transfer curves of Fig. 6.21. The input back-off is the difference in decibels between the carrier input at the operating point, and the saturation input which would be required for single-carrier operation. The output back-off is the corresponding drop in output power. Back-off values are always stated in decibels relative to the saturation point. As a rule of thumb, output back-off is about 5 dB less than input back-off. The need to incorporate back-off significantly reduces the channel capacity of a satellite link because of the reduced carrier-to-noise ratio received at the ground station. Allowance for back-off in the link budget calculations is dealt with in Secs. 10.7.2 and 10.8.1.

6.8   The Antenna Subsystem

The antennas carried aboard a satellite provide the dual functions of receiving the uplink and transmitting the downlink signals. They range from dipole-type antennas where omnidirectional characteristics are required to the highly directional antennas required for telecommunications purposes and TV relay and broadcast. Parts of the antenna structures for the HS 376 and HS 601 satellites can be seen in Figs. 6.1, 6.2, and 6.7.

Directional beams are usually produced by means of reflector-type antennas, the paraboloidal reflector being the most common. As shown in Chap. 5, the gain of the paraboloidal reflector, relative to an isotropic radiator, is given by Eq. (5.29)

$$G = \eta_I \left(\frac{\pi D}{\lambda} \right)^2$$

where λ is the wavelength of the signal, D is the reflector diameter, and η_I is the aperture efficiency. A typical value for η_I is 0.55. The -3-dB beamwidth is given approximately by Eq. (5.29) as

$$\theta_{3dB} \cong 70 \frac{\lambda}{D} \quad \text{degrees}$$

The ratio D/λ is seen to be the key factor in these equations, the gain being directly proportional to $(D/\lambda)^2$ and the beamwidth inversely proportional to D/λ. Hence the gain can be increased, and the beamwidth made narrower, by increasing the reflector size or decreasing the wavelength. The largest reflectors are those for the 6/4-GHz band. Comparable performance can be obtained with considerably smaller reflectors in the 14/12-GHz band.

Figure 6.22 shows the antenna subsystem for the Intelsat VI satellite (Johnston and Thompson, 1982). This provides a good illustration of the level of complexity which has been reached in large communications satellites. The largest reflectors are for the 6/4-GHz hemisphere and zone coverages as illustrated in Fig. 6.23. These are fed from horn arrays, and various groups of horns can be excited to produce the beam shape required. As can be seen, separate arrays are used for transmit and receive. Each array has 146 dual-polarization horns. In the 14/11-GHz band, circular reflectors are used to provide spot beams, one for east and one for west, also shown in Fig. 6.23. These beams are fully steerable. Each spot is fed by a single horn which is used for both transmit and receive.

Wide beams for global coverage are produced by simple horn antennas at 6/4 GHz. These horns beam the signal directly to the earth without the use of reflectors. Also as shown in Fig. 6.22, a simple biconical dipole antenna is used for the tracking and control signals. The complete antenna platform and the communications payload are despun as described in Sec. 6.3 to keep the antennas pointing to their correct locations on earth.

The same feed horn may be used to transmit and receive carriers with the same polarization. The transmit and receive signals are separated in a device known as a *diplexer*, and the separation is further

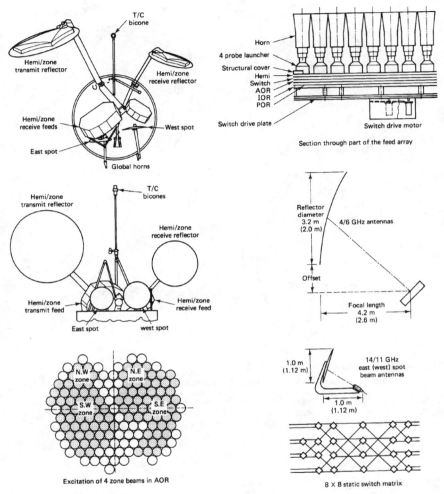

Figure 6.22 The antenna subsystem for the Intelsat VI satellite. (*From Johnston and Thompson, 1982. With permission.*)

aided by means of frequency filtering. Polarization discrimination may also be used to separate the transmit and receive signals using the same feed horn. For example, the horn may be used to transmit horizontally polarized waves in the downlink frequency band, while simultaneously receiving vertically polarized waves in the uplink frequency band. The polarization separation takes place in a device known as an *orthocoupler,* or orthogonal mode transducer (OMT). Separate horns may also be used for the transmit and receive functions, with both horns using the same reflector.

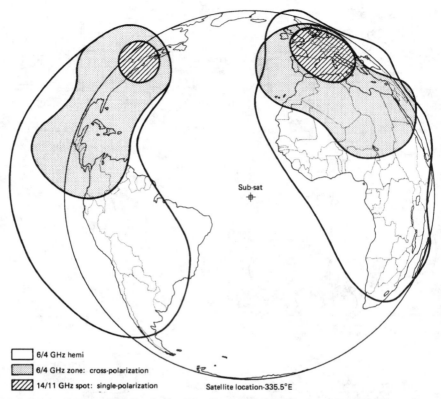

Figure 6.23   Intelsat V Atlantic Satellite Transmit Capabilities. (Note: The 14/11-GHz spot beams are steerable and may be moved to meet traffic requirements as they develop.) (*From Intelsat document BG-28-72E M/6/77. With permission.*)

6.9   Morelos

Figure 6.24 shows the communications subsystem of the Mexican satellite Morelos. Two such satellites were launched, Morelos A in June and Morelos B in November 1985. The satellites are from the Hughes 376 spacecraft series. Although these satellites had a predicted mission life of 9 years, the second Morelos is now scheduled to remain in operation until 1998. The payload carried on Morelos is referred to as a *hybrid,* or *dual-band,* payload because it carries C-band and K-band transponders. In the C band it provides 12 narrowband channels, each 36-MHz wide, and 6 wideband channels, each 72-MHz wide. In the K band it provides four channels, each 108 MHz wide. The 36-MHz channels use 7-W TWTAs with 14-for-12 redundancy. This method of stating redundancy simply means that 12 redundant units are available for 14 in-service units. The 72 MHz channels use 10.5-W TWTAs with 8-for-6 redundancy.

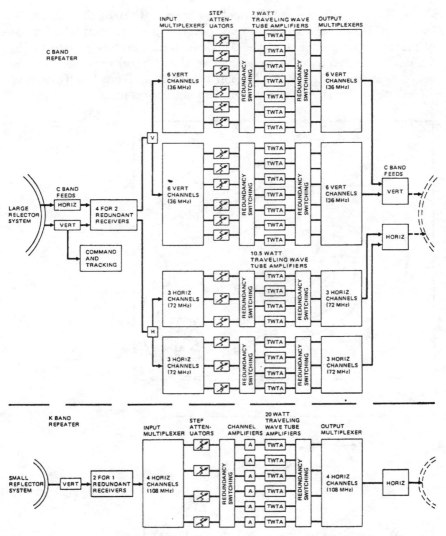

Figure 6.24 Communications subsystem functional diagram for Morelos. (*Courtesy Hughes Aircraft Company Space and Communications Group.*)

The four K-band repeaters use six 20-W TWTAs with 6-for-4 redundancy. The receivers are solid-state designs and there is a 4-for-2 redundancy for the C-band receivers and 2-for-1 redundancy for the K-band receivers.

As mentioned, the satellites are part of the Hughes 376 series, illustrated in Figs. 6.1 and 6.6. A 180-cm-diameter circular reflector is used for the C band. This forms part of a dual-polarization antenna, with separate C-band feeds for horizontal and vertical polarizations. The C-band footprints are shown in Fig. 6.25a.

The K-band reflector is elliptical in shape, with axes measuring 150 by 91 cm. It has its own feed array, producing a footprint which closely matches the contours of the Mexican land mass as shown in Fig. 6.25b. The K-band reflector is tied to the C-band reflector, and onboard tracking of a C-band beacon transmitted from the Tulancingo TT&C station ensures precise pointing of the antennas.

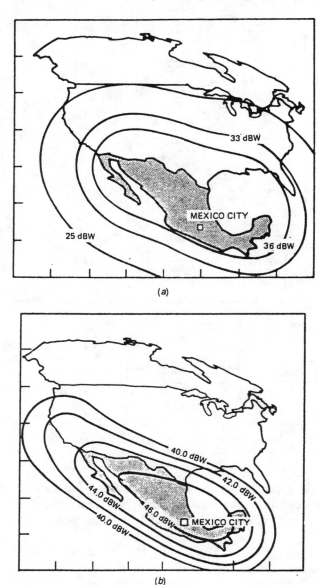

Figure 6.25 (a) C-band and (b) K-band transmit coverage for Morelos. (*Courtesy Hughes Aircraft Company Space and Communications Group.*)

6.10   Anik-E

The Anik-E satellites are the latest in the Canadian Anik series of satellites designed to provide communications services in Canada as well as cross-border services with the United States. The Anik-E is also a dual-band satellite which has an equivalent capacity of 56 television channels, or more than 50,000 telephone circuits. At the time of writing (1994), the Anik-E series (E1 and E2) were the largest domestic satellites ever built. Attitude control is of the momentum-bias, three-axes-stabilized type, and solar sails are used to provide power, the capacity being 3450 W at summer solstice and 3700 W at vernal equinox. Four NiH_2 batteries are provided for operation during eclipse. An exploded view of the Anik-E spacecraft configuration is shown in Fig. 6.26.

The C-band transponder functional block diagram is shown in Fig. 6.27. This is seen to use solid-state power amplifiers (SSPAs) which offer significant improvement in reliability and weight saving over traveling wave tube amplifiers. The antennas are fed through a broadband feeder network (BFN) to illuminate the large reflectors shown in Fig. 6.26. National, as distinct from regional, coverage is

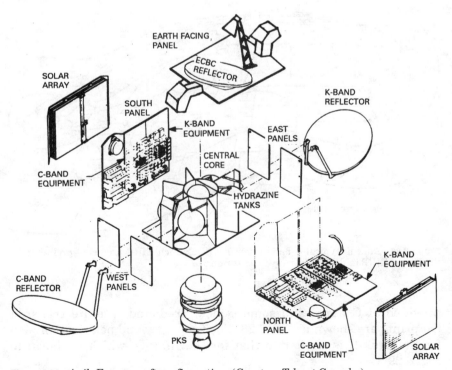

Figure 6.26   Anik-E spacecraft configuration. (*Courtesy Telesat Canada.*)

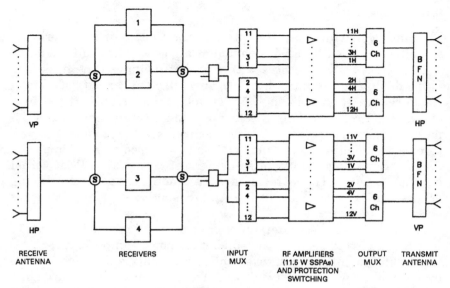

Figure 6.27 Anik-E C-band transponder functional block diagram. (*Courtesy Telesat Canada.*)

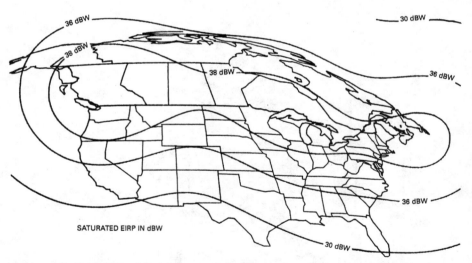

Figure 6.28 Anik-E Typical C-band coverage. (Prelaunch EIRP predicts—preliminary.) Saturated EIRP in dBW. (*Courtesy Telesat Canada.*)

provided at C band, and some typical predicted satellite transmit footprints are shown in Fig. 6.28. The frequency and polarization plan for the Anik-E is similar to that for the Anik-D, which is shown in Fig. 6.29.

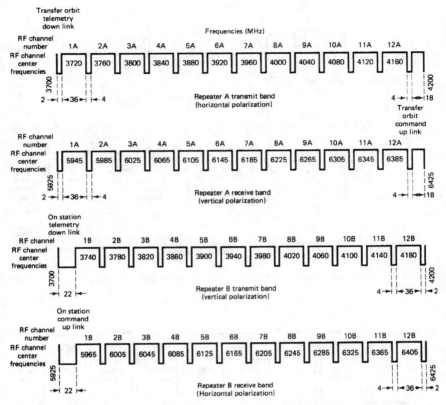

Figure 6.29 Anik-D 6/4 Ghz frequency and polarization plan. (*From Telesat Canada, 1985; courtesy Telesat Canada.*)

The Ku-band transponder functional block diagram is shown in Fig. 6.30. It will be noted that at these higher frequencies, traveling wave tube amplifiers are used. The Anik-E1 frequency and polarization plan is shown in Fig. 6.31. National and regional beams are provided for operation within Canada, as well as what is termed enhanced cross-border capability (ECBC), which provides services to customers with sites in the United States as well as Canada. The satellite transmit footprint for the ECBC is shown in Fig. 6.32, and the receive saturation flux density (SFD) contours in Fig. 6.33. The meaning and significance of the equivalent isotropic radiated power (EIRP) contours and the SFD contours will be explained in Chap. 10. The Anik-E system characteristics are summarized in Table 6.1.

The TWTAs aboard a satellite may also be switched to provide redundancy as illustrated in Fig. 6.34. The scheme shown is termed a four-for-two redundancy, meaning that four channels are provided

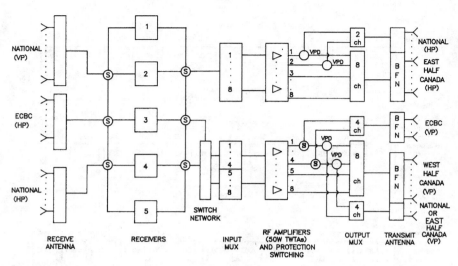

Figure 6.30 Ku-band transponder functional diagram for Anik-E. (*Courtesy Telesat Canada.*)

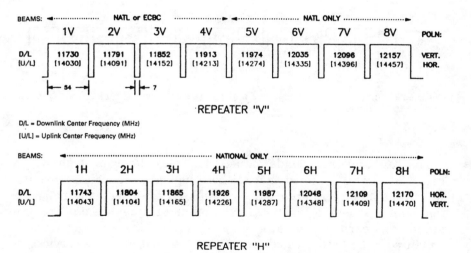

Figure 6.31 Anik-E1 frequency and polarization plan. Ku band. (*Courtesy Telesat Canada.*)

with two redundant amplifiers. For example, examination of the table in Fig. 6.34 shows that channel 1A has amplifier 2 as its primary amplifier, and amplifiers 1 and 3 can be switched in as backup amplifiers by ground command. In this system, 12 channels are designated as primary, and the remainder are either used with preemptible traffic or kept in reserve as backups.

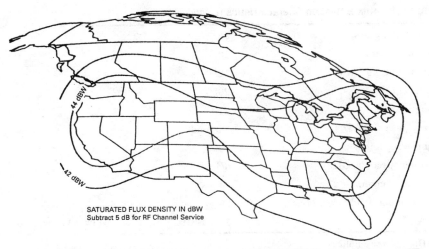

SATURATED FLUX DENSITY IN dBW
Subtract 5 dB for RF Channel Service

Figure 6.32 Anik-E enhanced cross-border capability (ECBC) Ku-band coverage: VP (prelaunch EIRP predicts—preliminary). Saturated EIRP in dBW. (*Courtesy Telesat Canada.*)

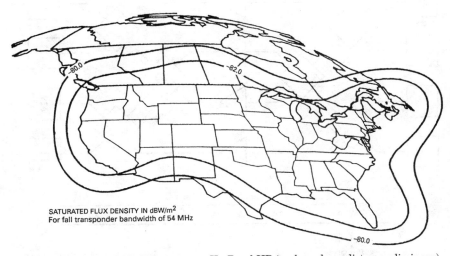

SATURATED FLUX DENSITY IN dBW/m^2
For fall transponder bandwidth of 54 MHz

Figure 6.33 Anik-E ECBC SFD contours: Ku-Band HP (prelaunch predicts—preliminary). (*Courtesy Telesat Canada.*)

TABLE 6.1   Anik-E System Characteristics

Prime contractor	Spar Aerospace
Type of satellite bus	GE Astro Series 5000
Number of satellites	2
Launch dates	E1: March 1990; E2: October 1990
Orbital range, ° W	104.5 to 117.5
Orbital position, ° W	E1: 107.5; E2: 110.5

TABLE 6.1   Anik-E System Characteristics (Continued)

Design life, years	12	
Fuel life, years	13.5	
Dry weight, kg	1280	
Transfer orbit mass, kg	2900	
Length, m	21.5	
Array power, kW	3.5 (end of life)	
Eclipse capability	100%	
Frequency bands, GHz	6/4	14/12
Number of channels	24	16
Transponder bandwidth, MHz	36	54
HPA* (W)	11.5 (SSPA)	50 (TWTA)

*High-power amplifier.

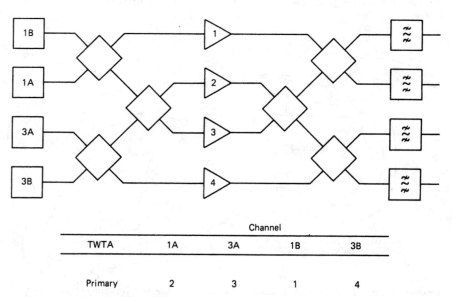

Channel				
TWTA	1A	3A	1B	3B
Primary	2	3	1	4

Figure 6.34 Anik D TWTA four-for-two redundancy switching arrangement. (*From Telesat Canada, 1983; courtesy Telesat Canada.*)

6.11 Advanced Tiros-N Spacecraft

Tiros is an acronym for Television and Infra-Red Observational Satellite. As described in Chap. 1, Tiros is a polar-orbiting satellite whose primary mission is to gather and transmit earth environmental data down to its earth stations. Although its payload differs fundamentally from the communications-relay-type payload, much of the bus equipment is similar. Table 1.2 lists the NOAA spacecraft used in the Advanced TIROS-N (ATN) program. The general features of these spacecraft are described in Schwalb, 1982*a*, 1982*b*, and a series of undated NASA/NOAA booklets are available describing the individual spacecraft. The main features of NOAA-J spacecraft are shown in Fig. 6.35, and the physical and orbital characteristics are given in Table 6.2.

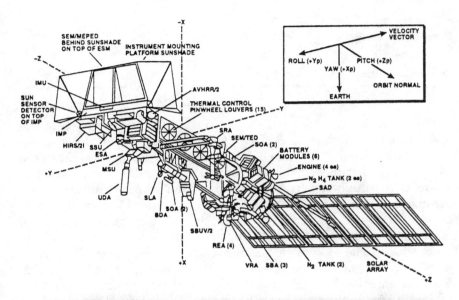

LEGEND			
AVHRR/2	Advanced Very High Resolution Radiometer	**SBUV/2**	Solar Backscatter Ultraviolet Spectral Radiometer
BDA	Beacon/Command Antenna	**SLA**	Search-and-Rescue Transmitting Antenna (L-Band)
ESA	Earth Sensor Assembly	**SOA**	S-Band Omni Antenna
HIRS/2I	High-resolution Infrared Sounder	**SRA**	Search-and-Rescue Receiving Antenna
IMP	Instrument Mounting Platform	**SSD**	Sun Sensor Detector
IMU	Inertial Measurement Unit	**SSU**	Stratospheric Sounding Unit
MSU	Microwave Sounding Unit	**UDA**	Ultra-High-Frequency Data Collection System Antenna
REA	Reaction Engine Assembly		
SAD	Solar-Array Drive	**VRA**	Very-High-Frequency Data Real-Time Antenna
SBA	S-Band Antenna	**SEM**	Space Environment Monitor

Figure 6.35 NOAA-J spacecraft with major features identified.

TABLE 6.2   Physical and Orbital Characteristics of the NOAA-J Spacecraft

Main body	4.18 m (13.7 ft) long, 1.88 m (6.2 ft) in diameter
Solar array	2.37 by 4.91 m (7.8 by 16.1 ft), 11.6 m^2 (125 ft^2)
Weight	At liftoff, 1712 kg (3775 lb); on orbit 1030 kg (2288 lb)
Power: orbit average, end of life	593 W for $\gamma = 0°$; 533 W for $\gamma = 80°$
Lifetime	Greater than 2 years
Apogee	870 km (470 nmi)
Perigee	870 km (470 nmi)
Minutes per orbit	102.12
Degrees inclination	98.86

Three nickel-cadmium (Ni-Cd) batteries supply power while the spacecraft is in darkness. The relatively short lifetime of these space-craft results largely from the effects of atmospheric drag, present at the low orbital altitudes. Attitude control of the NOAA spacecraft is achieved through the use of three reaction wheels similar to the arrangement shown in Fig. 6.8. A fourth, spare, wheel is carried, angled at 54.7° to each of the three orthogonal axes. The spare reaction wheel is normally idle, but is activated in the event of failure of any of the other wheels. The 54.7° angle permits its torque to be resolved into components along each of the three main axes. As can be seen from Fig. 6.35, the antennas are omnidirectional, but attitude control is needed to maintain directivity for the earth sensors. These must be maintained within ±0.2° of the local geographic reference (Schwalb, 1982*a*). A summary sheet for the ATN/NOAA E-J spacecraft is presented in Table 6.3.

TABLE 6.3   Advanced Tiros-N/NOAA E-J Summary Sheet

Spacecraft	Total weight 2220 lb (1009 kg) (excludes expendibles)
Payload weight	850 lb (386 kg) (including tape recorders)
Instrument complement	Advanced very high resolution radiometer (AVHRR/2) High-resolution infrared radiation sounder (HIRS/2) Stratospheric sounder unit (SSU) Microwave sounder unit (MSU) Data collection system—Argos (DCS) Space environment monitor (SEM) Search and rescue (SAR) satellite-aided tracking (Sarsat) Solar backscatter ultraviolet radiometer (SBUV/2), NOAA F and on Earth radiation budget experiment (ERBE), NOAA F and G only
Spacecraft size	165 in (3.71 m) in length 74 in (1.88 m) in diameter

TABLE 6.3   Advanced Tiros-N/NOAA E-J Summary Sheet (*Continued*)

Solar array	7.8×16.1 ft, 125 ft²
	(2.37×4.91 m, 11.6 m²)
	515 Watts, end of life at worst solar angle
	(violet, high-efficiency solar cells)
Power requirement	Full operation—475 W
	Reserved—40 W
Attitude control system	0.2° all axes
	0.14° determination
Communications	Command link—148.56 MHz
	Beacon—136.77, 137.77 MHz
	S band—1698, 1702.5, 1707 MHz
	APT*—137.50, 137.62 MHz
	DCS (uplink)—401.65 MHz
	SAR—1544.5 MHz
	SAR (uplink)—121.5, 243.0, 406 MHz
Data processing	All digital (APT translated to analog)
Orbit	833, 870-km nominal, sun-synchronous
Launch vehicle	Atlas E/F
Lifetime	2 years planned

SOURCE: Schwalb, 1982*b*.

*Automatic picture transmission.

6.12   Problems

6.1. Describe the *tracking, telemetry,* and *command* facilities of a satellite communications system. Are these facilities part of the space segment or part of the ground segment of the system?

6.2. Explain why some satellites employ cylindrical solar arrays, whereas others employ solar-sail arrays for the production of primary power. State the typical power output to be expected from each type. Why is it necessary for satellites to carry batteries in addition to solar-cell arrays?

6.3. Explain what is meant by satellite *attitude,* and briefly describe two forms of attitude control.

6.4. Define and explain the terms *roll, pitch,* and *yaw.*

6.5. Explain what is meant by the term *despun antenna,* and briefly describe one way in which the despinning is achieved.

6.6. Briefly describe the three-axis method of satellite stabilization.

6.7. Describe the east-west and north-south station-keeping maneuvers required in satellite station keeping. What are the angular tolerances in station keeping that must be achieved?

6.8. Referring to Fig. 6.10 and the accompanying text in Sec. 6.4, determine the minimum −3-dB beamwidth that will accommodate the tolerances in satellite position without the need for tracking.

6.9. Explain what is meant by *thermal control* and why this is necessary in a satellite.

6.10. Explain why an omnidirectional antenna must be used aboard a satellite for telemetry and command during the launch phase. How is the satellite powered during this phase?

6.11. Briefly describe the equipment sections making up a transponder channel.

6.12. Draw to scale the uplink and downlink channeling schemes for a 500-MHz-bandwidth C-band satellite, accommodating the full complement of 36-MHz-bandwidth transponders. Assume the use of 4-MHz guardbands.

6.13. Explain what is meant by *frequency reuse,* and describe briefly two methods by which this can be achieved.

6.14. Explain what is meant by a *redundant receiver* in connection with communication satellites.

6.15. Describe the function of the input demultiplexer used aboard a communications satellite.

6.16. Describe briefly the most common type of high-power amplifying device used aboard communications satellites.

6.17. What is the chief advantage of the traveling wave tube amplifier used aboard satellites compared to other types of high-power amplifying devices? What are the main disadvantages of the TWTA?

6.18. Define and explain the term *1-dB compression point.* What is the significance of this point in relation to the operating point of a TWT?

6.19. Explain why operation near the saturation point of a TWTA is to be avoided when multiple carriers are being amplified simultaneously.

6.20. State the type of satellite antenna normally used to produce a wide-beam radiation pattern, providing global coverage. How are spot beams produced?

6.21. Describe briefly how beam shaping of a satellite antenna radiation pattern may be achieved.

6.22. With reference to Fig. 6.34, explain what is meant by a *four-for-two redundancy switching arrangement.*

7

The Earth Segment

7.1 Introduction

The earth segment of a satellite communications system consists of the transmit and receive earth stations. The simplest of these are the home TV receive-only (TVRO) systems, and the most complex are the terminal stations used for international communications networks. Also included in the earth segment are those stations which are on ships at sea, and commercial and military land and aeronautical mobile stations.

As mentioned in Chap. 6, earth stations which are used exclusively for logistic support of satellites, such as those providing the telemetry, tracking, and command (TT&C) functions, are considered as part of the space segment.

7.2 Receive-Only Home TV Systems

Planned broadcasting directly to home TV receivers is scheduled to take place in the Ku (12-GHz) band. This service is known as direct broadcast satellite (DBS) service. There is some variation in the frequency bands assigned to different geographical regions. In the Americas for example, the downlink band is 12.2 to 12.7 GHz, as described in Sec. 1.4.

The comparatively large satellite receiving dishes (about 3-m diameter) which are a familiar sight around many homes, are used to receive downlink TV signals at C band (4 GHz). Such downlink signals were never intended for home reception but for network relay to commercial TV outlets (VHF and UHF TV broadcast stations and cable TV "headend" studios). Although the practice of intercepting these signals seems to be well-established at present, various technical and commercial and legal factors are combining to deter their direct reception. The major differences between the Ku-band and the C-band receive-only systems lies in the frequency of operation of the outdoor unit, and the fact that satellites intended for DBS have much higher EIRP as described in Table 1.4. For clarity, only the Ku-band system is described here.

Figure 7.1 shows the main units in a home terminal DBS TV receiving system. Although there will be variations from system to system,

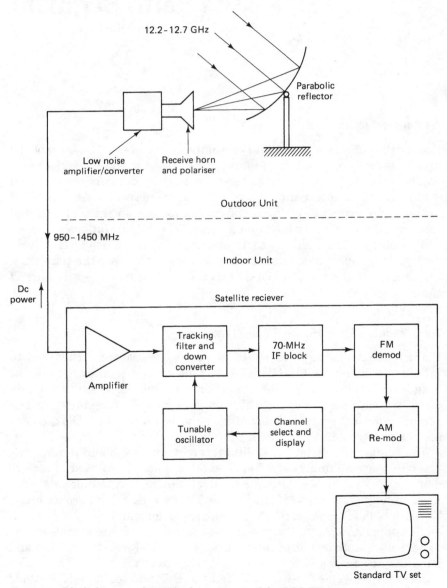

Figure 7.1   Block diagram showing a home terminal for DBS TV reception.

the diagram covers the basic concept. Cost estimates for a complete receive-only station for the home are in the range of $300 to $800, with some operators holding the view that the cost must be $350 or less for the DBS service to win public acceptance (Dement, 1984).

7.2.1 The Outdoor Unit

This consists of a receiving antenna feeding directly into a low-noise amplifier/converter combination. A parabolic reflector is generally used, with the receiving horn mounted at the focus. A common design is to have the focus directly in front of the reflector, but for better interference rejection, an offset feed may be used as shown.

Huck and Day, 1979, have shown that satisfactory reception can be achieved with reflector diameters in the range 0.6 m to 1.6 m (1.97–5.25 ft), and the two nominal sizes often quoted are 0.9 m (2.95 ft) and 1.2 m (3.94 ft). By contrast, the reflector diameter for 4-GHz reception is typically about 3 m (9.84 ft). As noted in Sec. 5.13, the gain of a parabolic dish is proportional to $(D/\lambda)^2$. Comparing the gain of a 3-m dish at 4 GHz with a 1-m dish at 12 GHz, the ratio $D/\lambda = 40$ in each case, so that the gains will be about equal. Although the free-space losses are much higher at 12 GHz compared to 4 GHz, as described in Chap. 10, a higher-gain receiving antenna is not needed because the direct broadcast satellites operate at a much higher equivalent isotropic radiated power (EIRP), as shown in Table 1.4.

The downlink frequency band of 12.2 to 12.7 GHz spans a range of 500 MHz, which accommodates 32 TV channels, each of which is 24 MHz wide. Obviously some overlap occurs between channels, but these are alternately polarized left-hand circular (LHC) and right-hand circular (RHC) to reduce interference to acceptable levels. This is referred to as *polarization interleaving*. A polarizer that may be switched to the desired polarization from the indoor control unit is required at the receiving horn.

The receiving horn feeds into a low-noise converter (LNC), or possibly a combination unit consisting of a low-noise amplifier (LNA) followed by a converter. The combination is referred to as an LNB, for low-noise block. The LNB provides gain for the broadband 12-GHz signal, and then converts the signal to a lower frequency range, so that a low-cost coaxial cable can be used as feeder to the indoor unit. The standard frequency range of this downconverted signal is 950 to 1450 MHz, as shown in Fig. 7.1. The coaxial cable, or an auxiliary wire pair, is used to carry dc power to the outdoor unit. Polarization-switching control wires are also required.

The low-noise amplification must be provided at the cable input in order to maintain a satisfactory signal-to-noise ratio. A low-noise amplifier at the indoor end of the cable would be of little use, as it would also amplify the cable thermal noise. Signal-to-noise ratio is considered in more detail in Sec. 10.5. Of course, having to mount the LNB outside means that it must be able to operate over a wide range of climatic conditions, and homeowners may have to contend with the added problems of vandalism and theft.

7.2.2 The Indoor Unit

The signal fed to the indoor unit is normally a wideband signal covering the range 950 to 1450 MHz. This is amplified and passed to a tracking filter which selects the desired channel, as shown in Fig. 7.1. As previously mentioned, polarization interleaving is used and only half the 32 channels will be present at the input of the indoor unit for any one setting of the antenna polarizer. This eases the job of the tracking filter, since alternate channels are well-separated in frequency.

The selected channel is again downconverted, this time from the 950- to 1450-MHz range to a fixed intermediate frequency, usually 70 MHz, although other values in the VHF range are also used. The 70-MHz amplifier amplifies the signal up to the levels required for demodulation. A major difference between DBS TV and conventional TV is that with DBS, frequency modulation is used, whereas with conventional TV amplitude modulation in the form of vestigial single sideband (VSSB) is used. The 70-MHz, frequency-modulated IF carrier must therefore be demodulated, and the baseband information used to generate a VSSB signal which is fed into one of the VHF/UHF channels of a standard TV set.

A DBS receiver provides a number of functions not shown on the simplified block diagram of Fig. 7.1. The demodulated video and audio signals are usually made available at output jacks. Also, as described in Sec. 11.3, an energy-dispersal waveform is applied to the satellite carrier to reduce interference, and this waveform has to be removed in the DBS receiver. Terminals may also be provided for the insertion of IF filters to reduce interference from terrestrial TV networks, and a descrambler may also be necessary for the reception of some programs.

7.3   Master Antenna TV System

A master antenna TV (MATV) system is used to provide reception of DBS TV channels to a small group of users, for example to the tenants in an apartment building. It consists of a single outdoor unit (antenna and LNA/C), feeding a number of indoor units as shown in Fig. 7.2. It is basically similar to the home system already described, but with each user having access to all of the channels independently of the other users. The advantage is that only one outdoor unit is required,

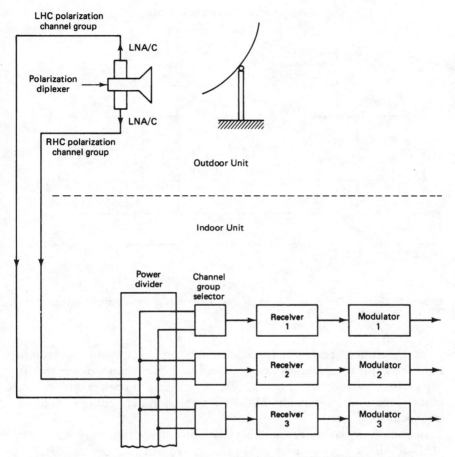

Figure 7.2 One possible arrangement for a master antenna TV (MATV) system.

but as shown, separate LNA/Cs and feeder cables are required for each sense of polarization. Compared to the single-user system, a larger antenna is also required (2- to 3-m diameter), in order to maintain a good signal-to-noise ratio at all of the indoor units.

Where more than a few subscribers are involved, the distribution system used is similar to the CATV system described in the next section.

7.4   Community Antenna TV System

The community antenna TV system employs a single outdoor unit, with separate feeds available for each sense of polarization, like the MATV system, so that all channels are made available simultaneous-

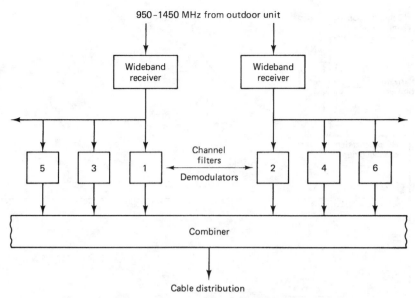

Figure 7.3 One possible arrangement for the indoor unit of a community antenna TV (CATV) system.

ly at the indoor receiver. Instead of having a separate receiver for each user, all the carriers are demodulated in a common receiver-filter system, as shown in Fig. 7.3. The channels are then combined into a standard multiplexed signal for transmission over cable to the subscribers.

In remote areas where a cable distribution system may not be installed, the signal can be rebroadcast from a low-power VHF TV transmitter. Figure 7.4 shows a remote TV station which employs an 8-m (26.2-ft), antenna for reception of the satellite TV signal in the C band.

With the CATV system, local programming material may also be distributed to subscribers, an option which is not permitted in the MATV system.

7.5 Transmit-Receive Earth Stations

In the previous sections, receive-only TV stations are described. Obviously, somewhere a transmit station must complete the uplink to the satellite. In some situations, a transmit-only station is required, for example, in relaying TV signals to the remote TV receive-only stations already described. Transmit-receive stations provide both functions, and are required for telecommunications traffic generally,

Figure 7.4   Remote television station. (*From Telesat Canada, 1983; courtesy Telesat Canada.*)

including network TV. There are considerable differences in detail between transmit-receive earth stations, depending on the types of service being provided. However, the basic elements are similar, and these are shown in Fig. 7.5 for a redundant earth station. As mentioned in connection with transponders in Sec. 6.7.1, redundancy means that certain units are duplicated. A duplicate, or redundant, unit is automatically switched into a circuit to replace a corresponding unit that has failed. Redundant units are shown by dashed lines in Fig. 7.5.

The block diagram is shown in more detail in Fig. 7.6, where for clarity, redundant units are not shown. Starting at the bottom of the diagram, the first block shows the interconnection equipment required between satellite station and the terrestrial network. For the purpose of explanation, telephone traffic will be assumed. This may consist of a number of telephone channels in a multiplexed format. Multiplexing is a method of grouping telephone channels together, usually in basic groups of 12, without mutual interference. It is described in detail in Chaps. 8 and 9.

It may be that groupings different from those used in the terrestrial network are required for satellite transmission, and the next block shows the multiplexing equipment in which the reformatting is carried out. Following along the transmit chain, the multiplexed signal is modulated onto a carrier wave at an intermediate frequency, usually 70 MHz. Parallel IF stages are required, one for each microwave

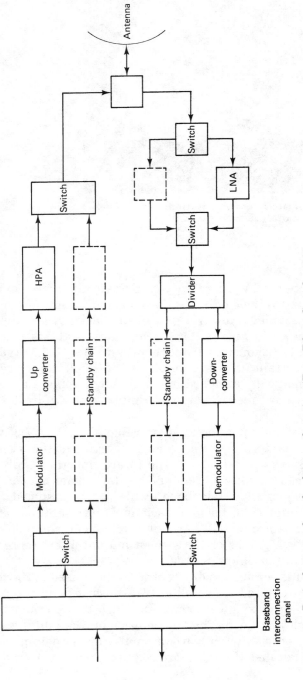

Figure 7.5 Basic elements of a redundant earth station. (*From Telesat Canada, 1983; courtesy Telesat Canada.*)

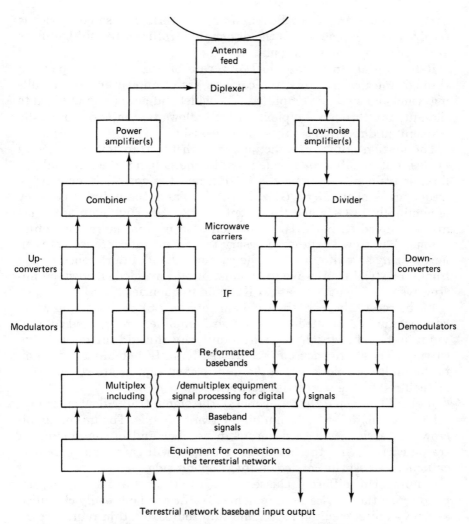

Figure 7.6 More detailed block diagram of a transmit-receive earth station.

carrier to be transmitted. After amplification at the 70-MHz IF, the modulated signal is then upconverted to the required microwave carrier frequency. A number of carriers may be transmitted simultaneously, and although these are at different frequencies they are generally specified by their nominal frequency, e.g., as 6-GHz or 14-GHz carriers.

It should be noted that the individual carriers may be multidestination carriers. This means that they carry traffic destined for different stations. For example, as part of its load, a microwave carrier may have telephone traffic for Boston and New York. The same carrier is received at both places, and the designated traffic sorted out by filters at the receiving earth station.

Referring again to the block diagram of Fig. 7.6, after passing through the upconverters, the carriers are combined and the resulting wideband signal is amplified. The wideband power signal is fed to the antenna through a diplexer, which allows the antenna to handle transmit and receive signals simultaneously.

The station's antenna functions in both the transmit and receive modes, but at different frequencies. In the C band, the nominal uplink, or transmit, frequency is 6 GHz, and the downlink, or receive, frequency is nominally 4 GHz. In the Ku band, the uplink frequency is nominally 14 GHz and the downlink, 12 GHz. High-gain antennas are employed in both bands, which also means narrow antenna beams. A narrow beam is necessary to prevent interference between neighboring satellite links. In the case of C band, interference to and from terrestrial microwave links must also be avoided. Terrestrial microwave links do not operate at Ku-band frequencies.

In the receive branch (the right side of Fig. 7.6), the incoming wideband signal is amplified in a low-noise amplifier and passed to a divider network, which separates out the individual microwave carriers. These are each downconverted to an IF band and passed on to the multiplex block where the multiplexed signals are reformatted as required by the terrestrial network.

It should be noted that, in general, the signal traffic flow on the receive side will differ from that on the transmit side. The incoming microwave carriers will be different in number and in the amount of traffic carried, and the multiplexed output will carry telephone circuits not necessarily carried on the transmit side.

A number of different classes of earth stations are available, depending on the service requirements. Traffic can be broadly classified as heavy route, medium route, and thin route. In a thin route circuit, a transponder channel (36 MHz) may be occupied by a number of single carriers, each associated with its own voice circuit. This mode of operation is known as single carrier per channel (SCPC), a multiple-access mode which is discussed further in Chap. 12. Antenna sizes range from 3.6 m (11.8 ft) for transportable stations up to 30 m (98.4 ft) for a main terminal.

A medium route circuit also provides multiple access, either on the basis of frequency-division multiple access (FDMA) or time-division multiple access (TDMA), multiplexed baseband signals being carried in either case. These access modes are also described in detail in Chap. 12. Antenna sizes range from 30 m (89.4 ft) for a main station to 10 m (32.8 ft) for a remote station.

In a 6/4-GHz heavy route system, each satellite channel (bandwidth 36 MHz) is capable of carrying over 960 one-way voice circuits simultaneously or a single color TV signal with associated audio (in some systems two TV signals can be accommodated). Thus, the transponder channel for a heavy route circuit carries one large bandwidth signal which may be TV or multiplexed telephony. The antenna diameter for a heavy route circuit is at least 30 m (98.4 ft). For international operation such antennas are designed to the Intelsat specifications for a Standard A earth station (Intelsat, 1982). Figure 7.7 shows a photograph of a 32-m (105-ft) Standard A earth station antenna.

It will be appreciated that for these large antennas, which may weigh in the order of 250 tons, the foundations must be very strong and stable. Such large diameters automatically mean very narrow beams, and therefore any movement which would deflect the beam unduly must be avoided. Where snow and ice conditions are likely to be encountered, built-in heaters are required. For the antenna shown in Fig. 7.7, deicing heaters provide reflector surface heat of 40W/ft^2 for the main reflectors and subreflectors, and 3000 W for the azimuth wheels.

Figure 7.7   Standard-A (C-band 6/4-GHz) 32-m antenna. (*Courtesy of TIW Systems, Inc., Sunnydale, Calif.*)

Although these antennas are used with geostationary satellites, some drift in the satellite position does occur, as shown in Chap. 2. This, combined with the very narrow beams of the larger earth station antennas, means that some provision must be made for a limited degree of tracking. Step adjustments in azimuth and elevation may be made under computer control, to maximize the received signal.

The continuity of the primary power supply is another important consideration in the design of transmit-receive earth stations. Apart from the smallest stations, power backup in the form of multiple feeds from the commercial power source and/or batteries and generators is provided. If the commercial power fails, batteries immediately take over with no interruption. At the same time, the standby generators start up, and once they are up to speed they automatically take over from the batteries.

7.6 Problems

7.1. Explain what is meant by *direct broadcast satellite* service. How does this differ from the home reception of satellite TV signals in the C band which is commonplace today?

7.2. Explain what is meant by *polarization interleaving*. On a frequency axis, draw to scale the channel allocations for the 32 TV channels in the Ku band, showing how polarization interleaving is used in this.

7.3. Why is it desirable to downconvert the satellite TV signal received at the antenna?

7.4. Explain why the low-noise amplifier in a satellite receiving system is placed at the antenna end of the feeder cable.

7.5. With the aid of a block schematic, briefly describe the functioning of the indoor receiving unit of a satellite TV receiving system intended for home reception.

7.6. In most satellite TV receivers the first IF band is converted to a second, fixed IF. Why is this second frequency conversion required?

7.7. For the standard home television set to function in a satellite receiving system, a demodulator/remodulator unit is needed. Explain why.

7.8. Describe and compare the master antenna TV and the community antenna TV systems.

7.9. Explain what is meant by the term *redundant earth station*.

7.10. With the aid of a block schematic, describe the functioning of a transmit-receive earth station used for telephone traffic. Describe a multidestination carrier.

8

Analog Signals

8.1 Introduction

Analog signals are electrical replicas of the original signals such as audio and video. *Baseband signals* are those signals which occupy the lowest, or base, band of frequencies, in the frequency spectrum used by the telecommunications network. A baseband signal may consist of one or more information signals. For example, a number of analog telephony signals may be combined into one baseband signal by the process known as frequency-division multiplexing. Other common types of baseband signals are the multiplexed video and audio signals which originate in the TV studio. In forming the multiplexed baseband signals, the information signals are *modulated* onto subcarriers. This modulation step must be distinguished from the modulation process which places the multiplexed signal onto the microwave carrier for transmission to the satellite.

In this chapter, the characteristics of the more common types of analog baseband signals are described, along with representative methods of analog modulation.

8.2 The Telephone Channel

Natural speech, including that of female and male voices, covers a frequency range of about 80 to 8000 Hz. The somewhat unnatural quality associated with telephone speech results from the fact that a considerably smaller band of frequencies is used for normal telephone transmission. The range of 300 to 3400 Hz is accepted internationally as the standard for "telephone quality" speech, and this is termed the *speech baseband.* In practice, some variations occur in the basebands used by different telephone companies. The telephone channel is often referred to as a voice frequency (VF) channel, and in this book this will be taken to mean the frequency range of 300 to 3400 Hz.

There are good reasons for limiting the frequency range. Noise, which covers a very wide frequency spectrum, is reduced by reducing the bandwidth. Also, reducing the bandwidth allows more telephone channels to be carried over a given type of circuit, as will be described in Sec. 8.4.

The signal levels encountered within telephone networks vary considerably. Audio signal levels are often measured in volume units (VU). For a sinusoidal signal within the voice frequency range, 0 VU corresponds to 1 mW of power, or 0 dBm. No simple relationship exists between volume units and power for speech signals, but as a rough guide, the power level in dBm of normal speech is given by −1.4 VU. As a rule of thumb, the average voice level on a telephone circuit (or mean talker level) is defined as −13 VU (see Freeman, 1981).

8.3   Single-Sideband Telephony

Figure 8.1a shows how the VF baseband may be represented in the frequency domain. In some cases the triangular representation has the small end of the triangle at 0 Hz, even though frequency components below 300 Hz may not actually be present. Also, in some cases the upper end is set at 4 kHz to indicate allowance for a guard band, the need for which will be described later.

When the telephone signal is multiplied in the time domain with a sinusoidal carrier of frequency f_c, a new spectrum results, in which the original baseband appears on either side of the carrier frequency. This is illustrated in Fig. 8.1b for a carrier of 20 kHz, where the band of frequencies below the carrier is referred to as the *lower sideband* and the band above the carrier as the *upper sideband*. To avoid distortion which would occur with sideband overlap, the carrier frequency must be greater than the highest frequency in the baseband.

The result of this multiplication process is referred to as *double-sideband suppressed-carrier* (DSBSC) *modulation,* as only the sidebands, and not the carrier, appear in the spectrum. Now, all of the

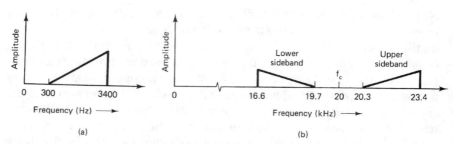

Figure 8.1   Frequency-domain representation of (a) a telephone baseband signal and (b) the double-sideband suppressed carrier (DSBSC) modulated version of (a).

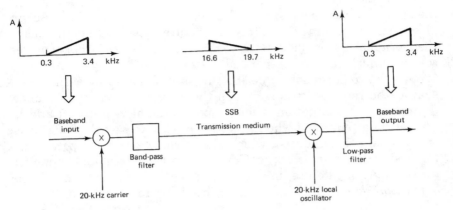

Figure 8.2 A basic SSB transmission scheme.

information in the original telephone signal is contained in either of the two sidebands, and therefore it is necessary to transmit only one of these. A filter may be used to select either one and reject the other. The resulting output is termed a *single-sideband* (SSB) *signal.*

The SSB process utilizing the lower sideband is illustrated in Fig. 8.2 where a 20-kHz carrier is used as an example. It will be seen that for the lower sideband, the frequencies have been inverted, the highest baseband frequency being translated to the lowest transmission frequency at 16.6 kHz, and the lowest baseband frequency to the highest transmission frequency at 19.7 kHz. This inversion does not affect the final baseband output, as the demodulation process reinverts the spectrum. At the receiver, the SSB signal is demodulated (that is, the baseband signal is recovered), by being passed through an identical multiplying process. In this case the multiplying sinusoid, termed the *local oscillator signal,* must have the same frequency as the original carrier. A low-pass filter is required at the output to select the baseband signal and reject other, unwanted frequency components generated in the demodulation process. This single-sideband modulation/demodulation process is illustrated in Prob. 8.2.

The way in which SSB signals are used for the simultaneous transmission of a number of telephone signals is described in the next section. It should be noted at this point that a number of different carriers are likely to be used in a satellite link. The radio-frequency carrier used in transmission to and from the satellite will be much higher in frequency than those used for the generation of the set of SSB signals. These latter carriers are sometimes referred to as a *voice-frequency carriers.* The term *subcarrier* is also used, and this practice will be followed here. Thus, the 20-kHz carrier shown in Fig. 8.2 is a subcarrier.

Companded single sideband (CSSB) refers to a technique in which the speech signal levels are compressed before transmission as a single sideband, and, at the receiver they are expanded again back to their original levels. (The term *compander* is derived from *compressor-expander*). In one companded system described by Campanella, 1983, a 2:1 compression in decibels is used, followed by a 1:2 expansion at the receiver. It is shown in the reference, that the expander decreases its attenuation when a speech signal is present, and increases its attenuation when it is absent. In this way the "idle" noise on the channel is reduced, which allows the channel to operate at a reduced carrier-to-noise ratio. This in turn permits more channels to occupy a given satellite link, a topic which comes under the heading of *multiple access* and which is described more fully in Chap. 12.

8.4 FDM Telephony

Frequency-division multiplexing (FDM) provides a way of keeping a number of individual telephone signals separate while transmitting them simultaneously over a common transmission link circuit. Each telephone baseband signal is modulated onto a separate subcarrier, and all the upper or all the lower sidebands are combined to form the frequency-multiplexed signal. Figure 8.3a shows how three voice channels may be frequency-division-multiplexed. Each voice channel occupies the range 300 to 3400 Hz, and each is modulated onto its own subcarrier. The subcarrier frequency separation is 4 kHz, this allowing for the basic voice bandwidth of 3.1 kHz plus an adequate guard band for filtering. The upper sidebands are selected by means of filters, and then combined to form the final three-channel multiplexed signal. The three-channel FDM *pregroup* signal can be represented by a single triangle, as shown in Fig. 8.3b.

To facilitate interconnection among the different telecommunications systems in use worldwide, the Comité Consultatif Internationale de Télégraphique et Téléphonique (CCITT) has recommended a standard modulation plan for FDM (CCITT G322 and G423, 1976). The standard *group* in the plan consists of 12 voice channels. One way to create such groups is to use an arrangement similar to that shown in Fig. 8.3a, except of course that 12 multipliers and 12 sideband filters are required. In the standard plan, the lower sidebands are selected by the filters, and the *group* bandwidth extends from 60 to 108 kHz.

As an alternative to forming a 12-channel group directly, the VF channels may be frequency-division-multiplexed in threes by using the arrangement shown in Fig. 8.3a. The four 3-channel-multiplexed signals, termed *pregroups,* are then combined to form the 12-channel group. This approach eases the filtering requirements, but does require an additional mixer stage which adds noise to the process.

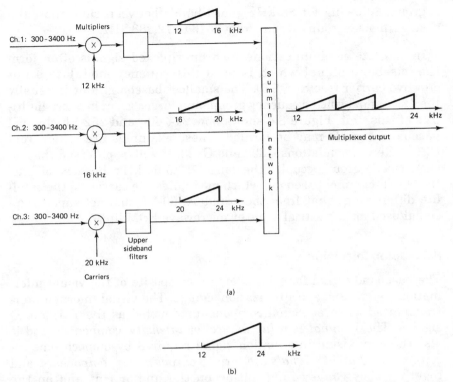

Figure 8.3 (*a*) Three-channel frequency-division multiplex scheme; (*b*) Simplified representation of the multiplexed signal.

The main group designations in the CCITT modulation plan are

Group.   As already described, this consists of 12 voice-frequency channels, each occupying a 4-kHz bandwidth in the multiplexed output. The overall bandwidth of a group extends from 60 to 108 kHz.

Supergroup.   A supergroup is formed by frequency-division multiplexing five groups together. The lower sidebands are combined to form a 60-VF-channel supergroup extending from 312 to 552 kHz.

Basic mastergroup.   A basic mastergroup is formed by frequency-division multiplexing five supergroups together. The lower sidebands are combined to form a 300-VF-channel basic mastergroup. Allowing for 8-kHz guard bands between sidebands, the basic mastergroup extends from 812 to 2044 kHz.

Super mastergroup.   A super mastergroup is formed by frequency-division multiplexing three basic mastergroups together. The lower sidebands are combined to form the 900-VF-channel super mastergroup. Allowing for 88-kHz guard bands between sidebands, the super mastergroup extends from 8516 to 12,388 kHz.

In satellite communications such multiplexed signals often form the baseband signal which is used to frequency-modulate a microwave carrier (see Sec. 8.6). The smallest baseband unit is usually the 12-channel voice-frequency group, and larger groupings are multiples of this unit. Figure 8.4 shows how 24-, 60-, and 252-VF-channel baseband signals may be formed. These examples are taken from CCITT Recommendations G322 and G423. It will be observed that, in each case, a group occupies the range 12 to 60 kHz. Because of this, the 60-VF-channel baseband which modulates the carrier to the satellite differs somewhat from the standard 60-VF-channel supergroup signal used for terrestrial cable or microwave FDM links.

8.5   Color Television

The baseband signal for television is a composite of the visual information signals and synchronization signals. The visual information is transmitted as three signal components, denoted as the Y, I, and Q signals. The *Y signal* is a *luminance,* or *intensity,* component and is also the only visual information signal required by monochrome receivers. The *I* and *Q signals* are termed *chrominance components,* and together they convey information on the hue or tint and on the amount of saturation of the coloring which is present.

The synchronization signal consists of narrow pulses at the end of each line scan for horizontal synchronization, and a sequence of narrower and wider pulses at the end of each field scan for vertical synchronization. Additional synchronization for the color information demodulation in the receiver is superimposed on the horizontal pulses as described below.

The luminance signal and the synchronization pulses require a base bandwidth of 4.2 MHz for North American standards. The baseband extends down to, and includes, a dc component. The composite signal containing the luminance and synchronization information is designed to be fully compatible with the requirements of monochrome (black-and-white) receivers.

In transmitting the chrominance information, use is made of the fact that the eye cannot resolve colors to the extent that it can intensity detail; this allows the chrominance signal bandwidth to be less than that of the luminance signal. The I and Q chrominance signals are transmitted within the luminance bandwidth by quadrature DSBSC (see below), modulating them onto a subcarrier which places them at the upper end of the luminance signal spectrum. Use is made of the

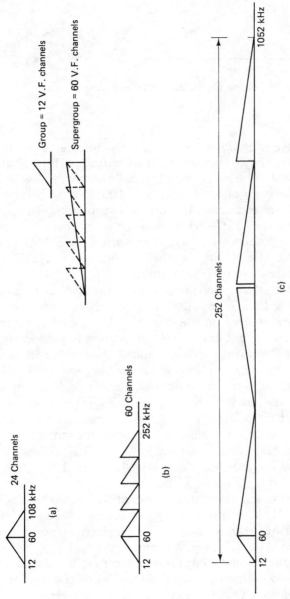

Figure 8.4 Examples of baseband signals for FDM telephony: (*a*) 24 channels; (*b*) 60 channels; (*c*) 252 channels.

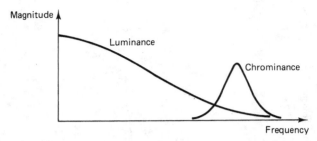

Figure 8.5 Frequency spectra for the luminance and chrominance signals.

fact that the eye cannot readily perceive the interference which results when the chrominance signals are transmitted within the luminance-signal bandwidth. The baseband response is shown in Fig. 8.5.

Different methods of chrominance subcarrier modulation are employed in different countries. In France, a system known as *Sequential Couleur a Mémoire* (SECAM) is used. In most other European countries, a system known as *Phase Alternation Line* (PAL) is used. In North America, the NTSC system is used, where NTSC stands for *National Television System Committee.*

In the NTSC system, each chrominance signal is modulated onto its subcarrier using DSBSC modulation as described in Sec. 8.3. A single oscillator source is used so that the I and Q signal subcarriers have the same frequency, but one of the subcarriers is shifted 90° in phase to preserve the separate chrominance information in the I and Q baseband signals. This method is known as *quadrature modulation* (QM). The I signal is the chrominance signal which modulates the in-phase carrier. Its bandwidth in the NTSC system is restricted to 1.5 MHz, and after modulation onto the subcarrier, a single-sideband filter removes the upper sideband components more than 0.5 MHz above the carrier. This is referred to as a vestigial sideband (VSB). The modulated I signal therefore consists of the 1.5-MHz lower sideband plus the 0.5 MHz upper VSB.

The Q signal is the chrominance signal which modulates the quadrature carrier. Its bandwidth is restricted to 0.5 MHz, and after modulation a DSBSC signal results. The spectrum magnitude of the combined I and Q signals is shown in Fig. 8.5.

The magnitude of the QM envelope contains the color saturation information, and the phase angle of the QM envelope contains the hue, or tint, information. The chrominance signal subcarrier frequency has to be precisely controlled, and in the NTSC system it is held at 3.579545 MHz ± 10 Hz, which places the subcarrier frequency midway between the 227th and the 228th harmonics of the horizontal scanning rate (frequency). The luminance and chrominance signals are both characterized by spectra wherein the power spectral density occurs in groups which are centered about the harmonics of the horizontal scan frequency. Placing the chrominance subcarrier midway between the 227th and 228th horizontal-scan harmonics of the luminance-plus-synchronization signals causes the luminance and the chrominance signals to be interleaved in the spectrum of the composite NTSC signal. This interleaving is most apparent in the range from about 3.0 to 4.1 MHz. The presence of the chrominance signal causes high-frequency modulation of the luminance signal and produces a very fine stationary dot-matrix pattern in the picture areas of high color saturation. To prevent this, most of the cheaper TV receivers limit the luminance channel video bandwidth to about 2.8 to 3.1 MHz. More expensive "high-resolution" receivers employ a *comb filter* to remove most of the chrominance signal from the luminance-channel signal while still maintaining about a 4-MHz luminance-channel video bandwidth.

Because the subcarrier is suppressed in the modulation process, a subcarrier frequency and phase reference carrier must be transmitted to allow the I and Q baseband chrominance signals to be demodulated at the receiver. This reference signal is transmitted in the form of bursts of eight to eleven cycles of the phase-shifted subcarrier, transmitted on the "backporch" of the horizontal blanking pulse. These bursts are transmitted toward the end of each line sync period, part of the line sync pulse being suppressed to accommodate them. One line waveform including the synchronization signals is shown in Fig. 8.6.

Figure 8.7 shows in block schematic form the NTSC system. The TV camera contains three separate camera tubes, one for each of the colors red, blue, and green. It is known that colored light can be syn-

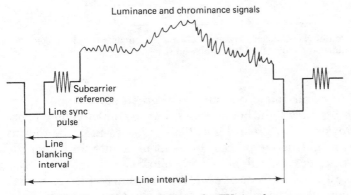

Figure 8.6 One line of waveform for a color TV signal.

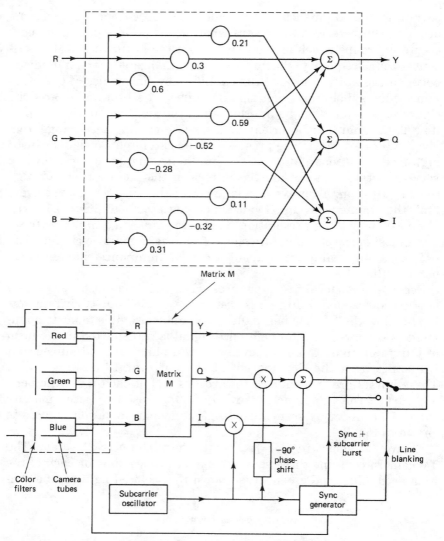

Figure 8.7 Generation of NTSC color TV signal. Matrix M converts the three color signals into the luminance and chrominance signals.

thesized by *additive* mixing of red, blue, and green light beams, these being the three primary light beam colors. For example, yellow is obtained by adding red and green light. (This process must be distinguished from the subtractive process of paint pigments, in which the primary pigment colors are red, blue, and yellow.)

Color filters are used in front of each tube to sharpen its response. In principle it would be possible to transmit the three color signals, and at the receiver reconstruct the color scene from them. However, this is not the best technical approach because such signals would not be compatible with monochrome television, and would require extra bandwidth. Instead, three new signals are generated which do provide compatibility and do not require extra bandwidth. These are the luminance signal and the two chrominance signals which have been already described. The process of generating the new signals from the color signals is mathematically equivalent to having three equations in three variables and rearranging these in terms of three new variables which are linear combinations of the original three. The details are shown in the matrix M block of Fig. 8.7, and derivation of the equations from this is left as Prob. 8.9.

At the receiver, the three color signals can be synthesized from the luminance and chrominance components. Again, this is mathematically equivalent to rearranging the three equations into their original form. The three color signals then modulate the electron beams which excite the corresponding color phosphors in the TV tube. The complete video signal is therefore a multiplexed baseband signal which extends from dc up to 4.2 MHz, and which contains all of the visual information plus synchronization signals.

In conventional TV broadcasting the aural signal is transmitted by a separate transmitter as shown in Fig. 8.8a. The aural information is received by stereo microphones, split into (L + R) and (L − R) signals, where L stands for left and R for right. The (L − R) signal is used to DSBSC-modulate a subcarrier at $2f_h$ (31.468 kHz). This DSBSC signal is then added to the (L + R) signal and used to frequency-modulate a separate transmitter whose rf carrier frequency is 4.5 MHz above the rf carrier frequency of the video transmitter. The outputs of these two transmitters may go to separate antennas or may be combined and fed into a single antenna as is shown in Fig. 8.8a.

The signal format for satellite TV differs from that of conventional TV as shown in Fig. 8.8. To generate the uplink microwave TV signal to a communications satellite transponder channel, the composite video signal (going from 0 Hz to about 4.2 MHz for the North American NTSC standard) is added to two or three FM carriers at frequencies of 6.2, 6.8, and/or 7.4 MHz, which carry audio information. This composite FDM signal is then, in turn, used to frequency-modulate the uplink microwave carrier signal, producing a signal with an rf bandwidth of about 36 MHz. The availability of three possible audio signal carriers permits the transmission of stereo and/or multilingual audio over the satellite link. Figure 8.8b shows a block diagram of this system.

As mentioned previously, three color TV systems, NTSC, PAL, and SECAM, are in widespread use. In addition, different countries use

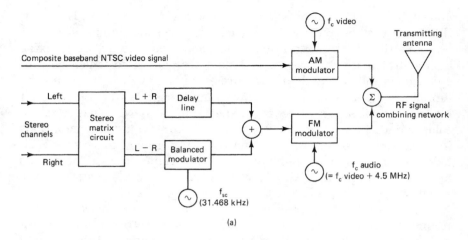

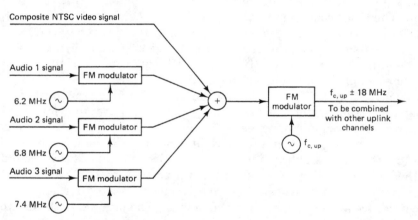

Figure 8.8 (*a*) Conventional TV broadcasting of the video and aural signals; (*b*) generation of a TV channel satellite uplink signal.

different line frequencies (determined by the frequency of the domestic power supply) and different numbers of lines per scan. Broadcasting between countries utilizing different standards requires the use of a converter. Transmission takes place using the standards of the country originating the broadcast, and conversion to the standards of the receiving country takes place at the receiving station. The conversion may take place through optical image processing, or by conversion of the electronic signal format. The latter can be further subdivided into analog and digital techniques. The digital converter, referred to as digital intercontinental conversion equipment (DICE) is favored because of its good performance and lower cost (see Miya, 1981).

8.6 Frequency Modulation

The analog signals discussed in the previous sections are transferred to the microwave carrier by means of frequency modulation (FM). Instead of being done in one step as shown in Fig. 8.8*b*, this modulation usually takes place at an intermediate frequency as is shown in Fig. 7.6. This signal is then frequency-multiplied up to the required uplink microwave frequency. In the receive branch of Fig. 7.6, the incoming (downlink) FM microwave signal is downconverted to an intermediate frequency and the "baseband" signal is recovered from the IF carrier in the demodulator. The actual baseband video signal is now available directly via a low-pass filter, but the audio channels must each undergo an additional step of FM demodulation to recover the baseband audio signals.

A major advantage associated with frequency modulation is the improvement in the postdetection signal-to-noise ratio at the receiver output, compared with other analog-modulation methods. This improvement can be attributed to three factors: (1) amplitude limiting, (2) a property of FM which allows an exchange between signal-to-noise ratio and bandwidth, and (3) a noise reduction inherent in the way noise phase-modulates a carrier. These factors are discussed in more detail in the following sections.

Figure 8.9 shows the basic circuit blocks of an FM receiver. The receiver noise, including that from the antenna, can be lumped into one equivalent noise source at the receiver input, as described in Sec. 10.5. It is emphasized at this point that thermal-like noise only is being considered, the main characteristic of which is that the spectral density of the noise power is constant, as given by Eq. (10.15). This is referred to as a *flat spectrum*. (This type of noise is also referred to as *white noise* in analogy to white light, which contains a uniform spectrum of colors.) Both the signal spectrum and the noise spectrum are converted to the intermediate frequency bands, with the bandwidth of the i-f stage determining the total noise power at the input to the demodulator. The IF bandwidth has to be wide enough to accommodate the FM signal as described in Sec. 8.6.2, but should be no wider.

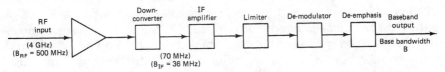

Figure 8.9 Elements of an FM receiver. Figures shown in parentheses are typical.

8.6.1 Limiters

The total thermal noise referred to the receiver input modulates the incoming carrier in amplitude and in phase. The rf limiter circuit (often referred to as an instantaneous or "hard" limiter) following the IF amplifier removes the amplitude modulation, leaving only the phase-modulation component of the noise. The limiter is an amplifier designed to operate as a class A amplifier for small signals. With large signals, positive excursions are limited by the saturation characteristics of the transistor (which is operated at a low collector voltage), and negative excursions generate a self-bias which drive the transistor into cutoff. Although the signal is severely distorted by this action, a tuned circuit in the output selects the FM carrier and its sidebands from the distorted signal spectrum, and thus the constant amplitude characteristic of the FM signal is restored. This is the amplitude-limiting improvement referred to previously. Only the noise phase modulation contributes to the noise at the output of the demodulator.

Amplitude limiting is also effective in reducing the interference produced by impulse-type noise, such as that generated by certain types of electrical machinery. Noise of this nature may be picked up by the antenna and superimposed as large amplitude excursions on the carrier, which the limiter removes. Limiting can also greatly alleviate the interference caused by other, weaker signals which occur within the IF bandwidth. When the limiter is either saturated or cut off by the larger signal, the weaker signal has no effect. This is known as *limiter capture* (see, for example, Young, 1990).

8.6.2 Bandwidth

When considering bandwidth it should be kept in mind that the word is used in a number of contexts. *Signal bandwidth* is a measure of the frequency spectrum occupied by the signal. *Filter bandwidth* is the frequency range passed by circuit filters. *Channel bandwidth* refers to the overall bandwidth of the transmission channel, which in general will include a number of filters at different stages. In a well-designed system the channel bandwidth will match the signal bandwidth.

Bandwidth requirements will be different at different points in the system. For example, at the receiver inputs for C-band and Ku-band satellite systems, the bandwidth typically is 500 MHz, this accommodating 12 transponders as described in Sec. 6.7. The individual transponder bandwidth is typically 36 MHz. In contrast, the baseband bandwidth for a telephony channel is typically 3.1 kHz.

In theory, the spectrum of a frequency-modulated carrier extends to infinity. In a practical satellite system the bandwidth of the transmitted FM signal is limited by the intermediate-frequency amplifiers. The IF bandwidth, denoted by B_{IF}, must be wide enough to pass all the significant components in the FM signal spectrum that is generated. The required bandwidth is usually estimated by *Carson's rule* as

$$B_{IF} = 2(\Delta F + F_M) \qquad (8.1)$$

where ΔF is the peak carrier deviation produced by the modulating baseband signal and F_M is the highest frequency component in the baseband signal. These maximum values, ΔF and F_M, are specified in the regulations governing the type of service. For example, for commercial FM sound broadcasting in North America, $\Delta F = 75$ kHz and $F_M = 15$ kHz.

The *deviation ratio D* is defined as the ratio

$$D = \frac{\Delta F}{F_M} \qquad (8.2)$$

Example 8.1. A video signal of bandwidth 4.2 MHz is used to frequency-modulate a carrier, the deviation ratio being 2.56. Calculate the peak deviation and the signal bandwidth.

solution

$$\Delta F = 2.56 \times 4.2 = 10.752 \text{ MHz}$$
$$B_{IF} = 2(10.752 + 4.2) = 29.9 \text{ MHz}$$

A similar ratio, known as the *modulation index,* is defined for sinusoidal modulation. This is usually denoted by ß in the literature. Letting Δf represent the peak deviation for sinusoidal modulation and f_m the sinusoidal modulating frequency gives

$$ß = \frac{\Delta f}{f_m} \qquad (8.3)$$

The difference between ß and D is that D applies for an arbitrary modulating signal and is the ratio of the maximum permitted values of deviation and baseband frequency, whereas ß applies only for sinusoidal modulation (or what is often termed *tone modulation*). Very often the analysis of an FM system will be carried out for tone modulation rather than for an arbitrary signal, as the mathematics are easier and the results usually give a good indication of what to expect with an arbitrary signal.

Example 8.2.   A test tone of frequency 800 Hz is used to frequency-modulate a carrier, the peak deviation being 200 kHz. Calculate the modulation index and the bandwidth.

solution

$$ß = \frac{200}{0.8} = 250$$

$$B = 2(200 + 0.8) = 401.6 \text{ kHz}$$

Carson's rule is widely used in practice, even though it tends to give an underestimate of the bandwidth required for deviation ratios in the range $2 < D < 10$, which is the range most often encountered in practice. For this range a better estimate of bandwidth is given by

$$B_{IF} = 2(\Delta F + 2F_M) \tag{8.4}$$

Example 8.3.   Recalculate the bandwidths for Examples 8.1 and 8.2.

solution

For the video signal

$$B_{IF} = 2(10.75 + 8.4) = 38.3 \text{ MHz}$$

For the 800 Hz tone:

$$B_{IF} = 2(200 + 1.6) = 403.2 \text{ kHz}$$

In Examples 8.1 through 8.3 it will be seen that, when the deviation ratio (or modulation index) is large, the bandwidth is determined mainly by the peak deviation and is given by either Eq. (8.1) or Eq. (8.4). However, for the video signal, for which the deviation ratio is relatively low, the two estimates of bandwidth are 29.9 and 38.3 MHz. In practice, the standard bandwidth of a satellite transponder required to handle this signal is 36 MHz.

The peak frequency deviation of an FM signal is proportional to the peak amplitude of the baseband signal. Increasing the peak amplitude results in increased signal power and hence a larger signal-to-noise ratio. At the same time ΔF, and hence the FM signal bandwidth, will increase as shown previously. Although the noise power at the demodulator input is proportional to the IF filter bandwidth, the noise power output after the demodulator is determined by the bandwidth of the baseband filters, and therefore an increase in IF filter bandwidth does not increase output noise. Thus an improvement in signal-to-noise ratio is possible but at the expense of an increase in the IF bandwidth. This is the large-amplitude signal improvement referred to in Sec. 8.6 and considered further in the next section.

8.6.3 FM detector noise and processing gain

At the input to the FM detector, the thermal noise is spread over the IF bandwidth, as shown in Fig. 8.10a. The noise is represented by the system noise temperature T_s as will be described in Sec. 10.5. At the input to the detector, the quantity of interest is the carrier-to-noise ratio. Since both the carrier and the noise are amplified equally by the receiver gain following the antenna input, this gain may be ignored in the carrier-to-noise ratio calculation, and the input to the detector represented by the voltage source shown in Fig. 8.10b. The carrier root-mean-square (rms) voltage is shown as E_c.

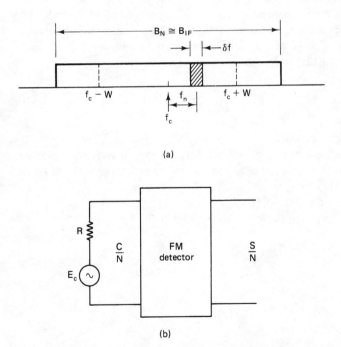

(a)

(b)

Figure 8.10 (a) The predetector noise bandwidth B_N is approximately equal to the IF bandwidth B_{IF}. The LF bandwidth W fixes the equivalent postdetector noise bandwidth at $2W$. f is an infinitesimally small noise bandwidth. (b) Receiving system, including antenna represented as a voltage source up to the FM detector.

The available carrier power at the input to the FM detector is $E_c/4R$, and the available noise power at the FM detector input is kT_sB_N (as explained in Sec. 10.5), so that the input carrier-to-noise ratio, denoted by C/N, is

$$\frac{C}{N} = \frac{E_c^2}{4RkT_sB_N} \tag{8.5}$$

The receiver noise voltage phase-modulates the incoming carrier. (It cannot directly frequency-modulate the carrier, whose frequency is determined at the transmitter, which is at a great distance from the receiver and which may be crystal controlled.) A characteristic of phase modulation is that, for the condition that the noise voltage is much less than the carrier voltage, the voltage spectral density of the noise at the detector output increases in direct proportion to the noise frequency, where the noise frequency is f_n as shown in Fig. 8.10a. The noise voltage spectral density (measured in volts per hertz) is as shown in Fig. 8.11.

The noise occurs on both sides of the carrier, but because the postdetector filters limit the frequency to some upper value W, in calculations of the output noise it is necessary only to consider the spectrum between the limits $f_c + W$ and $f_c - W$ as shown in Fig. 8.10a. The *average* noise power output is given by

$$P_n = \frac{AW^3}{1.5B_N C/N} \tag{8.6}$$

where A = detector constant
W = low-frequency bandwidth (that is, the low-frequency or postdetector output is assumed to cover the spectrum from 0 to W Hz)
B_N = noise bandwidth at IF
C/N = carrier-to-noise ratio given by Eq. (8.5)

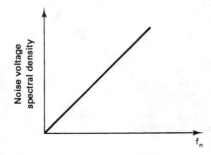

Figure 8.11 Noise voltage spectral density for FM.

When a sinusoidal signal of peak deviation Δf is present, the average signal power output is proportional to Δf^2 and is given by

$$P_s = A\Delta f^2 \qquad (8.7)$$

where A is the same constant of proportionality as used for noise. The output signal-to-noise ratio, denoted by S/N, is

$$\frac{S}{N} = \frac{P_s}{P_n}$$

$$= 1.5 \frac{C}{N} \frac{B_N \Delta f^2}{W^3} \qquad (8.8)$$

The *processing gain* of the detector is the ratio of signal-to-noise ratio to carrier-to-noise ratio. Denoting this by K_R gives

$$K_R = \frac{S/N}{C/N}$$

$$= \frac{1.5\, B_N\, \Delta f^2}{W^3} \qquad (8.9)$$

Using Carson's rule for the i-f bandwidth, $B_{IF} = 2(\Delta f + W)$, and assuming $B_N \cong B_{IF}$, the processing gain for sinusoidal modulation becomes after some simplification

$$K_R = 3(\beta + 1)\beta^2 \qquad (8.10)$$

Here, $\beta = \Delta f/W$ is the modulation index for a sinusoidal modulation frequency at the highest value W. Equation (8.10) shows that a high modulation index results in a high processing gain, which means that the signal-to-noise ratio can be increased even though the carrier-to-noise ratio is constant.

8.6.4 Signal-to-noise ratio

The term *signal-to-noise ratio* introduced in the preceding section is used to refer to the ratio of signal power to noise power at the receiver output. This ratio is sometimes referred to as the *post detector* or *destination* signal-to-noise ratio. In general, it differs from the carrier-to-noise ratio at the detector input (the words *detector* and *demodulator* may be used interchangeably), the two ratios being related through the receiver processing gain as shown by Eq. (8.9). Equation (8.9) may be written in decibel form as

$$10 \log_{10} \frac{S}{N} = 10 \log_{10} \frac{C}{N} + 10 \log_{10} K_R \qquad (8.11)$$

As indicated in App. G, it is useful to use square brackets to denote decibel quantities where these occur frequently. Equation (8.11) may therefore be written as

$$\left[\frac{S}{N}\right] = \left[\frac{C}{N}\right] + [K_R] \tag{8.12}$$

This shows that the signal-to-noise in decibels is proportional to the carrier-to-noise in decibels. However, these equations were developed for the condition that the noise voltage should be much less than the carrier voltage. At low carrier-to-noise ratios this assumption no longer holds, and the detector exhibits a *threshold effect*. This is a threshold level in the carrier-to-noise ratio below which the signal-to-noise degrades very rapidly. The threshold level is shown in Fig. 8.12, and is defined as the carrier-to-noise ratio at which the signal-to-noise ratio is 1 dB below the straight-line plot of Eq. (8.12). For con-

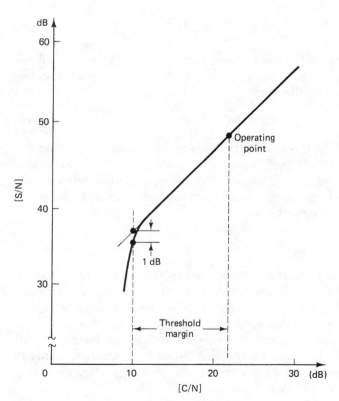

Figure 8.12 Output signal-to-noise ratio S/N versus input carrier-to-noise ratio C/N for a modulating index = 5. The straight-line section is a plot of Eq. (8.12).

ventional FM detectors (such as the Foster Seeley detector), the threshold level may be taken as 10 dB. *Threshold extension* detector circuits are available which can provide a reduction in the threshold level of between 3 and 7 dB (Fthenakis, 1984).

In normal operation, the operating point will always be above threshold, the difference between the operating carrier-to-noise ratio and the threshold level being referred to as the *threshold margin*. This is also illustrated in Fig. 8.12.

Example 8.4. A 1-kHz test tone is used to produce a peak deviation of 5 kHz in an FM system. Given that the received $[C/N]$ is 30 dB, calculate the receiver processing gain and the postdetector $[S/N]$.

solution   Since the $[C/N]$ is above threshold, Eq. (8.12) may be used. The modulation index is ß = 5 kHz/1 kHz = 5. Hence

$$K_R = 3 \times 5^2 \times (5 + 1) = 450$$

and $[K_R] = 26.5$ dB. From Eq. (8.12),

$$[S/N] = 30 + 26.5 = 56.5 \text{ dB}.$$

8.6.5   Preemphasis and deemphasis

As shown in Fig. 8.11, the noise voltage spectral density increases in direct proportion to the demodulated noise frequency. As a result the signal-to-noise ratio is worse at the high-frequency end of the baseband, a fact which is not apparent from the equation for signal-to-noise ratio, which uses average values of signal and noise power. For example, if a test tone is used to measure the signal-to-noise ratio in a TV baseband channel, the result will depend on the position of the test tone within the baseband, a better result being obtained at lower test tone frequencies. For FDM/FM telephony, the telephone channels at the low end of the FDM baseband would have better signal-to-noise ratios than those at the high end.

To equalize the performance over the baseband, a deemphasis network is introduced after the demodulator to attenuate the high-frequency components of noise. Over most of the baseband, the attenuation-frequency curve of the deemphasis network is the inverse of the rising noise-frequency characteristic shown in Fig. 8.11 (for practical reasons it is not feasible to have exact compensation over the complete frequency range). Thus, after deemphasis, the noise-frequency characteristic is flat, as shown in Fig. 8.13*d*. Of course, the deemphasis network will also attenuate the signal, and to correct for this, a complementary preemphasis characteristic is intro-

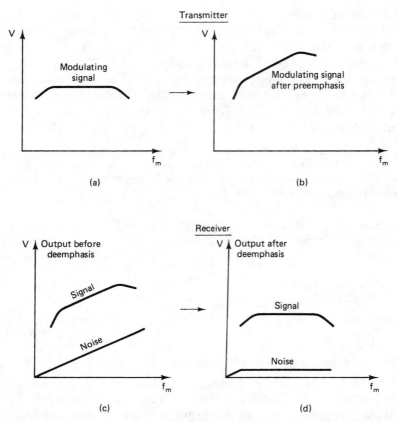

Figure 8.13 (a) and (b) Effect of preemphasis on the modulating signal frequency response at the transmitter. (c) and (d) Effect of deemphasis on the modulating signal and noise at the receiver output. The deemphasis cancels out the preemphasis for the signal while attenuating the noise at the receiver.

duced prior to the modulator at the transmitter. The overall effect is to leave the postdetection signal levels unchanged while the high-frequency noise is attenuated. The preemphasis, deemphasis sequence is illustrated in Fig. 8.13.

The resulting improvement in the signal-to-noise ratio is referred to variously as preemphasis improvement, deemphasis improvement, or simply as emphasis improvement. It is usually denoted by P, or $[P]$ decibels, and gives the reduction in the total postdetection noise power. Preemphasis curves for FDM/FM telephony are given in CCIR Recommendation 275-2, 1978 and for TV/FM in CCIR Recommendation 405-1, 1982. CCIR values for $[P]$ are 4 dB for the top channel in multichannel telephony; 13.1 dB for 525-line TV, and 13.0 dB for 625-line TV. Taking into account the emphasis improvement, Eq. (8.12) becomes

$$\left[\frac{S}{N}\right] = \left[\frac{C}{N}\right] + [K_R] + [P] \tag{8.13}$$

8.6.6 Noise weighting

Another factor that generally improves the postdetection signal-to-noise ratio is referred to as *noise weighting*. This is the way in which the flat noise spectrum has to be modified to take into account the frequency response of the output device and the subjective effect of noise as perceived by the observer. For example, human hearing is less sensitive to a given noise power density at low and high audio frequencies than at the middle frequency range.

Weighting curves have been established for various telephone handsets in use by different telephone administrations. One of these, the CCIR curve, is referred to as the psophometric weighting curve. When this is applied to the flat noise density spectrum, the noise power is reduced by 2.5 dB for a 3.1-kHz bandwidth (300 to 3400 Hz) compared to flat noise over the same bandwidth. The weighting improvement factor is denoted by [W], and hence for the CCIR curve [W] = 2.5 dB. (Do not confuse the symbol W here with that used for bandwidth earlier.) For a bandwidth of b kHz, a simple adjustment gives

$$[W] = 2.5 + 10 \log \frac{b}{3.1} \tag{8.14}$$

$$= -2.41 + [b]$$

Here, b is in kHz. A noise weighting factor can also be applied to TV viewing. The CCIR weighting factors are 11.7 dB for 525-line TV and 11.2 dB for 625-line TV. Taking weighting into account, Eq. (8.13) becomes

$$\left[\frac{S}{N}\right] = \left[\frac{C}{N}\right] + [K_R] + [P] + [W] \tag{8.15}$$

8.6.7 *S/N* and bandwidth for FDM/FM telephony

In the case of FDM/FM, the receiver processing gain, excluding emphasis and noise weighting, is given by (see, e.g., Miya, 1980, and Halliwell, 1974)

$$K_R = \frac{B_{IF}}{b} \left(\frac{\Delta F_{rms}}{f_m}\right)^2 \tag{8.16}$$

Here f_m is a specified baseband frequency in the channel of interest, at which K_R is to be evaluated. For example, f_m may be the center frequency of a given channel, or it may be the top frequency of the baseband signal. The channel bandwidth is b (usually 3.1 kHz), and ΔF_{rms} is the root-mean-square deviation per channel of the signal. The rms deviation is determined under specified test conditions, details of which will be found in CCIR Recommendation 404-2, 1982. Some values are shown in Table 8.1.

The peak deviation of the full baseband signal can be determined from a knowledge of the rms deviation per channel, and, again, recourse must be made to the relevant CCITT formulas. These are discussed for example, in Freeman, 1981, and are reproduced here:

$$\Delta F = g \times \Delta f_{rms} \times L \qquad (8.17)$$

Here g is a peak/rms ratio for the test tone deviation, a number which must be specified for the particular signal being considered. For a small number of channels g may be as high as 18.6 dB (Fthenakis, 1984), but typically g is around 13 dB. It is important to note that g is a voltage ratio, and when expressed in decibels the corresponding voltage ratio is obtained from

$$g = 10^{gdB/20} \qquad (8.18)$$

Here, gdB is the value specified in decibels.

L is a loading factor that depends on the number of channels n in the FDM signal.

TABLE 8.1   FDM/FM RMS Deviations

Maximum number of channels	RMS deviations per channel, kHz
12	35
24	35
60	50, 100, 200
120	50, 100, 200
300	200
600	200
960	200
1260	140, 200
1800	140
2700	140

For $n \geq 240$

$$20 \log L = -15 + 10 \log n \tag{8.19}$$

For $12 \leq n \leq 240$

$$20 \log L = -1 + 4 \log n \tag{8.20}$$

Once ΔF is determined from Eq. (8.17), the IF bandwidth can be found from Carson's rule and the receiver processing gain from Eq. (8.16). The following example illustrates the procedure.

Example 8.5

The carrier-to-noise ratio at the input to the demodulator of an FDM/FM receiver is 25 dB. Calculate the signal-to-noise ratio for the top channel in a 24-channel FDM baseband signal, evaluated under test conditions for which Table 8.1 applies. The emphasis improvement is 4 dB, noise weighting improvement is 2.5 dB, and the peak/rms factor is 13.57 dB. The audio channel bandwidth may be taken as 3.1 kHz.

solution

Given data:

$n : = 24$ $gdB : = 13.57$ $b : = 3.1 \cdot 10^3 \cdot Hz$ $PdB : = 4$ $WdB : = 2.5$

$CNRdB : = 25$

$L : = 10^{\frac{-1+4 \cdot \log(n)}{20}}$ $L = 1.683$ ...eq. (8.20)

$g : = 10^{\frac{gdB}{20}}$ $g = 4.77$ ...eq. (8.18)

From Table 8.1, for 24 channels

$$\Delta F_{rms} : = 35 \cdot 10^3 \cdot Hz$$

Using eq. (8.17)

$$\Delta F : = g \cdot \Delta F_{rms} \cdot L \qquad \Delta F = 2.809 \cdot 10^5 \cdot Hz$$

Assuming that the baseband spectrum is as shown in Fig. 8.4a, the top frequency is 108 kHz

$$f_m : = 108 \cdot 10^3 \cdot Hz$$

Carson's rule gives:

$$B_{IF} : = 2 \cdot (\Delta F + f_m)$$

From eq. (8.16)

$$K_R : = \frac{B_{IF}}{b} \cdot \left(\frac{\Delta F_{rms}}{f_m} \right)^2 \qquad K_R = 26.353$$

From eq. (8.15)

$$\text{SNRdB} := \text{CNRdB} + 10 \cdot \log(K_R) + \text{PdB} + \text{WdB} \qquad \text{SNRdB} = 45.7$$

$$=========$$

8.6.8 Signal-to-noise ratio for TV/FM

Television performance is measured in terms of the postdetector video signal-to-noise ratio, defined as (CCIR Recommendation 567-2, 1986)

$$\left(\frac{S}{N}\right)_V = \frac{\text{peak-to-peak video voltage}}{\text{rms noise voltage}} \qquad (8.21)$$

Because peak-to-peak video voltage is used, $2\,\Delta F$ replaces ΔF in Eq. (8.8). Also, since power is proportional to voltage squared, $(S/N)_V^2$ replaces S/N, and Eq. (8.8) becomes

$$\left(\frac{S}{N}\right)_V^2 = 1.5\,\frac{C}{N}\,\frac{B_N(2\,\Delta F)^2}{W^3} \qquad (8.22)$$

where W is the highest video frequency. With the deviation ratio $D = \Delta F/W$, and the processing gain for TV denoted as K_{RV},

$$K_{RV} = \frac{(S/N)_V^2}{C/N} = 12D^2(D+1) \qquad (8.23)$$

Some workers include an implementation margin to allow for non-ideal performance of filters and demodulators (Bischof et al., 1981). With the implementation margin denoted by [IMP], Eq. (8.15) becomes

$$\left(\frac{S}{N}\right)_V^2 = \left[\frac{C}{N}\right] + [K_R] + [P] + [W] - [IMP] \qquad (8.24)$$

Recall that this equation is in decibels. This is illustrated in the following example.

Example 8.6

A satellite TV link is designed to provide a video signal-to-noise ratio of 62 dB. The peak deviation is 9 MHz, and the highest video baseband frequency is 4.2 MHz. Calculate the carrier-to-noise ratio required at the input to the FM detector, given that the combined noise weighting, emphasis improvement, and implementation margin is 11.8 dB.

solution

$$D = \frac{9}{4.2} = 2.143$$

Equation (8.23) gives

$$K_{RV} = 12 \times 2.143^2 \times (2.142 + 1) = 173.2$$

Therefore

$$[K_{RV}] = 10 \log K_{RV} = 22.4 \text{ dB}$$

Since the required signal-to-noise ratio is 62 dB, eq. (8.24) can be written as

$$62 = \left[\frac{C}{N}\right] + 22.4 + 11.8$$

from which $[C/N] = 27.8$ dB.

8.7 Problems

8.1. State the frequency limits generally accepted for telephone transmission of speech, and typical signal levels encountered in the telephone network.

8.2. Show that when two sinusoids of different frequencies are multiplied together, the resultant product contains sinusoids at the sum and difference frequencies only. Hence show how a multiplier circuit may be used to produce a DSBSC signal.

8.3. Explain how a DSBSC signal differs from a conventional amplitude-modulated signal such as used in the medium-wave (broadcast) radio band. Describe one method by which an SSB signal may be obtained from a DSBSC signal.

8.4. Explain what is meant by *FDM telephony*. Sketch the frequency plans for the CCITT designations of group, supergroup, basic mastergroup, and super mastergroup.

8.5. With the aid of a block schematic, show how 12 VF channels could be frequency-division multiplexed.

8.6. Explain how a 252-VF-channel group is formed for satellite transmission.

8.7. Describe the essential features of the video signal used in the NTSC color TV scheme. How is the system made compatible with monochrome reception?

8.8. Explain how the sound information is added to the video information in a color TV transmission.

8.9. For the matrix network M shown in Fig. 8.7, derive the equations for the Y, Q, and I signals in terms of the input signals R, G, and B.

8.10.   Explain what is meant by *frequency modulation*. A 70-MHz carrier is frequency-modulated by a 1-kHz tone of 5-V peak amplitude. The frequency-deviation constant is 15 kHz/V. Write down the expression for instantaneous frequency.

8.11.   An angle-modulated wave may be written as sin $\theta(t)$, where the argument $\theta(t)$ is a function of the modulating signal. Given that the instantaneous angular frequency is $\omega_i = d\theta(t)/dt$, derive the expression for the FM carrier in Prob. 8.10.

8.12.   (*a*) Explain what is meant by *phase modulation*. (*b*) A 70-MHz carrier is phase modulated by a 1-kHz tone of 5-V peak amplitude. The phase modulation constant is 0.1 rad/V. Write down the expression for the argument $\theta(t)$ of the modulated wave.

8.13.   Determine the equivalent peak frequency deviation for the phase-modulated signal of Prob. 8.12.

8.14.   Show that when a carrier is phase-modulated with a sinusoid, the equivalent peak frequency deviation is proportional to the modulating frequency. Explain the significance of this on the output of an FM receiver used to receive the PM wave.

8.15.   In the early days of FM it was thought that the bandwidth could be limited to twice the peak deviation irrespective of the modulating frequency. Explain the fallacy behind this reasoning.

8.16.   A 10-kHz tone is used to frequency-modulate a carrier, the peak deviation being 75 kHz. Use Carson's rule to estimate the bandwidth required.

8.17.   Use Carson's rule to estimate the bandwidth required for the FM signal of Prob. 8.10.

8.18.   Use Carson's rule to estimate the bandwidth required for the PM signal of Prob. 8.12.

8.19.   Explain what is meant by preemphasis and deemphasis and why these are effective in improving signal-to-noise ratio in FM transmission. State typical improvement levels expected for both telephony and TV transmissions.

8.20.   Explain what is meant by *noise weighting*. State typical improvement levels in signal-to-noise ratios which result from the introduction of noise weighting, for both telephony and TV transmissions.

8.21.   Calculate the loading factor L for (*a*) a 12-channel, (*b*) a 120-channel, and (*c*) an 1800-channel FDM/FM telephony signal.

8.22.   Calculate the IF bandwidth required for (*a*) a 12-channel, (*b*) a 300-channel, and (*c*) a 900-channel FDM/FM telephony signal. Assume that the peak/rms factor 2[*g*] is equal to 10 dB for parts (*a*) and (*b*), and equal to 18 dB for part (*c*).

8.23.   Calculate the receiver processing gain for each of the signals given in Prob. 8.22.

8.24. A video signal has a peak deviation of 9 MHz and a video bandwidth of 4.2 MHz. Using Carson's rule, calculate the i-f bandwidth required and the receiver processing gain.

8.25. For the video signal of Prob. 8.24, the emphasis improvement figure is 13 dB, and the noise weighting improvement figure is 11.2 dB. Calculate in decibels (a) the signal-to-noise power ratio and (b) the video signal-to-noise ratio as given by Eq. (8.24). The $[C/N]$ value is 22 dB. Assume a sinusoidal video signal.

Chapter

9

Digital Signals

9.1 Introduction

As already mentioned in connection with analog signals, baseband signals are those signals which occupy the lowest, or base, frequency band in the frequency spectrum used by the telecommunications network. A baseband signal may consist of one or more information signals. For example, a number of telephony signals in digital form may be combined into one baseband signal by the process known as *time-division multiplexing*.

Analog signals may be converted into digital signals for transmission. Digital signals also originate in the form of computer and other data. In general, a digital signal is a coded version of the original data or analog signal. In this chapter, the characteristics of the more common types of digital baseband signals are described, along with representative methods of digital modulation.

9.2 Digital Baseband Signals

Digital signals are coded representations of information. Keyboard characters, for example, are usually encoded in binary digital code. A *binary code* has two symbols, usually denoted as 0 and 1, and these are combined to form binary words to represent the characters. For example, a teleprinter code may use the combination 11000 to represent the letter A.

Analog signals such as speech and video may be converted to a digital form through an analog-to-digital (A/D) converter. A particular form of A/D conversion is employed, known as *pulse code modulation,* which will be described in detail later. Some of these sources are illustrated diagrammatically in Fig. 9.1.

In digital terminology, a binary symbol is known as a *binit* from *bin*ary di*git*. The *information* carried by a binit is, in most practical

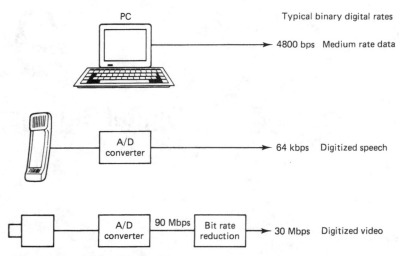

PC

Typical binary digital rates

→ 4800 bps   Medium rate data

A/D converter

→ 64 kbps   Digitized speech

A/D converter | 90 Mbps | Bit rate reduction

→ 30 Mbps   Digitized video

Figure 9.1   Examples of binary data sources.

situations, equal to a unit of information known as a *bit*. Thus it has become common practice to refer to binary symbols as bits rather than binits, and this practice will be followed here.

The digital information is transmitted as a waveform, some of the more common waveforms used for binary encoding being shown in Fig. 9.2. These will be referred to as *digital waveforms,* although strictly speaking they are analog representations of the digital information being transmitted. The binary sequence shown in Fig. 9.2 is 1010111. Detailed reasons for the use of different waveforms will be found in most books on digital communications (see, for example, Bellamy, 1982).

The duration of a bit is referred to as the *bit period* and is shown as T_b. The bit rate is given by

$$R_b = \frac{1}{T_b} \tag{9.1}$$

With T_b in seconds the bit rate will be in bits per second, usually denoted by b/s.

Figure 9.2a shows a *unipolar* waveform, meaning that the waveform excursions from zero are always in the same direction, either positive or negative. They are shown as positive A in Fig. 9.2a. Because it has a dc component, the unipolar waveform is unsuitable for use on telephone lines and radio networks, including satellite links.

Figure 9.2b shows a *polar* waveform, which utilizes positive and negative polarities. (In Europe this is referred to as a *bipolar* wave-

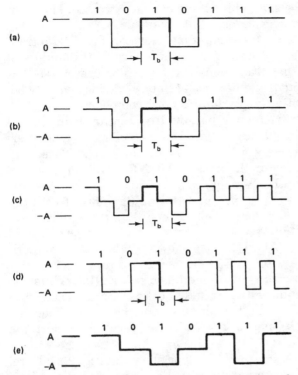

Figure 9.2 Examples of binary waveforms used for encoding digital data: (*a*) unipolar NRZ; (*b*) polar NRZ; (*c*) polar RZ; (*d*) split phase or Manchester, (*e*) alternate mark inversion.

form, but the term *bipolar* in North American usage is reserved for a specific waveform, described later). For a long, random sequence of 1's and 0's the dc component would average out to zero. However, long sequences of like symbols result in a gradual drift in the dc level, which creates problems at the receiver decoder. Also the decoding process requires knowledge of the bit timing, which is derived from the zero crossovers in the waveform, and these are obviously absent in long strings of like symbols. Both the unipolar and polar waveforms shown in Fig. 9.2*a* and *b* are known as *nonreturn to zero* (NRZ) waveforms. This is because the waveform does not return to the zero baseline at any point during the bit period.

Figure 9.2*c* shows an example of a polar *return to zero* (RZ) waveform. Here the waveform does return to the zero baseline in the middle of the bit period, so transitions will always occur even within a long string of like symbols, and bit timing can be extracted. However, dc drift still occurs with long strings of like symbols.

In the *split-phase* or *Manchester* encoding shown in Fig. 9.2*d* a transition between positive and negative levels occurs in the middle of each bit. This ensures that transitions will always be present so that bit timing can be extracted, and because each bit is divided equally between positive and negative levels there is no dc component.

A comparison of the frequency bandwidths required for digital waveforms can be obtained by considering the waveforms which alternate at the highest rate between the two extreme levels. These will appear as square waves. For the basic polar NRZ waveform of Fig. 9.2*b*, this happens when the sequence is . . . 101010. . . . The periodic time of such a square wave is $2T_b$ and the fundamental frequency component is $f = 1/2T_b$. For the split-phase encoding the square wave with the highest repetition frequency occurs with a long sequence of like symbols such as . . . 1111111 . . . as shown in Fig. 9.2*d*. The periodic time of this square wave is T_b, and hence the fundamental frequency component is twice that of the basic polar NRZ. Thus the split-phase encoding requires twice the bandwidth compared to that for the basic polar NRZ, while the bit rate remains unchanged. The utilization of bandwidth, measured in bits per second per hertz, is therefore less efficient.

An *alternate mark inversion* (AMI) code is shown in Fig. 9.2*e*. Here, the binary 0's are at the zero baseline level and the binary 1's alternate in polarity. In this way, the dc level is removed, while bit timing can be extracted easily, except when a long string of zeros occurs. Special techniques are available to counter this last problem. The highest pulse repetition frequency occurs with a long string of . . . 111111 . . . , the periodic time of which is $2T_b$, the same as the waveform of Fig. 9.2*b*. The AMI waveform is also referred to as a *bipolar* waveform in North America.

Bandwidth requirements may be reduced by utilizing multilevel digital waveforms. Figure 9.3*a* shows a polar NRZ signal for the sequence 11010010. By arranging the bits in groups of two, four levels can be used. For example, these may be

$$11 \quad 3A$$

$$10 \quad A$$

$$01 \quad -A$$

$$00 \quad -3A$$

This is referred to as *quarternary* encoding, and the waveform is shown in Fig. 9.3*b*. The encoding is symmetrical about the zero axis,

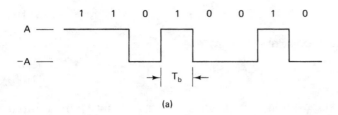

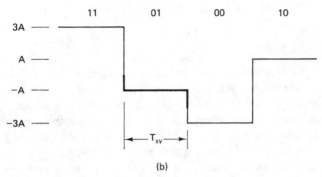

Figure 9.3 Encoding of 11010010 in (*a*) binary polar NRZ and (*b*) quarternary polar NRZ.

the spacing between adjacent levels being 2A. Each level represents a *symbol,* the duration of which is the *symbol period.* For the quarternary waveform the symbol period is seen to be equal to twice the bit period, and the symbol rate is

$$R_{sym} = \frac{1}{T_{sym}} \tag{9.2}$$

The symbol rate is measured in units of *bauds,* where one baud is one symbol per second.

The periodic time of the square wave having the greatest symbol repetition frequency is $2T_{sym}$, which is equal to $4T_b$, and hence the bandwidth, compared to the basic binary waveform, is halved. The bit rate (as distinct from the symbol rate) remains unchanged, and hence the bandwidth utilization in terms of bits per second per hertz is doubled.

In general, a waveform may have *M* levels (sometimes referred to as an *M-ary waveform*), where each symbol represents *m* bits and

$$m = \log_2 M \tag{9.3}$$

The symbol period is therefore

$$T_{sym} = T_b m \qquad (9.4)$$

and the symbol rate in terms of bit rate is

$$R_{sym} = \frac{R_b}{m} \qquad (9.5)$$

For satellite transmission, the encoded message must be modulated onto the microwave carrier. Before examining the modulation process, we describe the way in which speech signals are converted to a digital format through pulse code modulation.

9.3 Pulse Code Modulation

In the previous section describing baseband digital signals, the information was assumed to be encoded in one of the digital waveforms shown in Figs. 9.2 and 9.3. Speech and video appear naturally as analog signals and these must be converted to digital form for transmission over a digital link. In Fig. 9.1 the speech and video analog signals are shown converted to digital form through the use of analog-to-digital converters. The particular form of A/D conversion used is known as *pulse-code modulation* (PCM). Commercially available integrated circuits known as PCM *codecs* (for coder-decoder) are used to implement PCM. Figure 9.4a shows a block schematic for the Motorola MC145500 series of codecs. The analog signal enters at the Tx terminals and passes through a low-pass filter, followed by a high-pass filter to remove any 50/60-Hz interference which may appear on the line. The low-pass filter has a cutoff frequency of about 4 kHz, which allows for the filter roll-off above the audio limit of 3400 Hz. As shown in connection with single-sideband systems, a voice channel bandwidth extending from 300 Hz to 3400 Hz is considered satisfactory for speech. Band limiting the audio signal in this way reduces noise. It has another important consequence associated with the analog-to-digital conversion process. The analog signal is digitized by taking samples at periodic intervals. A theorem, known as the *sampling theorem*, states in part that the *sampling frequency* must be at least twice the highest frequency in the spectrum of the signal being sampled. With the upper cutoff frequency of the audio filter at 4 kHz, the sampling frequency can be standardized at 8 kHz.

The sampled voltage levels are encoded as binary digital numbers in the A/D converter following the high-pass filter. The binary number which is transmitted actually represents a range of voltages, and all samples which fall within this range are encoded as the same number. This process, referred to as *quantization*, obviously will introduce some distortion (termed *quantization noise*) into the signal. In a properly designed system the quantization noise is kept well within

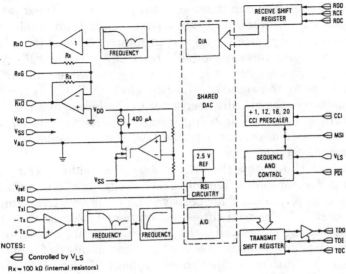

(a)

Chord Number	Number of Steps	Step Size	Normalized Encode Decision Levels	Digital Code								Normalized Decode Levels
				1 Sign	2 Chord	3 Chord	4 Chord	5 Step	6 Step	7 Step	8 Step	
			8159	1	0	0	0	0	0	0	0	8031
8	16	256	7903 ⋮					⋮				⋮
			4319	1	0	0	0	1	1	1	1	4191
			4063 ⋮					⋮				⋮
7	16	128	2143	1	0	0	1	1	1	1	1	2079
			2015 ⋮					⋮				⋮
6	16	64	1055	1	0	1	0	1	1	1	1	1023
			991 ⋮					⋮				⋮
5	16	32	511	1	0	1	1	1	1	1	1	495
			479 ⋮					⋮				⋮
4	16	16	239	1	1	0	0	1	1	1	1	231
			223 ⋮					⋮				⋮
3	16	8	103	1	1	0	1	1	1	1	1	99
			95 ⋮					⋮				⋮
2	16	4	35	1	1	1	0	1	1	1	1	33
			31 ⋮					⋮				⋮
1	15	2	3	1	1	1	1	1	1	1	0	2
	1	1	1	1	1	1	1	1	1	1	1	0
			0									

NOTES:
1. Characteristics are symmetrical about analog zero with sign bit = 0 for negative analog values.
2. Digital code includes inversion of all magnitude bits.

(b)

Figure 9.4 (a) MC145500/01/02/03/05 PCM CODEC/filter monocircuit block diagram. (b) Mu-law encode-decode characteristics. (*Courtesy Motorola Inc.*)

acceptable limits. The quantization steps follow a nonlinear law, with large signals being quantized into coarser steps than small signals. This is termed *compression,* and it is introduced to keep the signal-to-quantization noise ratio reasonably constant over the full dynamic range of the input signal, while maintaining the same number of bits per codeword. At the receiver (the D/A block in Fig. 9.4) the binary codewords are automatically decoded into the larger quantized steps for the larger signals, this being termed *expansion.* The expansion law is the inverse of the compression law, and the combined processing is termed *companding.*

Figure 9.4b shows how the MC145500 codec achieves compression by using a *chorded approximation.* The leading bit of the digital codeword is a *sign* bit, being 1 for positive and 0 for negative samples of the analog signal. The next three bits are used to encode the chord in which the analog signal falls, the three bits giving a total of eight chords. Each chord is made to cover the same number of input steps, but the step size increases from chord to chord. The chord bits are followed by four bits indicating the step in which the analog value lies. The normalized decision levels shown in Fig. 9.4b are the analog levels at which the comparator circuits change from one chord to the next, and from one step to the next. These are normalized to a value 8159 for convenience in presentation. For example, the maximum value may be considered to be 8159 mV and then the smallest step would be 1 mV. The first step is shown as 1 (mV), but it should be kept in mind that the first quantized level spans the analog zero so that 0^+ must be distinguished from 0^-. Thus the level representing zero has in fact a step size of ± 1 mV.

As an example, suppose the sampled analog signal has a value + 500 mV. This falls within the normalized range 479 to 511 mV, and therefore the binary code is 10111111. It should be mentioned that normally the first step in a chord would be encoded 0000 but the bits are inverted, as noted in Fig. 9.4b. This is because low values are more likely than high values, and inversion increases the 1-bit density, which helps in maintaining synchronization.

The table in Fig. 9.4b shows mu-law encode-decode characteristics. The term mu law, usually written as μ law, originated with older analog compressors, where μ was a parameter in the equation describing the compression characteristic. The μ-law characteristic is standard in North America and Japan, while in Europe and many other parts of the world, a similar law known as the A law is in use. Figure 9.5 shows the curves for $\mu = 255$ and $A = 87.6$, which are the standard values in use. These are shown as smooth curves, which could be approached with the older analog compression circuits. The chorded approximation approaches these in straight line segments, or *chords,* for each step.

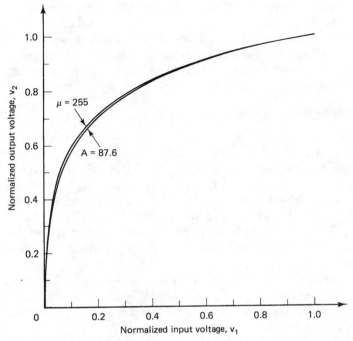

Figure 9.5 Compressor characteristics. Input and output voltage scales are normalized to the maximum values.

Because of the similarity of the A-law and μ-law curves, the speech quality, as affected by companding, will be similar in both systems, but otherwise the systems are incompatible, and conversion circuitry is required for interconnections such as might occur with international traffic. The MC145500 can be configured for use with either law through appropriate pin selections, but of course the transmitting and receiving functions must be configured for the same law.

In the receiver, the output from the D/A converter is passed through a low-pass filter which selects the original analog spectrum from the quantized signal. Its characteristics are similar to those of the low-pass filter used in the transmitter. Apart from the quantization noise (which should be negligible), the final output is a replica of the filtered analog signal at the transmitter.

With a sampling rate of 8 kHz or 8000 samples per second, and 8 bits for each sample codeword, the bit rate for a single-channel PCM signal is

$$R_b = 8000 \times 8 = 64 \text{ kb/s} \qquad (9.6)$$

The frequency spectrum occupied by a digital signal is proportional to the bit rate, and in order to conserve bandwidth it may be necessary to reduce the bit rate. For example, if 7-bit codewords were to be used, the bit rate would be 56 kb/s. Various *data reduction* schemes are in use which give much greater reductions, and some of these can achieve bit rates as low as 2400 b/s (Hassanein et al., 1989 and 1992).

9.4 Time-Division Multiplexing

A number of signals in binary digital form can be transmitted through a common channel by interleaving the pulses in time, this being referred to as *time-division multiplexing* (TDM). For speech signals, a separate codec may be used for each voice channel, the outputs from these being combined to form a TDM baseband signal as shown in Fig. 9.6. At the baseband level in the receiver the TDM signal is demultiplexed, the PCM signals being routed to separate codecs for decoding. In satellite systems the TDM waveform is used to modulate the carrier wave as described later.

The time-division multiplexed signal format is best described with reference to the widely used Bell T1 system. The signal format is illustrated in Fig. 9.7a. Each PCM word contains 8 bits, and a *frame* contains 24 PCM channels. In addition, a periodic *frame synchronizing* signal must be transmitted, and this is achieved by inserting a bit from the frame synchronizing codeword at the beginning of every frame. At the receiver, a special detector termed a *correlator* is used to detect the frame synchronizing codeword in the bit stream, which enables the frame timing to be established. The total number of bits

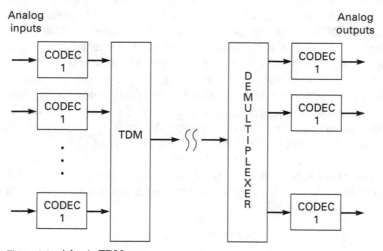

Figure 9.6 A basic TDM system.

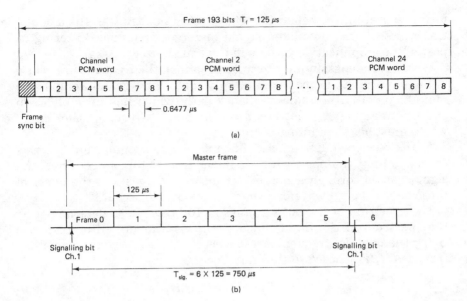

Figure 9.7 Bell T1 PCM format.

in a frame is therefore $24 \times 8 + 1 = 193$. Now, as established earlier, the sampling frequency for voice is 8 kHz, and so the interval between PCM words for a given channel is $1/8000 = 125$ μs. For example the leading bit in the PCM codewords for a given channel must be separated in time by no more than 125 μs. As can be seen from Fig. 9.7a, this is also the frame period and therefore the bit rate for the T1 system is

$$R_b = \frac{193}{125 \text{ μs}} = 1.544 \text{ Mb/s} \qquad (9.7)$$

Signaling information is also carried as part of the digital stream. By *signaling* is meant such data as number dialed, busy signals, and billing information. Signaling can take place at a lower bit rate and in the T1 system the eighth bit for every channel, in every sixth frame, is replaced by a signaling bit. This is illustrated in Fig. 9.7b. The time separation between signaling bits is 6×125 μs $= 750$ μs, and the signaling bit rate is therefore $1/750$ μs $= 1.333$ kb/s.

9.5   Bandwidth Requirements

In a satellite transmission system, the baseband signal is modulated onto a carrier for transmission. Filtering of the signals takes place at a number of stages. The baseband signal itself is band-limited by filtering to prevent the generation of excessive sidebands in the modulation process. The modulated signal undergoes bandpass filtering as part of the amplification process in the transmitter.

Where transmission lines form the channel, the frequency response of the lines must also be taken into account. With a satellite link, the main channel is the radio-frequency path, which has little effect on the frequency spectrum, but does introduce a propagation delay which must be taken into account.

At the receive end, bandpass filtering of the incoming signal is necessary to limit the noise which is introduced at this stage. Thus, the signal passes through a number of filtering stages and the effect of these on the digital waveform must be taken into account.

The spectrum of the output pulse at the receiver is determined by the spectrum of the input pulse $V_i(f)$, the transmit filter response $H_T(f)$, the channel frequency response $H_{CH}(f)$, and the receiver filter response $H_R(f)$. These are shown in Fig. 9.8. Thus

$$V(f) = V_i(f)H_T(f)H_{CH}(f)H_R(f) \tag{9.8}$$

Inductive and capacitive elements are an inherent part of the filtering process. These do not dissipate power, but energy is periodically cycled between the magnetic and electric fields and the signal. The time required for this energy exchange results in part of the signal being delayed so that a square pulse entering at the transmitting end may exhibit "ringing" as it exits at the receiving end. This is illustrated in Fig. 9.9a.

Because the information is digitally encoded in the waveform, the distortion apparent in the pulse shape is not important, as long as the receiver can distinguish the binary 1 pulse from the binary 0 pulse. This requires the waveform to be sampled at the correct instants in order to determine its polarity. With a continuous waveform, the "tails" which result from the "ringing" of all the preceding pulses can combine to interfere with the particular pulse being sampled. This is known as *intersymbol interference* (ISI), and it can be severe enough to produce an error in the detected signal polarity.

The ringing cannot be removed, but the pulses can be shaped, such that the sampling of a given pulse occurs when the tails are at zero crossover points. This is illustrated in Fig. 9.9b, where two tails are

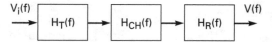

Figure 9.8 Frequency spectrum components of Eq. (9.8).

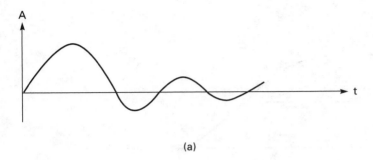

(a)

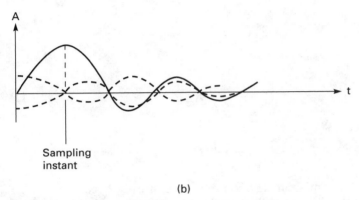

Sampling
instant

(b)

Figure 9.9 (a) Pulse ringing. (b) Sampling to avoid ISI.

shown overlapping the pulse being sampled. In practice, perfect pulse shaping cannot be achieved so that some ISI occurs, but it can be reduced to negligible proportions.

The pulse shaping is carried out by controlling the spectrum of the received pulse as given by Eq. (9.8). One theoretical model for the spectrum is known as the *raised cosine response,* which is shown in Fig. 9.10. Although a theoretical model, it can be approached closely with practical designs. The raised cosine spectrum is described by

$$V(f) = \begin{cases} 1 & f < f_1 \\ 0.5 \left(1 + \cos \dfrac{\pi(f - f_1)}{B - f_1} \right) & f_1 < f < B \\ 0 & B < f \end{cases} \quad (9.9)$$

The frequencies f_1 and B are determined by the symbol rate, and a design parameter known as the *roll-off factor,* denoted here by the symbol ρ. The roll-off factor is a specified parameter in the range

$$0 \le \rho \le 1 \quad (9.10)$$

In terms of ρ and the symbol rate, the bandwidth B is given by

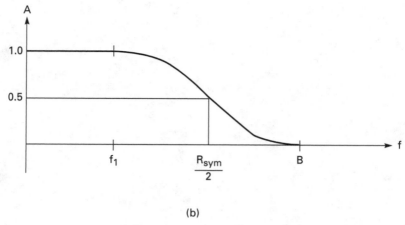

Figure 9.10 The raised cosine reponse.

$$B = \frac{(1 + \rho)}{2}R_{sym} \qquad (9.11)$$

and

$$f_1 = \frac{(1 - \rho)}{2}R_{sym} \qquad (9.12)$$

For binary transmission the symbol rate simply becomes the bit rate. Thus, for the T1 signal the required baseband bandwidth is

$$B = \frac{1 + \rho}{2} \times 1.544 \times 10^6 = 0.772(1 + \rho) \qquad \text{MHz} \qquad (9.13)$$

For a roll-off factor of unity, the bandwidth for the T1 system becomes 1.544 MHz.

Although a satellite link requires the use of a modulated carrier wave, the same overall baseband response is needed for the avoidance of ISI. Fortunately, the channel for a satellite link does not introduce frequency distortion, so the pulse shaping can take place in the transmit and receive filters. The modulation of the baseband signal onto a carrier is discussed in the next section.

9.6   Digital Carrier Systems

For transmission to and from a satellite, the baseband digital signal must be modulated onto a microwave carrier. In general, the digital baseband signals may be multilevel (*M*-ary), requiring multilevel modulation methods. The main binary modulation methods are illustrated in Fig. 9.11. They are defined as follows.

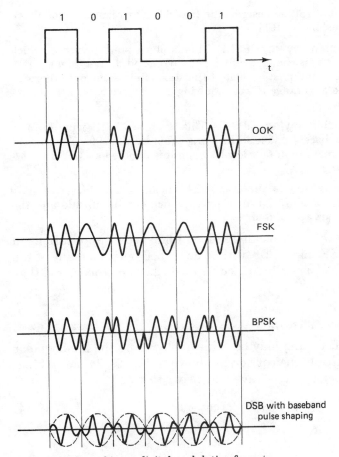

Figure 9.11   Some binary digital modulation formats.

On-off keying (OOK), also known as amplitude-shift keying (ASK)   The binary signal in this case is unipolar and is used to switch the carrier on and off.

Frequency-shift keying (FSK)   The binary signal is used to frequency-modulate the carrier, one frequency being used for a binary 1, and another for a binary 0. These are also referred to as the mark-space frequencies.

Binary phase-shift keying (BPSK)   Polarity changes in the binary signal are used to produce 180° changes in the carrier phase. This may be achieved through the use of double-sideband, suppressed-carrier modulation (DSBSC), with the binary signal as a polar NRZ waveform. In effect, the carrier amplitude is multiplied by a ±1 pulsed waveform. When the binary signal is +1, the carrier sinusoid is unchanged, and when it is −1, the carrier sinusoid is changed in phase by 180°. Binary phase-shift keying is also known as phase-reversal keying (PRK). The binary signal may be filtered at baseband before modulation, to limit the sidebands produced, and as part of the filtering needed for the reduction of ISI, as described in Sec. 9.5. The resulting modulated waveform is sketched in Fig. 9.11.

Differential phase-shift keying (DPSK) This is phase-shift keying in which the phase of the carrier is changed only if the current bit differs from the previous one. A reference bit must be sent at the start of message, but otherwise the method has the advantage of not requiring a reference carrier at the receiver for demodulation.

Quadrature phase-shift keying (QPSK) This is phase-shift keying for a 4-symbol waveform, adjacent phase shifts being equispaced by 90°. The concept can be extended to more than four levels, when it is denoted as *MPSK* for *M-ary phase-shift keying*.

Quadrature amplitude modulation (QAM) This is also a multilevel (meaning higher than binary) modulation method in which the amplitude and the phase of the carrier are modulated.

Although all of the methods mentioned find specific applications in practice, only BPSK and QPSK will be described here, as many of the general properties can be illustrated through these methods, and they are widely used.

9.6.1 Binary Phase-Shift Keying

Binary phase-shift keying may be achieved by using the binary polar NRZ signal to multiply the carrier, as shown in Fig. 9.12a. For a binary signal $p(t)$, the modulated wave may be written as

$$e(t) = p(t) \cos \omega_0 t \qquad (9.14)$$

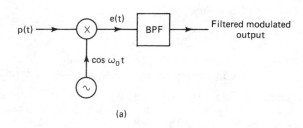

(a)

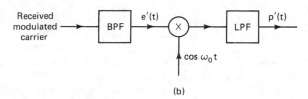

(b)

Figure 9.12 (*a*) BPSK modulator; (*b*) coherent detection of a BPSK signal.

When $p(t) = +1$, $e(t) = \cos \omega_0 t$, and when $p(t) = -1$, $e(t) = -\cos \omega_0 t$, which is equivalent to $\cos (\omega_0 t \pm 180°)$. Bandpass filtering (BPF) of the modulated wave may be used instead of baseband filtering to limit the radiated spectrum. The bandpass filter may also incorporate the square root of the raised-cosine roll-off, described in Sec. 9.5, required to reduce ISI (see, for example, Pratt and Bostian, 1986).

At the receiver, Fig. 9.12b, the received modulated carrier will undergo further bandpass filtering, to complete the raised-cosine response and to limit input noise. The filtered modulated wave, $e'(t) = p'(t) \cos \omega_0 t$, is passed into another multiplier circuit, where it is multiplied by a replica of the carrier wave $\cos \omega_0 t$. The output from the multiplier is therefore equal to $p'(t) \cos^2 \omega_0 t$. This can be expanded as $p'(t)(0.5 + 0.5 \cos 2\omega_0 t)$. The low-pass filter is used to remove the second harmonic component of the carrier, leaving the low-frequency output, which is $0.5p'(t)$, where $p'(t)$ is the filtered version of the input binary wave $p(t)$. It will be seen that the modulator is basically the same as that used to produce the DSBSC signal described in Sec. 8.3. In the present instance, the bandpass filter following the modulator is used to select the complete DSBSC signal rather than a single sideband.

The receiver is shown in more detail in Fig. 9.13. As shown, a locally generated version of the unmodulated carrier wave is required as one of the inputs to the multiplier. The locally generated carrier has to be exactly in phase with the incoming carrier, and hence this type of detection is termed *coherent detection*. Coherent detection necessitates recovering the unmodulated carrier phase information from the incoming modulated wave, and this is achieved in the carrier recovery (CR) section shown in Fig. 9.13.

As discussed in Sec. 9.5, to avoid ISI, sampling must be carried out at the bit rate and at the peaks of the output pulses. This requires the

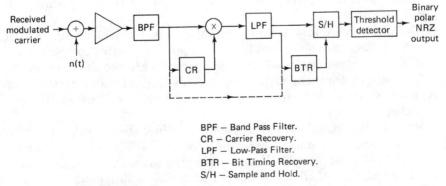

BPF — Band Pass Filter.
CR — Carrier Recovery.
LPF — Low-Pass Filter.
BTR — Bit Timing Recovery.
S/H — Sample and Hold.

Figure 9.13 Block schematic of a coherent detector showing the carrier recovery section and the bit timing recovery.

sample-and-hold circuit to be accurately synchronized to the bit rate, which necessitates a *bit timing recovery* (BTR) section as shown in Fig. 9.13.

Thermal noise at the receiver will result in noise phase modulation of the carrier, and so the demodulated waveform $p'(t)$ will be accompanied by noise. The noisy $p'(t)$ signal is passed into the threshold detector which regenerates a noise-free output but one containing some bit errors as a result of the noise already present on the waveform.

The QPSK signal has many features in common with BPSK, and will be examined before describing in detail the carrier and bit timing recovery circuits and the effects of noise.

9.6.2   Quadrature Phase-Shift Keying

With quadrature phase-shift keying, the binary data are converted into 2-bit symbols which are then used to phase-modulate the carrier. Since four combinations containing 2 bits are possible from a binary alphabet (logical 1's and 0's), the carrier phase can be shifted to one of four states.

Figure 9.14a shows one way in which QPSK modulation can be achieved. The incoming bit stream $p(t)$ is converted in the serial-to-parallel converter into two binary streams. The conversion is illustrated by the waveforms of Fig. 9.14b. For illustration purposes the bits in the $p(t)$ waveform are labeled a, b, c, d, e, and f. The serial-to-parallel converter switches bit a to the I port, and at the same time switches bit b to the Q port. In the process, each bit duration is doubled, and so the bit rates at the I and Q outputs are half that of the input bit rate.

The $p_i(t)$ bit stream is combined with a carrier cos $\omega_0 t$ in a BPSK modulator, while the $p_q(t)$ bit stream is combined with a carrier sin $\omega_0 t$, also in a BPSK modulator. These two BPSK waveforms are added to give the QPSK wave, the various combinations being shown in Table 9.1.

The phase modulation angles are shown in the phasor diagram of Fig. 9.15. Because the output from the I port modulates the carrier directly, it is termed the *in-phase* component, and hence the designation I. The output from the Q port modulates a quadrature carrier, one which is shifted by 90° from the reference carrier, and hence the designation Q.

Because the modulation is carried out at half the bit rate of the incoming data, the bandwidth required by the QPSK signal is exactly half that required by a BPSK signal carrying the same input data. This is the advantage of QPSK compared to BPSK modulation. The disadvantage is that the modulator and demodulator circuits are more complicated, being equivalent essentially to two BPSK systems in parallel.

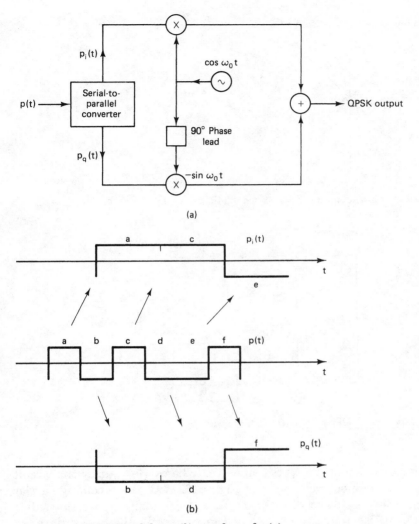

Figure 9.14 (a) QPSK modulator; (b) waveforms for (a).

TABLE 9.1 QPSK Modulator States

$p_i(t)$	$p_q(t)$	QPSK
1	1	$\cos \omega_0 t - \sin \omega_0 t = \sqrt{2} \cos (\omega_0 t + 45°)$
1	-1	$\cos \omega_0 t + \sin \omega_0 t = \sqrt{2} \cos (\omega_0 t - 45°)$
-1	1	$-\cos \omega_0 t - \sin \omega_0 t = \sqrt{2} \cos (\omega_0 t + 135°)$
-1	-1	$-\cos \omega_0 t + \sin \omega_0 t = \sqrt{2} \cos (\omega_0 t - 135°)$

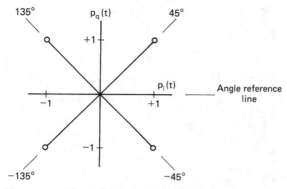

Figure 9.15 Phase diagram for QPSK modulation.

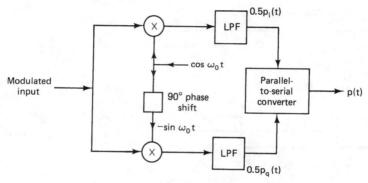

Figure 9.16 Demodulator circuit for QPSK modulation.

Demodulation of the QPSK signal may be carried out by the circuit shown in the block schematic of Fig. 9.16. With the incoming carrier represented as $p_i(t) \cos \omega_0 t - p_q(t) \sin \omega_0 t$, it is easily shown that, after low-pass filtering, the output of the upper BPSK demodulator is $0.5 p_i(t)$ and the output of the lower BPSK demodulator is $0.5 p_q(t)$. These two signals are combined in the parallel-to-serial converter to yield the desired output $p(t)$. As with the BPSK signal, noise will create errors in the demodulated output of the QPSK signal.

9.6.3 Transmission Rate and Bandwidth for PSK Modulation

Equation (9.14), which shows the baseband signal $p(t)$ multiplied onto the carrier $\cos \omega_0 t$, is equivalent to double-sideband, suppressed-carrier modulation. The digital modulator circuit of Fig. 9.12*a* is similar to the single-sideband modulator circuit shown in Fig. 8.2, the difference being that, after the multiplier, the digital modulator requires a bandpass filter while the analog modulator requires a single-sideband filter. As shown in Fig. 8.1, the DSBSC spectrum extends to twice the highest frequency in the baseband spectrum. For BPSK modulation the latter is given by Eq. (9.11) with R_{sym} replaced with R_b:

$$B_{IF} = 2B = (1 + \rho)R_b \qquad (9.15)$$

Thus, for BPSK with a roll-off factor of unity, the IF bandwidth in Hz is equal to twice the bit rate in bits per second.

As shown in the previous section, QPSK is equivalent to the sum of two orthogonal BPSK carriers, each modulated at a rate $R_b/2$, and therefore the symbol rate is $R_{sym} = R_b/2$. The spectra of the two BPSK modulated waves overlap exactly, but interference is avoided at the receiver because of the coherent detection using quadrature carriers. Equation (9.15) is modified for QPSK to

$$B_{IF} = (1 + \rho)R_{sym} \qquad (9.16)$$

$$= \frac{(1 + \rho)}{2}R_b$$

An important characteristic of any digital modulation scheme is the ratio of data bit rate to transmission bandwidth. The units for this ratio are usually quoted as bits per second per hertz (a dimensionless ratio in fact, since it is equivalent to bits per cycle). Note that it is the data bit rate R_b and not the symbol rate R_{sym} which is used.

For BPSK, Eq. (9.15) gives an R_b/B_{IF} ratio of $1/(1 + \rho)$, and for QPSK, Eq. (9.16) gives an R_b/B_{IF} ratio of $2/(1 + \rho)$. Thus, QPSK is twice as efficient as BPSK in this respect. However, more complex equipment is required to generate and detect the QSPK modulated signal.

9.6.4 Bit Error Rate for PSK Modulation

Referring back to Fig. 9.13, the noise at the input to the receiver can cause errors in the detected signal. The noise voltage, which adds to the signal, fluctuates randomly between positive and negative values, and thus the sampled value of signal plus noise may have the opposite polarity to that of the signal alone. This would constitute an error in the received pulse. The noise can be represented by a source at the front of the receiver, shown in Fig. 9.13 (this is considered in more detail in Chap. 10). It is seen that the noise is filtered by the receiver input filter. Thus, the receive filter, in addition to contributing to minimizing the ISI, must minimize noise while maximizing the received signal. In short, it must maximize the received *signal-to-noise ratio*.

In practice for satellite links (or radio links), this can usually be achieved by making the transmit and receive filters identical, each having a frequency response which is the square root of the raised-cosine response. Having identical filters is an advantage from the point of view of manufacturing.

The most commonly encountered type of noise has a flat frequency spectrum, meaning that the noise power spectrum density, measured in joules (or watts per hertz) is constant. The noise spectrum density will be denoted by N_o. When the filtering is designed to maximize the received signal-to-noise ratio, the maximum signal-to-noise voltage ratio is found to be equal to $\sqrt{2E_b/N_o}$, where E_b is the average bit energy. The average bit energy can be calculated knowing the average received power P_R and the bit period T_b:

$$E_b = P_R T_b \qquad (9.17)$$

The probability of the detector making an error as a result of noise is given by

$$P_e = \frac{1}{2}\,\text{erfc}\,\sqrt{\frac{E_b}{N_o}} \qquad (9.18)$$

where erfc stands for *complementary error function,* whose value is available in tabular or graphical form in books of mathematical tables and as built-in functions in computational packages such as Mathcad. In Mathcad, the *error function* denoted by erf (x) is provided, and

$$\text{erfc}\,(x) = 1 - \text{erf}\,(x) \qquad (9.19)$$

Equation (9.18) applies for polar NRZ baseband signals and for BPSK and QPSK modulation systems. The probability of bit error is also referred to as the *bit error rate* (BER). A $P_e = 10^{-6}$ signifies a BER of 1 bit in a million, on average. The graph of P_e vs. E_b/N_o in decibels is shown in Fig. 9.17. Note carefully that the energy ratio, not the decibel value, of E_b/N_o must be used in Eq. (9.18). This is illustrated in the following Mathcad example.

Example 9.1

The average power received in a binary polar transmission is 10 mW and the bit period is 100 μs. If the noise power spectral density is 0.1μJ, and optimum filtering is used, determine the bit error rate.

solution

Given data:

$$P_R := 10^{-2} \cdot \text{watt} \qquad T_b := 10^{-4} \cdot \text{sec} \qquad N_o := 10^{-7} \cdot \text{joule}$$

Computations:

$$E_b := P_R \cdot T_b$$

$$BER := .5 \cdot \left(1 - \text{erf}\left(\sqrt{\frac{E_b}{N_o}}\right)\right) \qquad BER = 3.9 \cdot 10^{-6}$$

Equation (9.18) is sometimes expressed in the alternative form

$$P_e = Q\left(\sqrt{\frac{2E_b}{N_o}}\right) \qquad\qquad (9.20)$$

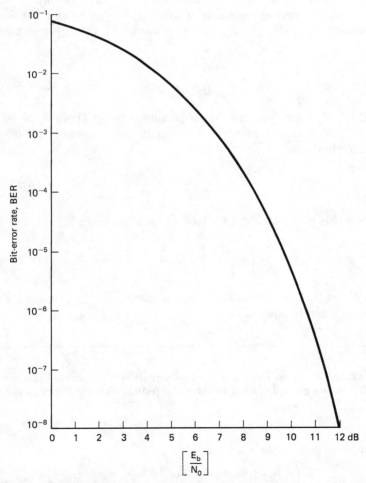

Figure 9.17 BER versus (E_b/N_o) for baseband signaling using a binary polar NRZ waveform. The curve also applies for BPSK and QPSK modulated signals.

Here, the $Q(.)$ function is simply an alternative way of expressing the complementary error function, and in general

$$\text{erfc}\,(x) = 2Q\left(\sqrt{2}x\right) \tag{9.21}$$

These relationships are given for reference only and will not be used further in this book.

An important parameter for carrier systems is the ratio of the average carrier power to the noise power density, usually denoted by $[C/N_o]$. The $[E_b/N_o]$ and $[C/N_o]$ ratios can be related as follows. The average carrier power at the receiver is P_R watts. The energy per symbol is therefore P_R/R_{sym} joules, with R_{sym} in symbols per second. Since each symbol contains m bits, the energy per bit is P_R/mR_{sym} joules. But $mR_{\text{sym}} = R_b$ and therefore the energy per bit E_b is

$$E_b = \frac{P_R}{R_b} \tag{9.22}$$

As before, let N_o represent the noise power density. Then $E_b/N_o = P_R/R_bN_o$. But P_R/N_o is the carrier-to-noise density ratio, usually denoted by C/N_o and therefore

$$\frac{E_b}{N_o} = \frac{C/N_o}{R_b} \tag{9.23}$$

Rearranging this and putting it in decibel notation gives

$$\left[\frac{C}{N_o}\right] = \left[\frac{E_b}{N_o}\right] + [R_b] \tag{9.24}$$

It should be noted that, whereas E_b/N_o has units of dB, C/N_o has units of dBHz as explained in App. G.

Example 9.2

The downlink transmission rate in a satellite circuit is 61 Mb/s and the required $[E_b/N_o]$ at the ground station receiver is 9.5 dB. Calculate the required $[C/N_o]$.

solution

The transmission rate in decibels is

$$[R_b] = 10 \log 61 \times 10^6 = 77.85 \text{ dBb/s}$$

Hence, $[C/N_o] = 77.85 + 9.5 = 87.35 \text{ dBHz}$

The equations giving the probability of bit error are derived on the basis that the filtering provides maximum signal-to-noise ratio. In practice there are a number of reasons why the optimum filtering may not be achieved. The raised cosine response is a theoretical model that can only be approximated in practice. Also, for economic reasons it is desirable to use production filters manufactured to the same specifications for the transmit and receive filter functions, and this may result in some deviation from the desired theoretical response. The usual approach in practice is that one knows the BER that is acceptable for a given application. The corresponding ratio of bit energy to noise density can then be found from Eq. (9.18) or from a graph such as that shown in Fig. 9.17. Once the theoretical value of E_b/N_o is found, an *implementation margin,* amounting to a few decibels at most, is added to allow for imperfections in the filtering. This is illustrated in the following example.

Example 9.3

A BPSK satellite digital link is required to operate with a bit error rate of no more than 10^{-5}, the implementation margin being 2 dB. Calculate the required E_b/N_o ratio in decibels.

solution

The graph of Fig. 9.17 shows that E_b/N_o is around 9 dB for a BER of 10^{-5}. By plotting this region to an expanded scale a more accurate value of E_b/N_o can be obtained. For ease of presentation the E_b/N_o ratio will be denoted by x. A suitable range for x is

$$x : = 8, 8.2 .. 10$$

In decibels this is:

$$xdB(x) : = 10 \cdot \log(x)$$

Eq. (9.18) gives:

$$P_e(x) : = .5 \cdot \left(1 - \mathrm{erf}\left(\sqrt{x}\right)\right)$$

From Fig. 9.18, E_b/N_o is seen to be 9.6 dB. This is without an implementation margin. The required value, including an implementation margin is 9.6 + 2 = 11.6 dB.

To summarize, BER is a specified requirement, which enables E_b/N_o to be determined by using Eq. (9.18) or Fig. 9.17. The rate R_b will also be specified, and hence the $[C/N_o]$ ratio can be found by using Eq. (9.24). The $[C/N_o]$ ratio is then used in the link budget calculations, as described in Chap. 10.

With purely digital systems, the BER will be directly reflected in errors in the data being transmitted. With analog signals which have

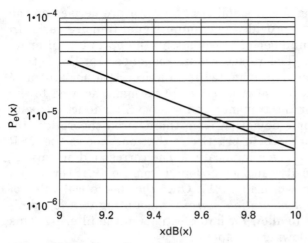

Figure 9.18 Solution to Example 9.3.

been converted to digital form through PCM, the BER contributes to the output signal-to-noise ratio, along with the quantization noise, as described in Sec. 9.3. Curves showing the contributions of thermal noise and quantization noise to the signal-to-noise output for analog systems are in Fig. 9.19. The signal-to-noise power ratio is given by (Taub and Schilling, 1986):

$$\frac{S}{N} = \frac{Q^2}{1 + 4Q^2 P_e} \qquad (9.25)$$

where $Q = 2^n$ is the number of quantized steps and n is the number of bits per sample.

9.6.5 Improved BER through Error Control Coding

As shown by Fig. 9.17, the BER can be reduced by increasing $[E_b/N_o]$, but there are practical limits to this approach. Equation (9.24) shows that for a given bit rate R_b, $[E_b/N_o]$ is directly proportional to $[C/N_o]$. An increase in $[C/N_o]$ can be achieved by increasing transmitted power and/or reducing the system noise temperature (to reduce N_o). Both of these measures are limited by cost, and, in the case of the on-board satellite equipment, size. In practical terms, a BER of about 10^{-4}, which is satisfactory for voice transmissions, can be achieved with off-the-shelf equipment. For lower BER values, such as required for some data, error control coding must be used.

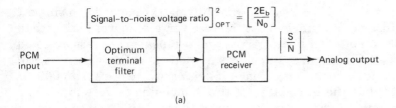

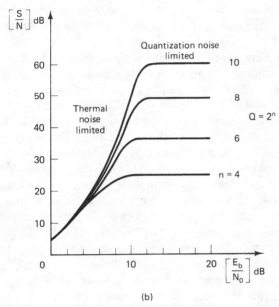

Figure 9.19 (a) Use of optimum terminal filter to maximize the signal-to-noise voltage ratio; (b) plot of Eq. (9.19).

Encoding in extra bits, termed *redundant bits,* in a predetermined and ordered fashion, enables errors to be detected and corrected in the decoding process. A code which allows for the detection of errors is termed an *error detecting code.* For example, it may be possible to detect the fact that one bit in a code word is wrong, but it may not be possible to locate which bit. In this case retransmission may be requested by an automatic repeat request (ARQ). With more complex codes, the location of the bit error can also be found and corrected. Where correction takes place without the need for retransmission of data, the code is termed a *forward error correcting* (FEC) code. The difference between error detecting and error correcting codes can be illustrated by means of a simple repetitive encoding scheme. Suppose the message consists of single-bit data words 1 and 0. *Triple-redun-*

dancy encoding would consist in transmitting 111 instead of a single 1 and 000 instead of a single 0. At the receiver, a code word other than 111 or 000 would be a known error, and a repeat transmission could be requested. This is an example of error detection. Of course it is still possible for an error to get through if noise converted a 111 to 000, or 000 to 111, but the probability of this happening is very much less than an error occurring in a single bit. If the probability of a single bit error is 10^{-2}, then, assuming this remains unchanged for the coded data stream (which implies that $[E_b/N_o]$ is not changed by the coding), the probability of an error getting through with triple redundancy encoding is $(10^{-2})^3 = 10^{-6}$.

An error correcting arrangement could consist of a logic circuit, arranged to produce an output to agree with the majority bits in a received code word. Thus, if 101 is received, the logic circuit generates a 1 on the assumption that the original word was 111. This is an example of forward error correction, since a retransmission is not needed. In this case, probability calculations shows that the probability of an error slipping through is about 3×10^{-4} compared to 10^{-2} without FEC.

In the example given above it was assumed that $[E_b/N_o]$ remained unchanged between the uncoded and coded messages, and in the case of the triple redundancy coding this would require that the transmission time be tripled. In more complex, but more efficient, coding schemes, the transmission time is not increased, but because of the additional bits, the bit rate is increased. The new bit rate is usually referred to as the data rate or the transmission rate R_T. It must be understood, however, that the message rate remains unaltered. The ratio of message bit rate to transmission bit rate is termed the *code rate r*. Thus

$$r = \frac{R_b}{R_T} \tag{9.26}$$

This can also be stated as

$$r = \frac{k}{n} \tag{9.27}$$

where k is the number of bits in the original message or data word and n is the number of bits in the corresponding code word. Where error control coding is employed, the transmission rate R_T must be used in place of R_b in Eqs. (9.15) and (9.16) to determine the IF bandwidth.

Example 9.4

A T1 digital signal is transmitted using a 7/8 FEC code. Calculate the transmitted bit rate. Given that the signal is transmitted by QPSK and assuming a roll-off response factor of 0.2, calculate the bandwidth required.

solution

As shown in Sec. 9.4, the T1 message rate is 1.544 Mb/s. When 7/8 FEC is applied, the transmission rate becomes $1.544 \times 8/7 = 1.765$ Mb/s. From Eq. (9.16)

$$B_{IF} = 1.765 \times (1.2)/2 = 1.06 \text{ MHz}$$

Certain codes, which fall into the general classification of *block codes,* can be used to correct all errors up to some maximum number, denoted here by t. If n is the number of bits in each code word, and P_e the probability of bit error occurring in transmission, then the decoded bit error rate is given to a good approximation by

$$\text{BER} \cong \frac{(n-1)!}{t!(n-1-t)!} P_e^{\,t+1} \tag{9.28}$$

Because the transmission bit rate is greater than the message rate, while the message time remains fixed, the transmit bit period, and hence the energy in a bit is reduced by a factor r. This leads to a received value of $[rE_b/N_o]$ which in turn leads to a worse P_e. To be beneficial, therefore, the FEC must give an overall improvement in BER. This is illustrated in Example 9.5.

Example 9.5

For the link specified in Example 9.3 determine (a) the probability of bit error on the transmission path, and (b) the decoded bit error rate, when a FEC code of rate 7/8, capable of correcting 1 error is used.

solution

Given data: $\qquad t := 1 \qquad r := \dfrac{7}{8} \qquad n := 8$

Computations
From Example 9.3, the basic E_b/N_o ratio is 9.6 dB (i.e., without the implementation margin) for a probability of bit error of 10^{-5} without error correction coding. Denoting the received ratio as x, then with FEC, the value of x is

$$x := 10^{\frac{9.6}{10}} \cdot r$$

(a) The probability of bit error in transmission with the FEC encoded signal is

$$P_e := .5 \cdot \left(1 - \text{erf}\left(\sqrt{x}\right)\right) \qquad P_e = 3.2 \cdot 10^{-5}$$
$$\texttt{=========}$$

This should be compared with the value of 10^{-5} for the signal without FEC.
(b) The decoded bit error rate is

$$\text{BER} := \frac{(n-1)!}{t! \cdot (n-1-t)!} \cdot P_e^{\,(t+1)} \qquad \text{BER} = 7.3 \cdot 10^{-9}$$
$$\texttt{===========}$$

Typical error curves with and without coding are shown in Fig. 9.20. The curves have a crossover point, labeled *A,* and the operating $[E_b/N_o]$ should be greater than the crossover value. The *coding gain* is the difference in the $[E_b/N_o]$ values with and without coding, for a given BER, as shown in Fig. 9.20. Note that with FEC the received ratio is rE_b/N_o but both curves are plotted against E_b/N_o.

The encoder is inserted between the digital source and the digital modulator as shown in Fig. 9.21, and the decoder follows the demodulator in the receiver. The coding is chosen in conjunction with the type of modulation used. When the decoder in Fig. 9.21 has to choose be-

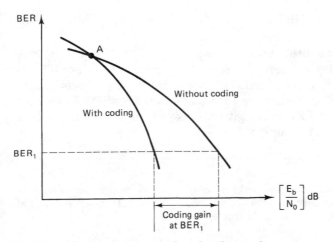

Figure 9.20 Error-rate curves with and without coding.

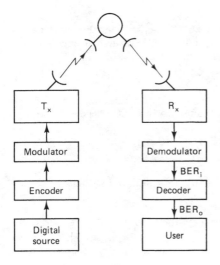

Figure 9.21 Satellite circuit with encoder and decoder blocks.

tween symbols (the output from the demodulator) which are assumed equiprobable, it is called a *hard decision* decoder. Other codes have been devised for which all symbols are not equiprobable. When such statistical information is taken into account in the decoding, the decoder is called a *soft decision* decoder. Soft decision decoding improves the coding gain but is more complex to implement than hard decision decoding. Some typical values taken from Freeman, 1981, for a BER of 10^{-5}, achievable coding gains range from 2.1 to 3.6 dB for hard decision decoding and from 3.6 to 5.2 dB for soft decision decoding.

Block codes have been designed for specific purposes, and are usually named after their discoverers. The code illustrated in the above example is an example of a *Hamming* code for which the number of bits in the data word is $n = 2^m - 1$ and the number of bits in the corresponding code word is $k = n - m$, where m is an integer. The Hamming code can correct a single error. The *Golay* code, for which $n = 23$ and $k = 12$, can correct up to 3 errors. The *Bose-Chaudhuri-Hocquenghem* (BCH) code is a multiple error correction code, and the *Reed Solomon* (RS) code is designed to cope with errors which occur in bursts.

Another form of encoding known as *convolution encoding* enables error control coding to be applied to a continuous stream of data, without the need for partitioning it into blocks of data. Data are moved into a shift register k bits at a time, where the bits are combined in a known manner with message bits from the same stream using modulo-2 adders. For each k bits of input, n bits of encoded data are generated. (For details of block and convolution codes see, for example, Taub and Schilling, 1986.)

9.7 Carrier Recovery Circuits

To implement coherent detection a local oscillator which is exactly synchronized to the carrier must be provided at the receiver. The BPSK signal is a DSBSC type signal which has no carrier component directly available, and a nonlinear circuit is used to recover the carrier from the received signal. A circuit which is widely used for carrier recovery is the *squaring loop* shown in Fig. 9.22.

Consider first the operation of the squaring loop with a BPSK signal. Let the incoming signal be represented as $\sin(\omega_c t + d(t)\pi/2)$ where $d(t) = \pm 1$ is random binary modulation. Using the trigonometric identity $\sin^2 \theta = [\frac{1}{2}(1 - \cos 2\theta)]$ shows that the output following the squaring circuit contains the term $\cos 2\theta = \cos(2\omega_c t + d(t)\pi)$. The $d(t)$ term therefore produces a $\pm \pi$ phase shift which has no effect on the cosine term. The resultant double-frequency signal is selected by the bandpass filter and passed to the phase detector of the phase-

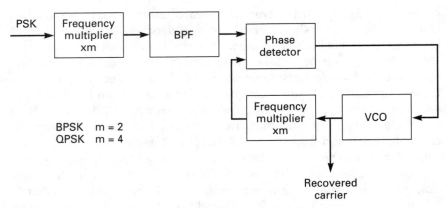

Figure 9.22   Functional block diagram for carrier recovery.

locked loop. The voltage-controlled oscillator (VCO) in the phase-locked loop (PLL) operates at or near the carrier frequency, and the second harmonic of the VCO output provides the second input to the phase detector. The voltage output from the phase detector, determined by the phase difference of the two input signals, controls the VCO frequency in such a way that its second harmonic locks onto the second harmonic of the carrier. The VCO output is used as the recovered carrier for the coherent detection process.

The phase changes in the QPSK signal are integral multiples of $\pi/2$. Frequency quadrupling changes these to integer multiples of π, and the PLL operates as for the BPSK signal.

Frequency multiplication can be avoided by use of a method known as the *Costas loop*. Details of this, along with an analysis of the effects of noise on the squaring loop and the Costas loop methods will be found in Gagliardi, 1991. Other methods are also described in detail in Franks, 1980.

9.8   Bit Timing Recovery

Accurate bit timing is needed at the receiver in order to be able to sample the received waveform at the optimum points. In the most common arrangements, the clocking signal is recovered from the demodulated waveform, these being known as *self-clocking* or *self-synchronizing* systems. Where the waveform has a high density of zero crossings, a zero-crossing detector can be used to recover the clocking signal. In practice the received waveform is often badly distorted by the frequency response of the transmission link and by noise, and the design of the bit timing recovery circuit is quite complicated. In most instances the spectrum of the received waveform will not contain a

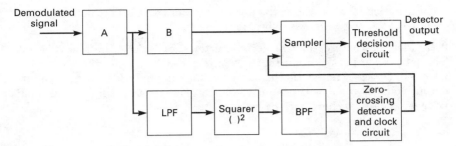

Figure 9.23 Functional block diagram for bit-timing recovery.

discrete component at the clock frequency. However, it can be shown that a periodic component at the clocking frequency is present in the squared waveform for digital signals (unless the received pulses are exactly rectangular, in which case squaring simply produces a dc level for a binary waveform). A commonly used baseband scheme is shown in block schematic form in Fig. 9.23 (Franks, 1980). The filters A and B form part of the normal signal filtering (e.g., raised-cosine filtering). The signal for the bit timing recovery is tapped from the junction between A and B and passed along a separate branch which consists of a filter, a squaring circuit, and a bandpass filter which is sharply tuned to the clock frequency component present in the spectrum of the squared signal. This is then used to synchronize the clocking circuit, the output of which clocks the sampler in the detector branch.

The *early-late gate circuit* provides a method of recovering bit timing which does not rely on a clocking component in the spectrum of the received waveform. The circuit utilizes a feedback loop in which the magnitude changes in the outputs from matched filters control the frequency of a local clocking circuit (for an elementary description see, for example, Roddy and Coolen, 1995). Detailed analyses of these and other methods will be found in Franks, 1980, and Gagliardi, 1991.

9.9 Problems

9.1. For a test pattern consisting of alternating binary 1's and 0's, determine the frequency spectra in terms of the bit period T_b for the following signal formats: (*a*) unipolar; (*b*) polar NRZ; (*c*) polar RZ; (*d*) Manchester.

9.2. The raised-cosine frequency response may be written as

$$V(f) = \frac{1}{2}\left[1 + \cos\frac{\pi(f - f_1)}{B - f_1}\right]$$

where $B = (1 + \rho)R_b/2$; $f_1 = (1 - \rho)R_b/2$; and ρ is the roll-off factor. Plot the characteristic for $\rho = 1$. Use the inverse Fourier transform to determine the shape of the pulse time waveform for $\rho = 1$.

9.3. Plot the compressor transfer characteristics for $\mu = 100$ and $A = 100$. The μ-law compression characteristic is given by $v_0 = \text{sign}\,(v_i)\,\ln\,(1 + \mu|v_i|)/\ln\,(1 + \mu)$ where v_0 is the output voltage normalized to the maximum output voltage, and v_i is the input voltage normalized to the maximum input voltage. The A-law characteristic is given by

$$v_0 = \text{sign}\,(v_i)\,\frac{A|v_i|}{1 + \ln A} \qquad \text{for } |v_i| \leq \frac{1}{A}$$

and

$$v_0 = \text{sign}\,(v_i)\left[\frac{1 + \ln\,(A|v_i|)}{1 + \ln A}\right] \qquad \text{for } \frac{1}{A} \leq |v_i| \leq 1.$$

9.4. Write down the expander transfer characteristics corresponding to the compressor characteristics given in Prob. 9.3.

9.5. Assuming that the normalized levels shown in Fig. 9.4b represent millivolts, write out the digitally encoded words for input levels of (a) ± 90 mV, (b) ± 100 mV, (c) ± 190 mV, (d) ± 3000 mV.

9.6. Determine the decoded output voltage levels for the input levels given in Prob. 9.5. Determine also the quantization error in each case.

9.7. (a) A test tone having the full peak-to-peak range is applied to a PCM system. If the number of bits per sample is 8, determine the quantization S/N. Assume uniform sampling of step size ΔV, for which the mean square noise voltage is $(\Delta V)^2/12$. (b) Given that a raised-cosine filter is used with $\rho = 1$, determine the bandwidth expansion factor B/W, where B is the PCM bandwidth and W is the upper cutoff frequency of the input.

9.8. A PCM signal uses the polar NRZ format. Following optimum filtering, the $[E_b/N_o]$ at the input to the receiver decision detector is 10 dB. Determine the bit error rate (BER) at the output of the decision detector.

9.9 Using Equation (9.18), calculate the probability of bit error for E_b/N_o values of (a) 0 dB, (b) 10 dB, and (c) 40 dB.

9.10. A PCM system uses 8 bits per sample and polar NRZ transmission. Determine the output $[S/N]$ for the signal-to-noise voltage ratio specified in Prob. 9.9, at the input to the decision detector.

9.11. A binary periodic waveform of period $3T_b$ is low-pass filtered before being applied to a BPSK modulator. The low-pass filter cuts off at $B = T_b$. Derive the trigonometric expansion for the modulated wave, showing that only side frequencies and no carrier are present. Given that the bit period is 100 μs and the carrier frequency is 100 kHz, sketch the spectrum, showing the frequencies to scale.

9.12. Explain what is meant by *coherent detection* as used for the demodulation of PSK bandpass signals. An envelope detector is an example of a *noncoherent detector.* Can such a detector be used for BPSK? Give reasons for your answer.

9.13. Explain how a quadrature phase-shift keyed (QPSK) signal can be represented by two binary phase-shift keyed (BPSK) signals. Show that the bandwidth required for QPSK signal is one-half that required for a BPSK signal operating at the same data rate.

9.14. The input data rate on a satellite circuit is 1.544 Mbps. Calculate the bandwidths required for BPSK modulation and for QPSK modulation given that raised-cosine filtering is used with a roll-off factor of 0.2 in each case.

9.15. A QPSK system operates at a $[E_b/N_o]$ ratio of 8 dB. Determine the bit error rate.

9.16. A BPSK system operates at a $[E_b/N_o]$ ratio of 16 dB. Determine the bit error rate.

9.17. The received power in a satellite digital communications link is 0.5 pW. The carrier is BPSK modulated at a bit rate of 1.544 Mb/s. If the noise power density at the receiver is 0.5×10^{-19} J, determine the bit error rate.

9.18. The received $[C/N_o]$ ratio in a digital satellite communications link is 86.5 dBHz, and the data bit rate is 50 Mb/s. Calculate the $[E_b/N_o]$ ratio and the BER for the link.

9.19. For the link specified in Prob. 9.18, the $[C/N_o]$ ratio is improved to 87.5 dBHz. Determine the new BER.

9.20. Explain what is meant by a *forward error correcting* (FEC) *code.* FEC coding at a code rate of 3/4 is used in a digital system. Given that the message bit rate is 1.544 Mb/s, calculate the transmission rate.

9.21. FEC coding at a code rate of 7/8 is used with the signals in Prob. 9.14. Determine the new bandwidth required in each case.

9.22. FEC coding at a code rate of 3/4 is used on the satellite digital communications link specified in Prob. 9.17. Given that the received power, the noise power density, and the message bit rate remain unchanged, determine the received bit energy to noise density ratio.

9.23. (*a*) Explain what is meant by *coding gain* as applied to error correcting coding. (*b*) When FEC coding at a 3/4 code rate is used on a digital link, a coding gain of 3 dB is achieved. Calculate the decibel reduction in transmitted carrier power that this permits.

9.24. A digital message is 100 megabits long. If the BER is 10^{-5}, calculate the average number of errors expected.

9.25. The type of probability distribution that applies for digital messages is known as a Poisson distribution, for which the standard deviation is equal to the square root of the mean. There is about a 95.5 percent probability that the number of errors will be between the mean $\pm 2\sigma$, where σ is the standard deviation. Assuming that this gives the range of errors expected, calculate the minimum and maximum number of errors likely for a digital message 1000 megabits long for which the BER is 10^{-7}.

9.26. An error correcting code is used which has $n = 15$, $k = 11$ [written as a (15,11) block code]. The code is capable of correcting one error at most. On the same set of axes, plot the BER for the coded and uncoded cases for $[E_b/N_o]$ between 2 dB and 12 dB. What is the coding gain at a BER of 10^{-6}?

Chapter

10

The Space Link

10.1 Introduction

This chapter describes how the link power budget calculations are made. These calculations basically relate two quantities, the transmit power and the receive power, and show in detail how the difference between these two powers is accounted for.

Link budget calculations are usually made using decibel or decilog quantities. These are explained in App. G. In this text square brackets are used to denote decibel quantities using the basic power definition. Where no ambiguity arises regarding the units, the abbreviation dB is used. For example Boltzmann's constant is given as -228.6 dB, although strictly speaking this should be given as -228.6 decilogs relative to one joule per kelvin. Where it is desirable to show the reference unit, this is indicated in the abbreviation, for example, $dBHz$ means decibels relative to one hertz.

10.2 Equivalent Isotropic Radiated Power

A key parameter in link budget calculations is the equivalent isotropic radiated power, conventionally denoted as EIRP. From Eqs. (5.4) and (5.5) the maximum power flux density at some distance r from a transmitting antenna of gain G is

$$\Psi_M = \frac{GP_s}{4\pi r^2} \tag{10.1}$$

An isotropic radiator with an input power equal to GP_s would produce the same flux density. Hence this product is referred to as the equivalent isotropic radiated power, or

$$\text{EIRP} = GP_s \tag{10.2}$$

EIRP is often expressed in decibels relative to one watt, or *dBW*. Let P_s be in watts; then

$$[EIRP] = [P_s] + [G] \quad dBW \tag{10.3}$$

where $[P_s]$ is also in dBW and $[G]$ is in dB.

Example 10.1

A satellite downlink at 12 GHz operates with a transmit power of 6 W and an antenna gain of 48.2 dB. Calculate the EIRP in dBW.

solution

$$[EIRP] = 10 \log 6 + 48.2$$
$$= 56 \text{ dBW}$$

For a paraboloidal antenna the isotropic power gain is given by Eq. (5.29). This equation may be rewritten in terms of frequency, since this is the quantity which is usually known:

$$G = \eta(10.472fD)^2 \tag{10.4}$$

where f is the carrier frequency in gigahertz, D is the reflector diameter in meters, and η is the aperture efficiency. A typical value for aperture efficiency is 0.55, although values as high as 0.73 have been specified (Andrew Antenna, 1985).

With the diameter D in feet, and all other quantities as before, the equation for power gain becomes

$$G = \eta(3.192fD)^2 \tag{10.5}$$

Example 10.2

Calculate the gain of a 3-m paraboloidal antenna operating at a frequency of 12 GHz. Assume an aperture efficiency of 0.5.

solution

$$G = 0.55 \times (10.472 \times 12 \times 3)^2 \cong 78,168$$

Hence

$$[G] = 10 \log 78,168 = 48.9 \text{ dB}$$

10.3   Transmission Losses

The [EIRP] may be thought of as the power input to one end of the transmission link, and the problem is to find the power received at the other end. Losses will occur along the way, some of which are constant. Other losses can only be estimated from statistical data, and some of these are dependent on weather conditions, especially on rainfall.

The first step in the calculations is to determine the losses for *clear-weather*, or *clear-sky*, conditions. These calculations take into account the losses, including those calculated on a statistical basis, which do not vary significantly with time. Losses which are weather-related, and other losses which fluctuate with time, are then allowed for by introducing appropriate *fade margins* into the transmission equation.

10.3.1 Free-Space Transmission

As a first step in the loss calculations, the power loss resulting from the spreading of the signal in space must be determined. This calculation is similar for the uplink and the downlink of a satellite circuit. Using Eqs. (10.1) and (10.2) gives the power flux density at the receiving antenna as

$$\Psi_M = \frac{\text{EIRP}}{4\pi r^2} \tag{10.6}$$

The power delivered to a matched receiver is this power flux density multiplied by the effective aperture of the receiving antenna, given by Eq. (5.12). The received power is therefore

$$P_R = \Psi_M A_{\text{eff}}$$

$$= \frac{\text{EIRP}}{4\pi r^2} \frac{\lambda^2 G_R}{4\pi} \tag{10.7}$$

$$= (\text{EIRP})(G_R)\left(\frac{\lambda}{4\pi r}\right)^2$$

Recall that r is the distance, or range, between the transmit and receive antennas and G_R is the isotropic power gain of the receiving antenna. The subscript R is used to identify the receiving antenna.

The right-hand side of Eq. (10.7) is separated into three terms associated with the transmitter, receiver, and free space, respectively. In decibel notation, the equation becomes

$$[P_R] = [\text{EIRP}] + [G_R] - 10 \log\left(\frac{4\pi r}{\lambda}\right)^2 \tag{10.8}$$

The received power in dBW is therefore given as the sum of the transmitted EIRP in dBW, plus the receiver antenna gain in dB, minus a third term which represents the free-space loss in decibels. The free space loss component in decibels is given by

$$[FSL] = 10 \log \left(\frac{4\pi r}{\lambda} \right)^2 \qquad (10.9)$$

Normally the frequency rather than wavelength will be known, and the substitution $\lambda = c/f$ can be made, where $c = 10^8$ m/s. With frequency in megahertz and distance in kilometers, it is left as an exercise for the student to show that the free-space loss is given by

$$[FSL] = 32.4 + 20 \log r + 20 \log f \qquad (10.10)$$

Equation (10.8) can then be written as

$$[P_R] = [EIRP] + [G_R] - [FSL] \qquad (10.11)$$

The received power $[P_R]$ will be in dBW when the [EIRP] is in dBW and [FSL] in dB. Equation (10.9) is applicable to both the uplink and the downlink of a satellite circuit, as will be shown in more detail shortly.

Example 10.3

The range between a ground station and a satellite is 42,000 km. Calculate the free-space loss at a frequency of 6 GHz.

solution

$$[FSL] = (32.4 + 20 \log 42,000 + 20 \log 6000) = 200.4 \text{ dB}$$

This is a very large loss. Suppose the [EIRP] is 56 dBW (as calculated in Example 10.1 for a radiated power of 6 W) and the receive antenna gain is 50 dB. The receive power would be $56 + 50 - 200.4 = -94.4$ dBW. This is 355 pW. It may also be expressed as -64.4 dBm, which is 64.4 dB below the 1-mW reference level.

Equation (10.11) shows that the received power is increased by increasing antenna gain as expected, and Eq. (5.29) shows that antenna gain is inversely proportional to the square of the wavelength. Hence it might be thought that increasing the frequency of operation (and therefore decreasing wavelength) would increase the received power. However, Eq. (10.9) shows that the free-space loss is also inversely proportional to the square of the wavelength, and so these two effects cancel. It follows therefore that *for a constant EIRP* the received power is independent of frequency of operation.

If the transmit power is a specified constant, rather than the *EIRP*, then the received power will increase with increasing frequency for given antenna dish sizes at the transmitter and receiver. It is left as an exercise for the student to show that under these conditions the received power is directly proportional to the square of the frequency.

10.3.2 Feeder Losses

Losses will occur in the connection between the receive antenna and the receiver proper. Such losses will occur in the connecting waveguides, filters, and couplers. These will be denoted by RFL, or [RFL] dB, for *receiver feeder losses*. The [RFL] are added to [FSL] in Eq. (10.11). Similar losses will occur in the filters, couplers, and waveguides connecting the transmit antenna to the high-power amplifier (HPA) output. However, provided that the EIRP is stated, Eq. (10.11) can be used without knowing the transmitter feeder losses. These are needed only when it is desired to relate EIRP to the HPA output as described in Secs. 10.7.4 and 10.8.2.

10.3.3 Antenna Misalignment Losses

When a satellite link is established, the ideal situation is to have the earth station and satellite antennas aligned for maximum gain as shown in Fig. 10.1*a*. There are two possible sources of off-axis loss, one at the satellite and one at the earth station, as shown in Fig. 10.1*b*. The off-axis loss at the satellite is taken into account by designing the link for operation on the actual satellite antenna contour; this is described in more detail in later sections. The off-axis loss at

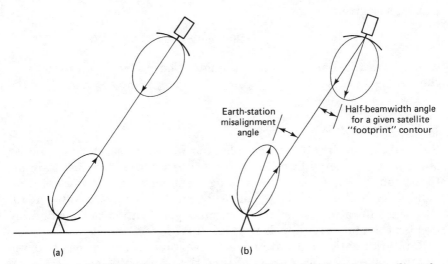

Figure 10.1 (*a*) Satellite and earth-station antennas aligned for maximum gain; (*b*) earth station situated on a given satellite "footprint," and earth-station antenna misaligned.

the earth station is referred to as the *antenna pointing loss*. Antenna pointing losses are usually only a few tenths of a decibel; typical values are given in Table 10.1.

In addition to pointing losses, losses may result at the antenna from misalignment of the polarization direction (these are in addition to the polarization losses described in Chap. 5). The polarization misalignment losses are usually small, and it will be assumed that the antenna misalignment losses, denoted by [AML], include both pointing and polarization losses resulting from antenna misalignment. It should be noted that the antenna misalignment losses have to be estimated from statistical data, based on the errors actually observed for a large number of earth stations, and of course the separate antenna misalignment losses for the uplink and the downlink must be taken into account.

10.3.4 Fixed Atmospheric and Ionospheric Losses

Atmospheric gases result in losses by absorption as described in Sec. 3.2. Equation (3.1) gives the losses in decibels as a function of the angle of elevation of the path as

$$[AA] = [AA]_{90} \csc \theta$$

where θ is the angle of elevation of the radio path and $[AA]_{90}$ is the absorption loss for vertical incidence, as shown in Fig. 3.2.

Also, as discussed in Sec. 4.5, the ionosphere introduces a depolarization loss given by

$$[PL] = 20 \log (\cos \theta_F)$$

where θ_F is the Faraday rotation angle. These losses add directly to the free-space losses.

10.3.5 Effects of Rain

Rainfall results in attenuation of radio waves by scattering, and by absorption, of energy from the wave as described in Sec. 3.4. Rain attenuation increases with increasing frequency, and is worse at Ku band than at C band. Studies have shown (CCIR Report 338-3, 1978) that the rain attenuation for horizontal polarization is considerably greater than for vertical polarization.

Rain attenuation data are usually available in the forms of curves or tables showing the fraction of time that a given attenuation is exceeded, or equivalently, the probability that a given attenuation will be exceeded (see Hogg et al., 1975; Lin et al., 1980; Webber et al., 1986). Some yearly average Ku-band values are shown in Table 10.1.

The percentage figures at the head of the first three columns give the percentage of time, averaged over any year, that the attenuation exceeds the dB values given in each column. For example, at Thunder Bay, the rain attenuation exceeds, on average throughout the year, 0.2 dB for 1 percent of the time, 0.3 dB for 0.5 percent of the time, and 1.3 dB for 0.1 percent of the time. Alternatively, one could say that, for 99 percent of the time, the attenuation will be equal to or less than 0.2 dB; for 99.5 percent of the time, it will be equal to or less than 0.3 dB; and for 99.9 percent of the time, it will be equal to or less than 1.3 dB.

Rain attenuation is accompanied by noise generation, and both the attenuation and the noise adversely affect satellite circuit performance, as described in Secs. 10.7.4 and 10.8.3.

As a result of falling through the atmosphere, raindrops are somewhat flattened in shape, becoming elliptical rather than spherical. When a radio wave with some arbitrary polarization passes through raindrops, the component of electric field in the direction of the major axes of the raindrops will be affected differently from the component

TABLE 10.1   Rain Attenuation, Atmospheric Absorption Loss, and Satellite Pointing Loss for Cities and Communities in the Province of Ontario

Location	Rain attenuation, dB			Atmospheric absorption, dB, summer	Satellite antenna pointing loss, dB	
	1%	0.5%	0.1%		1/4 Canada coverage	1/2 Canada coverage
Cat Lake	0.2	0.4	1.4	0.2	0.5	0.5
Fort Severn	0.0	0.1	0.4	0.2	0.9	0.9
Geraldton	0.1	0.2	0.9	0.2	0.2	0.1
Kingston	0.4	0.7	1.9	0.2	0.5	0.4
London	0.3	0.5	1.9	0.2	0.3	0.6
North Bay	0.3	0.4	1.9	0.2	0.3	0.2
Ogoki	0.1	0.2	0.9	0.2	0.4	0.3
Ottawa	0.3	0.5	1.9	0.2	0.6	0.2
Sault Ste. Marie	0.3	0.5	1.8	0.2	0.1	0.3
Sioux Lookout	0.2	0.4	1.3	0.2	0.4	0.3
Sudbury	0.3	0.6	2.0	0.2	0.3	0.2
Thunder Bay	0.2	0.3	1.3	0.2	0.3	0.2
Timmins	0.2	0.3	1.4	0.2	0.5	0.2
Toronto	0.2	0.6	1.8	0.2	0.3	0.4
Windsor	0.3	0.6	2.1	0.2	0.5	0.8

SOURCE: Telesat Canada Design Workbook.

along the minor axes. This produces a depolarization of the wave; in effect the wave becomes elliptically polarized (see Sec. 4.6). This is true for both linear and circular polarizations, and the effect seems to be much worse for circular polarization (Freeman, 1981). Where only a single polarization is involved, the effect is not serious, but where frequency reuse is achieved through the use of orthogonal polarization (as described in Chap. 3), depolarizing devices, which compensate for the rain depolarization, may have to be installed.

Where the earth station antenna is operated under cover of a radome, the effect of the rain on the radome must be taken into account. Rain falling on a hemispherical radome forms a water layer of constant thickness. Such a layer introduces losses both by absorption and reflection. Results presented by Hogg and Chu, 1975, show an attenuation of about 14 dB for a 1-mm-thick water layer. It is desirable therefore that earth station antennas be operated without radomes where possible. Without a radome, water will gather on the antenna reflector, but the attenuation produced by this is much less serious than that produced by the wet radome (Hogg and Chu, 1975).

10.4   The Link Power Budget Equation

As mentioned at the beginning of Sec. 10.3, the [EIRP] can be considered as the input power to a transmission link. Now that the losses for the link have been identified, the power at the receiver, which is the power output of the link, may be calculated simply as [EIRP] − [LOSSES] + $[G_R]$, where the last quantity is the receiver antenna gain. Note carefully that decibel addition must be used.

The major source of loss in any ground-satellite link is the free-space spreading loss [FSL], as shown in Sec. 10.3.1, where Eq. (10.13) is the basic link power-budget equation taking into account this loss only. However, the other losses must also be taken into account, and these are simply added to [FSL]. The losses for clear-sky conditions are

$$[LOSSES] = [FSL] + [RFL] + [AML] + [AA] + [PL] \quad (10.12)$$

The decibel equation for the received power is then

$$[P_R] = [EIRP] + [G_R] - [LOSSES] \quad (10.13)$$

where $[P_R]$ = received power, dBW
   [EIRP] = equivalent isotropic radiated power, dBW
   [FSL] = free-space spreading loss, dB
   [RFL] = receiver feeder loss, dB

[AML] = antenna misalignment loss, dB
  [AA] = atmospheric absorption loss, dB
  [PL] = polarization mismatch loss, dB

Example 10.4

A satellite link operating at 14 GHz has receiver feeder losses of 1.5 dB and a free-space loss of 207 dB. The atmospheric absorption loss is 0.5 dB, and the antenna pointing loss is 0.5 dB. Depolarization losses may be neglected. Calculate the total link loss for clear-sky conditions.

solution

The total link loss is the sum of all the losses;

[LOSSES] = [FSL] + [RFL] + [AA] + [AML] = 207 + 1.5 + 0.5 + 0.5 = 209.5 dB

10.5   System Noise

It is shown in Sec. 10.3 that the receiver power in a satellite link is very small, of the order of picowatts. This by itself would be no problem since amplification could be used to bring the signal strength up to an acceptable level. However, electrical noise is always present at the input, and unless the signal is significantly greater than the noise, amplification will be of no help, since it will amplify signal and noise to the same extent. In fact, the situation will be worsened by the noise added by the amplifier.

The major source of electrical noise in equipment is that which arises from the random thermal motion of electrons in various resistive and active devices in the receiver. Thermal noise is also generated in the lossy components of antennas, and thermal-like noise is picked up by the antennas as radiation.

The available noise power from a thermal noise source is given by

$$P_N = kT_N B_N \qquad (10.14)$$

Here, T_N is known as the equivalent noise temperature, B_N the equivalent noise bandwidth, and $k = 1.38 \times 10^{-23}$ J/K is Boltzmann's constant. With the temperature in kelvins and bandwidth in hertz, the noise power will be in watts. The noise power bandwidth is always wider than the -3-dB bandwidth determined from the amplitude-frequency response curve, and a useful rule of thumb is that the noise bandwidth is equal to 1.12 times the -3dB bandwidth, or $B_N \cong 1.12 \times B_{-3\text{dB}}$.

The main characteristic of thermal noise is that it has a *flat frequency spectrum*; that is, the noise power per unit bandwidth is a constant. The noise power per unit bandwidth is termed the *noise power spectral density*. Denoting this by N_0, then from Eq. (10.14),

$$N_0 = \frac{P_N}{B_N} = kT_N \quad \text{joules} \tag{10.15}$$

The noise temperature is directly related to the physical temperature of the noise source, but is not always equal to it. This is discussed more fully in the following sections. The noise temperatures of various sources which are connected together can be added directly to give the total noise.

Example 10.5

An antenna has a noise temperature of 35 K, and is matched into a receiver which has a noise temperature of 100 K. Calculate (*a*) the noise power density and (*b*) the noise power for a bandwidth of 36 MHz.

solution

(*a*) $N_0 = (35 + 100) \times 1.38 \times 10^{-23} = 1.86 \times 10^{-21}$ J

(*b*) $P_N = 1.86 \times 10^{-21} \times 36 \times 10^6 = 0.067$ pW

In addition to these thermal noise sources, intermodulation distortion in high power amplifiers (see, for example, Sec. 10.7.3) can result in signal products which appear as noise, and in fact is referred to as *intermodulation noise*. This is discussed in Sec. 10.10.

10.5.1   Antenna Noise

Antennas operating in the receiving mode introduce noise into the satellite circuit. Noise will therefore be introduced by the satellite antenna and the ground station receive antenna. Although the physical origins of the noise in either case are similar, the magnitudes of the effects differ significantly.

The antenna noise can be broadly classified into two groups: noise originating from antenna losses and *sky noise*. Sky noise is a term used to describe the microwave radiation which is present throughout the universe, and which appears to originate from matter in any form, at finite temperatures. Such radiation in fact covers a wider spectrum than just the microwave spectrum. The equivalent noise temperature of the sky, as seen by an earth station antenna, is shown in Fig. 10.2. The lower graph is for the antenna pointing directly overhead, while the upper graph is for the antenna pointing just above the horizon. The increased noise in the latter case results from the thermal radiation of the earth, and this in fact sets a lower limit of

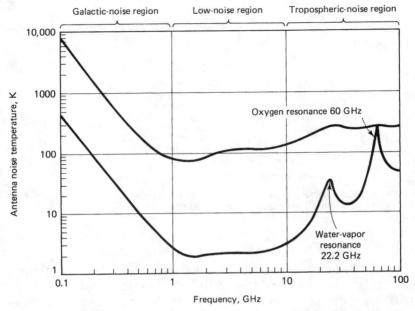

Figure 10.2 Irreducible noise temperature of an ideal, ground-based antenna. The antenna is assumed to have a very narrow beam without sidelobes or electrical losses. Below 1 GHz, the maximum values are for the beam pointed at the galactic poles. At higher frequencies, the maximum values are for the beam just above the horizon and the minimum values for zenith pointing. The low-noise region between 1 and 10 GHz is most amenable to application of special, low-noise antennas. (*From Philip F. Panter, "Communications Systems Design," McGraw-Hill Book Company, New York, 1972. With permission.*)

about 5° at C band and 10° at Ku band on the elevation angle which may be used with ground-based antennas.

The graphs show that, at the low-frequency end of the spectrum, the noise decreases with increasing frequency. Where the antenna is zenith-pointing, the noise temperature falls to about 3 K at frequencies between about 1 GHz and 10 GHz. This represents the residual background radiation in the universe. Above about 10 GHz, two peaks in temperature are observed, resulting from resonant losses in the earth's atmosphere. These are seen to coincide with the peaks in atmospheric absorption loss shown in Fig. 10.2.

Any absorptive loss mechanism generates thermal noise, there being a direct connection between the loss and the effective noise temperature as shown in Sec. 10.5.5. Rainfall introduces attenuation, and therefore it degrades transmissions in two ways: it attenuates the signal, and it introduces noise. The detrimental effects of rain are much worse at Ku-band frequencies than at C-band, and the downlink rain-fade margin discussed in Sec. 10.8.3 must also allow for the increased noise which is generated.

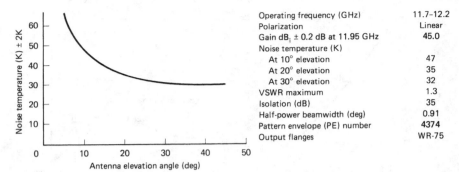

Figure 10.3 Antenna noise temperature as a function of elevation for 1.8-m antenna characteristics. (*Andrew Bulletin 1206; courtesy Andrew Antenna Company, Limited.*)

Figure 10.2 applies to ground-based antennas. Satellite antennas are generally pointed toward the earth, and therefore they receive the full thermal radiation from it. In this case the equivalent noise temperature of the antenna, excluding antenna losses, is approximately 290 K.

Antenna losses add to the noise received as radiation, and the total antenna noise temperature is in the sum of the equivalent noise temperature of all these sources. For large ground-based C-band antennas, the total antenna noise temperature is typically about 60 K, and for the Ku band, about 80 K under clear sky conditions. These values do not apply to any specific situation and are quoted merely to give some idea of the magnitudes involved. Figure 10.3 shows the noise temperature as a function of angle of elevation for a 1.8-m antenna operating in the Ku band.

10.5.2 Amplifier Noise Temperature

Consider first the noise representation of the antenna and the low-noise amplifier (LNA) shown in Fig. 10.4a. The available power gain of the amplifier is denoted as G, and the noise power output as P_{no}. For the moment we will work with the noise power per unit bandwidth, which is simply noise energy in joules as shown by Eq. (10.15). The input noise energy coming from the antenna is

$$N_{o,\mathrm{ant}} = kT_{\mathrm{ant}} \tag{10.16}$$

The output noise energy $N_{o,\mathrm{out}}$ will be $GN_{o,\mathrm{ant}}$ plus the contribution made by the amplifier. Now all the amplifier noise, wherever it occurs in the amplifier, may be *referred to the input* in terms of an equivalent

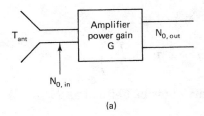

(a)

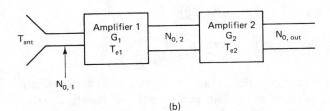

(b)

Figure 10.4 Circuit used in finding equivalent noise temperature of (*a*) an amplifier and (*b*) two amplifiers in cascade.

input noise temperature for the amplifier, T_e. This allows the output noise to be written as

$$N_{o,\text{out}} = Gk(T_{\text{ant}} + T_e) \tag{10.17}$$

The total noise referred to the input is simply $N_{o,\text{out}}/G$, or

$$N_{o,\text{in}} = k(T_{\text{ant}} + T_e) \tag{10.18}$$

T_e can be obtained by measurement, a typical value being in the range 35 to 100 K. Typical values for T_{ant} are given in Sec. 10.5.1.

10.5.3 Amplifiers in Cascade

The cascade connection is shown in Fig. 10.4*b*. For this arrangement the overall gain is

$$G = G_1 G_2 \tag{10.19}$$

The noise energy of amplifier 2 referred to its own input is simply kT_{e2}. The noise input to amplifier 2 from the preceding stages is $G_1 k(T_{\text{ant}} + T_{e2})$ and thus the total noise energy *referred to amplifier 2 input* is

$$N_{0,2} = G_1 k(T_{\text{ant}} + T_{e1}) + kT_{e2} \tag{10.20}$$

This noise energy may be referred to amplifier 1 input by dividing by the available power gain of amplifier 1

$$N_{0,1} = \frac{N_{0,2}}{G_1}$$

$$= k\left(T_{ant} + T_{e1} + \frac{T_{e2}}{G_1}\right) \tag{10.21}$$

A system noise temperature may now be defined as T_S by

$$N_{0,1} = kT_S \tag{10.22}$$

and hence it will be seen that T_S is given by

$$T_S = T_{ant} + T_{e1} + \frac{T_{e2}}{G_1} \tag{10.23}$$

This is a very important result. It shows that the noise temperature of the second stage is divided by the power gain of the first stage when referred to the input. Therefore, in order to keep the overall system noise as low as possible, the first stage (usually an LNA) should have high power gain as well as low noise temperature.

This result may be generalized to any number of stages in cascade, giving

$$T_S = T_{ant} + T_{e1} + \frac{T_{e2}}{G_1} + \frac{T_{e3}}{G_1 G_2} \tag{10.24}$$

10.5.4 Noise Factor

An alternative way of representing amplifier noise is by means of its *noise factor F.* In defining the noise factor of an amplifier, the source is taken to be at *room temperature,* denoted by T_0, usually taken as 290 K. The input noise from such a source is kT_0, and the output noise from the amplifier is

$$N_{0,out} = FGkT_0 \tag{10.25}$$

Here, G is the available power gain of the amplifier as before, and F is its noise factor.

A simple relationship between noise temperature and noise factor can be derived. Let T_e be the noise temperature of the amplifier, and let the source be at room temperature as required by the definition of F. This means that $T_{ant} = T_0$. Since the same noise output must be available whatever the representation, it follows that

$$Gk(T_0 + T_e) = FGkT_0$$

or

$$T_e = (F - 1)T_0 \qquad (10.26)$$

This shows the direct equivalence between noise factor and noise temperature. As a matter of convenience, in a practical satellite receiving system, noise temperature is specified for low-noise amplifiers and converters, while noise factor is specified for the main receiver unit.

The *noise figure* is simply F expressed in decibels:

$$\text{Noise figure} = [F] = 10 \log F \qquad (10.27)$$

Example 10.6

An LNA is connected to a receiver which has a noise figure of 12 dB. The gain of the LNA is 40 dB and its noise temperature is 120 K. Calculate the overall noise temperature referred to the LNA input.

solution

12 dB is a power ratio of 15.85:1, and therefore

$$T_{e2} = (15.85 - 1) \times 290 = 4306 \text{ K}$$

A gain of 40 dB is a power ratio of 10^4:1, and therefore

$$T_{in} = 120 + \frac{4306}{10^4} = 120.43 \text{ K}$$

In Example 10.6 it will be seen that the decibel quantities must be converted to power ratios. Also, even though the main receiver has a very high noise temperature, its effect is made negligible by the high gain of the LNA.

10.5.5   Noise Temperature of Absorptive Networks

An *absorptive network* is one which contains resistive elements. These introduce losses by absorbing energy from the signal and converting it to heat. Resistive attenuators, transmission lines, and waveguides are all examples of absorptive networks, and even rainfall, which absorbs energy from radio signals passing through it, can be considered a form of absorptive network. Because an absorptive network contains resistance it generates thermal noise.

Consider an absorptive network which has a power loss L (L is the ratio of available input power to available output power, and is greater than 1) and which is matched to an input source. Let the source be at some temperature T_x so that its available noise energy is kT_x. The network "power gain" is $1/L$ and therefore the source contribution to the output noise is kT_x/L. Let $T_{NW,o}$ represent the noise temperature of the network, referred to the output, so that the network contribution to the output noise is $kT_{NW,o}$. The total output noise is therefore

$$N_{0,\text{out}} = \frac{kT_x}{L} + kT_{NW,o} \qquad (10.28)$$

Let the network initially be at the same temperature T_x as the source. Because the network is matched to the source, the available noise energy at the output is also given by kT_x (just as it was for the source itself), and therefore

$$kT_x = \frac{kT_x}{L} + kT_{NW,o}$$

or

$$T_{NW,o} = T_x\left(1 - \frac{1}{L}\right) \qquad (10.29)$$

It follows that the noise contribution of the network alone to the noise output is $kT_x(1 - 1/L)$. This depends only on the network, and is quite independent of the source conditions. The source temperature may be changed to a new value T_i, but provided the network is kept at temperature T_x, its noise temperature referred to the output remains is given by Eq. (10.29).

The network noise can also be referred to the input of the network simply by dividing the output noise by the network power gain $1/L$. This is equivalent to dividing the output noise temperature by the power gain, or in effect, multiplying it by L. Thus, the noise temperature of the lossy network referred to its input is

$$T_{NW,i} = LT_{NW,o}$$

$$= T_x(L - 1) \qquad (10.30)$$

If the lossy network should happen to be at room temperature, that is, $T_x = T_0$, then a comparison of Eqs. (10.26) and (10.30) shows that

$$F = L \qquad (10.31)$$

This shows that at room temperature the noise factor of a lossy network is equal to its power loss.

10.5.6   Overall System Noise Temperature

Figure 10.5a shows a typical receiving system. Applying the results of the previous sections yields, for the system noise temperature referred to the input,

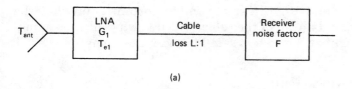

(a)

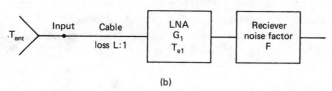

(b)

Figure 10.5 Connections used in examples illustrating overall noise temperature of system, Sec. 10.5.6).

$$T_S = T_{\text{ant}} + T_{e1} + \frac{(L-1)T_0}{G_1} + \frac{L(F-1)T_0}{G_1} \qquad (10.32)$$

The significance of the individual terms is illustrated in the following examples.

Example 10.7

For the system shown in Fig. 10.5a, the receiver noise figure is 12 dB, the cable loss is 5 dB, the LNA gain is 50 dB, and its noise temperature 100 K. The antenna noise temperature is 35 K. Calculate the noise temperature referred to the input.

solution

For the main receiver, $F = 10^{1.2} = 15.85$. For the cable, $L = 10^{0.5} = 3.16$. For the LNA, $G = 10^5$. Hence

$$T_S = 35 + 150 + \frac{(3.16 - 1) \times 290}{10^5} + \qquad\qquad \cong 185\ \text{K}$$

Example 10.8

Repeat the calculation when the system of Fig. 10.5a is arranged as shown in Fig. 10.5b.

solution

In this case the cable precedes the LNA and therefore the equivalent noise temperature referred to the cable input is

$$T_S = 35 + (3.16 - 1) \times 290 + 3.16 \times 150 + \qquad\qquad = 1136\ \text{K}$$

Examples 10.7 and 10.8 illustrate the important point that the LNA must be placed ahead of the cable, which is why one sees amplifiers mounted right at the dish in satellite receive systems.

10.6 Carrier-to-Noise Ratio

A measure of the performance of a satellite link is the ratio of carrier power to noise power at the receiver input, and link budget calculations are often concerned with determining this ratio. Conventionally the ratio is denoted by C/N (or CNR) which is equivalent to P_R/P_N. In terms of decibels,

$$\left[\frac{C}{N}\right] = [P_R] - [P_N] \tag{10.33}$$

Equations (10.17) and (10.18) may be used for $[P_R]$ and $[P_N]$, resulting in

$$\left[\frac{C}{N}\right] = [\text{EIRP}] + [G_R] - [\text{LOSSES}] - [k] - [T_s] - [B_N] \tag{10.34}$$

The G/T ratio is a key parameter in specifying the receiving system performance. The antenna gain G_R and the system noise temperature T_s can be combined in Eq. (10.34) as

$$[G/T] = [G_R] - [T_S] \qquad \text{dBK}^{-1} \tag{10.35}$$

Therefore the link equation, Eq. (10.34), becomes

$$\left[\frac{C}{N}\right] = [\text{EIRP}] + \left[\frac{G}{T}\right] - [\text{LOSSES}] - [k] - [B_N] \tag{10.36}$$

The ratio of carrier power to noise power density P_R/N_0 may be the quantity actually required. Since $P_N = kT_N B_N = N_0 B_N$ then

$$\left[\frac{C}{N}\right] = \left[\frac{C}{N_o B_N}\right]$$

$$= \left[\frac{C}{N_o}\right] - [B_N]$$

and therefore

$$\left[\frac{C}{N_0}\right] = \left[\frac{C}{N}\right] + [B_N] \tag{10.37}$$

$[C/N]$ is a true power ratio in units of decibels, and $[B_N]$ is in decibels relative to one hertz, or dBHz. Thus the units for $[C/N_0]$ are dBHz.

Substituting Eq. (10.37) for $[C/N]$ gives

$$\left[\frac{C}{N_0}\right] = [\text{EIRP}] + \left[\frac{G}{T}\right] - [\text{LOSSES}] - [k] \qquad \text{dBHz} \qquad (10.38)$$

Example 10.9

In a link budget calculation at 12 GHz, the free-space loss is 206 dB, the antenna pointing loss is 1 dB, and the atmospheric absorption is 2 dB. The receiver G/T ratio is 19.5 dB/K and receiver feeder losses are 1 dB. The EIRP is 48 dBW. Calculate the carrier-to-noise spectral density ratio.

solution

The data are best presented in tabular form, and in fact lend themselves readily to spreadsheet-type computations. For brevity the units are shown as decilogs and losses are entered as negative numbers to take account of the minus sign in Eq. (10.38). Recall that Boltzmann's constant equates to -228.6 decilogs, so that $-[k] = 228.6$ decilogs as shown in the table. Entering data in this way allows the final result to be entered in a table cell as the sum of the terms in the rows above the cell, a feature usually incorporated in spreadsheets and word processors. This is illustrated in the following table.

Quantity	Decilogs
Free-space loss	−206.00
Atmospheric absorption loss	−2.00
Antenna pointing loss	−1.00
Receiver feeder losses	−1.00
Polarization mismatch loss	0.00
Receiver G/T ratio	19.50
EIRP	48.00
$-[k]$	228.60
$[C/N_0]$, Eq. (10.38)	86.10

The final result, 86.10 dBHz, is the algebraic sum of the quantities as given in Eq. (10.38).

10.7 The Uplink

The uplink of a satellite circuit is the one in which the earth station is transmitting the signal and the satellite is receiving it. Equation (10.38) can be applied to the uplink, but subscript U will be used to denote specifically that the uplink is being considered. Thus Eq. (10.38) becomes

$$\left[\frac{C}{N_0}\right]_U = [\text{EIRP}]_U + \left[\frac{G}{T}\right]_U - [\text{LOSSES}]_U - [k] \qquad (10.39)$$

In Eq. (10.39) the values to be used are the earth station EIRP, the satellite receiver feeder losses, and satellite receiver G/T. The free-space loss, and other losses which are frequency-dependent, are calculated for the uplink frequency. The resulting carrier-to-noise density ratio given by Eq. (10.39) is that which appears at the satellite receiver.

In some situations the flux density appearing at the satellite receive antenna is specified rather than the earth station EIRP, and Eq. (10.39) is modified as explained next.

10.7.1 Saturation Flux Density

As explained in Sec. 6.7.3, the traveling wave tube amplifier (TWTA) in a satellite transponder exhibits power output saturation, shown in Fig. 6.21. The flux density required at the receiving antenna to produce saturation of the TWTA is termed the *saturation flux density*. The saturation flux density is a specified quantity in link budget calculations, and knowing it, one can calculate the required EIRP at the earth station. To show this, consider again Eq. (10.6) which gives the flux density in terms of EIRP, repeated here for convenience:

$$\Psi_M = \frac{\text{EIRP}}{4\pi r^2}$$

In decibel notation this is

$$[\Psi_M] = [\text{EIRP}] + 10 \log \frac{1}{4\pi r^2} \qquad (10.40)$$

But from Eq. (10.9) for free-space loss we have

$$-[\text{FSL}] = 10 \log \frac{\lambda^2}{4\pi} + 10 \log \frac{1}{4\pi r^2} \qquad (10.41)$$

Substituting this in Eq. (10.40) gives

$$[\Psi_M] = [\text{EIRP}] - [\text{FSL}] - 10 \log \frac{\lambda^2}{4\pi} \qquad (10.42)$$

The $\lambda^2/4\pi$ term has dimensions of area, and in fact from Eq. (5.12) it is the effective area of an isotropic antenna. Denoting this by A_0 gives

$$[A_0] = 10 \log \frac{\lambda^2}{4\pi} \qquad (10.43)$$

Since frequency rather than wavelength is normally known, it is left as an exercise for the student to show that, with frequency f in gigahertz, Eq. (10.43) can be rewritten as

$$[A_0] = -(21.45 + 20 \log f) \qquad (10.44)$$

Combining this with Eq. (10.42) and rearranging slightly gives the EIRP as

$$[EIRP] = [\Psi_M] + [A_0] + [FSL] \qquad (10.45)$$

Equation (10.45) was derived on the basis that the only loss present was the spreading loss, denoted by [FSL]. But as shown in the previous sections, the other propagation losses are the atmospheric absorption loss, the polarization mismatch loss, and the antenna misalignment loss. When allowance is made for these, Eq. (10.45) becomes

$$[EIRP] = [\Psi_M] + [A_0] + [FSL] + [AA] + [PL] + [AML] \quad (10.46)$$

In terms of the total losses given by Eq. (10.12), Eq. (10.46) becomes

$$[EIRP] = [\Psi_M] + [A_0] + [LOSSES] - [RFL] \qquad (10.47)$$

This is for clear sky conditions and gives the *minimum* value of [EIRP] which the earth station must provide to produce a given flux density at the satellite. Normally, the saturation flux density will be specified. With saturation values denoted by the subscript S, Eq. (10.47) is rewritten as

$$[EIRP_S]_U = [\Psi_S] + [A_0] + [LOSSES]_U - [RFL] \qquad (10.48)$$

Example 10.10

An uplink operates at 14 GHz, and the flux density required to saturate the transponder is $-120 dB(W/m^2)$. The free space loss is 207 dB, and the other propagation losses amount to 2 dB. Calculate the earth-station [EIRP] required for saturation, assuming clear sky conditions. Assume [RFL] is negligible.

solution

At 14 GHz,

$[A_0] = -(21.45 + 20 \log 14) = -44.37$ dB

The losses in the propagation path amount to $207 + 2 = 209$ dB. Hence, from Eq. (10.48),

$$[EIRP_S]_U = -120 - 44.37 + 209$$
$$= 44.63 \text{ dBW}$$

10.7.2 Input back-off

As described in Sec. 10.7.3, where a number of carriers are present simultaneously in a TWTA, the operating point must be backed off to a linear portion of the transfer characteristic to reduce the effects of intermodulation distortion. Such multiple carrier operation occurs with frequency-division multiple access (FDMA) and is described in Chap. 12. The point to be made here is that back-off must be allowed for in the link budget calculations.

Suppose that the saturation flux density for single-carrier operation is known. Input back-off will be specified for multiple-carrier operation, referred to the single-carrier saturation level. The earth station EIRP will have to be reduced by the specified back-off (BO), resulting in an uplink value of

$$[\text{EIRP}]_U = [\text{EIRP}_S]_U - [\text{BO}]_i \qquad (10.49)$$

Although some control of the input to the transponder power amplifier is possible through the ground TT&C station as described in Sec. 10.7.3, input back-off is normally achieved through reduction of the [EIRP] of the earth stations actually accessing the transponder.

Equations (10.48) and (10.49) may now be substituted in Eq. (10.39) to give

$$\left[\frac{C}{N_0}\right]_U = [\Psi_S] + [A_0] - [\text{BO}]_i + \left[\frac{G}{T}\right]_U - [k] - [\text{RFL}] \qquad (10.50)$$

Example 10.11

An uplink at 14 GHz requires a saturation flux density of -91.4 dBW/m^2 and an input back-off of 11 dB. The satellite G/T is -6.7 dBK^{-1} and receiver feeder losses amount to 0.6 dB. Calculate the carrier-to-noise density ratio.

solution

As in Example 10.9, the calculations are the best carried out in tabular form. $[A_0] = -44.37$ dBm^2 for a frequency of 14 GHz is calculated by using Eq. (10.44) as in Example 10.10.

Quantity	Decilogs
Saturation flux density	-91.4
$[A_0]$ at 14 GHz	-44.4
Input back-off	-11.0
Satellite saturation $[G/T]$	-6.7
$-[k]$	228.6
Receiver feeder loss	-0.6
Total	74.5

Note that $[k] = -228.6$ dB so that $-[k]$ in Eq. (10.50) becomes 228.6 dB. Also, [RFL] and $[BO]_i$ are entered as negative numbers to take account of the minus signs attached to them in Eq. (10.50). The total gives the carrier-to-noise density ratio at the satellite receiver as 74.5 dBHz.

Since fade margins have not been included at this stage, Eq. (10.50) applies for *clear sky* conditions. Usually, the most serious fading is caused by rainfall as described in Sec. 10.7.4.

10.7.3   The Earth Station HPA

The earth station high-power amplifier has to supply the radiated power, plus the transmit feeder losses, denoted here by TFL, or [TFL] dB. These include waveguide, filter, and coupler losses between the HPA output and the transmit antenna. Referring back to Eq. (10.3), the power output of the HPA is given by

$$[P_{HPA}] = [EIRP] - [G_T] + [TFL] \qquad (10.51)$$

The [EIRP] is that given by Eq. (10.49) and thus includes any input back-off that is required at the satellite.

The earth station itself may have to transmit multiple carriers, and its output will also require back-off, denoted by $[BO]_{HPA}$. The earth-station HPA must be rated for a saturation power output given by

$$[P_{HPA,sat}] = [P_{HPA} + [BO]_{HPA} \qquad (10.52)$$

Of course, the HPA will be operated at the backed-off power level so that it provides the required power output $[P_{HPA}]$. To ensure operation well into the linear region, an HPA with a comparatively high saturation level can be used, and a high degree of back-off introduced. The large physical size and high power consumption associated with larger tubes do not carry the same penalties they would if used aboard the satellite. Again, it is emphasized that back-off at the earth station may be required quite independently of any back-off requirements at the satellite transponder. The power rating of the earth station HPA should also be sufficient to provide a fade margin as discussed in the next section.

10.7.4   Uplink Rain-Fade Margin

Up to this point, the $[C/N_0]$ calculations have been made for clear sky conditions. As discussed in the previous sections, fading of the signal can arise in a number of ways. In the C band, and more especially in the Ku band, rainfall is the most significant of these. It results in attenuation of the signal, and an increase in noise temperature, thus degrading the $[C/N_0]$ at the satellite in two ways. The increase in noise, however, is not usually a major factor for the uplink. This is because the satellite antenna is pointed toward a "hot" earth, and this added to the satellite receiver noise temperature tends to mask any additional noise induced by rain attenuation. What is important is that the uplink carrier power at the satellite must be held within close limits for certain modes of operation, and some form of *uplink power control* is necessary to compensate for rain fades. The power output from the satellite may be monitored by a central control station, or in some cases by each earth station, and the power output from any given earth station increased if required to compensate for fading. Thus, the earth-station HPA must have sufficient reserve power to meet the fade margin requirement.

Some typical rain-fade margins are shown in Table 10.1. As an example, for Ottawa, the rain attenuation exceeds 1.9 dB for 0.1 percent of the time. This means that to meet the specified power requirements at the input to the satellite for 99.9 percent of the time, the earth station must be capable of providing a 1.9 dB margin over the clear sky conditions.

10.8   Downlink

The downlink of a satellite circuit is the one in which the satellite is transmitting the signal and the earth station is receiving it. Equation (10.38) can be applied to the downlink, but subscript D will be used to denote specifically that the downlink is being considered. Thus Eq. (10.38) becomes

$$\left[\frac{C}{N_0}\right]_D = [\text{EIRP}]_D + \left[\frac{G}{T}\right]_D - [\text{LOSSES}]_D - [k] \qquad (10.53)$$

In Eq. (10.53) the values to be used are the satellite EIRP, the earth station receiver feeder losses, and the earth station receiver G/T. The free-space and other losses are calculated for the downlink frequency. The resulting carrier-to-noise density ratio given by Eq. (10.53) is that which appears at the detector of the earth station receiver.

Where the carrier-to-noise ratio is the specified quantity rather than carrier-to-noise density ratio, Eq. (10.38) is used. This becomes, on assuming that the signal bandwidth B is equal to the noise bandwidth B_N:

$$\left[\frac{C}{N}\right]_D = [\text{EIRP}]_D + \left[\frac{G}{T}\right]_D - [\text{LOSSES}]_D - [k] - [B] \qquad (10.54)$$

Example 10.12

A satellite TV signal occupies the full transponder bandwidth of 36 MHz, and it must provide a C/N ratio at the destination earth station of 22 dB. Given that the total transmission losses are 200 dB and the destination earth station G/T ratio is 31 dB/K, calculate the satellite EIRP required.

solution

Equation (10.54) can be rearranged as

$$[\text{EIRP}]_D = \left[\frac{C}{N}\right]_D - \left[\frac{G}{T}\right]_D + [\text{LOSSES}]_D + [k] + [B]$$

Setting this up in tabular form, and keeping in mind that $+[k] = -228.6$ dB and that losses are numerically equal to $+200$ dB, we obtain

Quantity	Decilogs
$[C/N]$	22
$-[G/T]$	-31
[LOSSES]	200
$[k]$	-228.6
[B]	75.6
[EIRP]	38

The required EIRP is 38 dBW or equivalently 6.3 kW.

Example 10.12 illustrates the use of Eq. (10.54). Example 10.13 shows the use of Eq. (10.53) applied to a digital link.

Example 10.13

A QPSK signal is transmitted by satellite. Raised cosine filtering is used for which the roll-off factor is 0.2 and a BER of 10^{-5} is required. For the satellite downlink, the losses amount to 200 dB, the receiving earth station G/T ratio is 32 dBK^{-1}, and the transponder bandwidth is 36 MHz. Calculate (a) the bit rate which can be accommodated, and (b) the *eirp* required.

solution
Given data:

$$B := 36 \cdot \text{MHz} \quad \rho := 0.2 \quad \text{GTRdB} := 31 \quad \text{LOSSESdB} := 200 \quad \text{kdB} := -228.6$$

Given that $\rho = 0.2$, eq. (9.16) gives:

$$\text{Rb} := \frac{2 \cdot B}{1 + \rho} \qquad \text{Rb} = 6 \cdot 10^{7} \cdot \text{sec}^{-1}$$
$$\overline{\overline{=========}}$$

for BER = 10^{-5}, Fig. 9.17 gives an E_b/N_o ratio of:

$$\text{EbNoRdB} := 9.6$$

Converting Rb to decilogs:

$$RbdB := 10 \cdot \log\left(\frac{Rb}{\sec^{-1}}\right)$$

From eq. (8.48) the required C/N_0 ratio is:

$$CNoRdB := EbNoRdB + RbdB$$

From eq. (10.53)

$$eirpdBW := CNoRdB - GTRdB + LOSSESdB + kdB \qquad eirpdBW = 27.8$$
$$==========$$

10.8.1 Output Back-off

Where input back-off is employed as described in Sec. 10.7.2, a corresponding output back-off must be allowed for in the satellite EIRP. As the curve of Fig. 6.21 shows, output back-off is not linearly related to input back-off. A rule of thumb frequently used is to take the output back-off as the point on the curve which is 5 dB below the extrapolated linear portion, as shown in Fig. 10.6. Since the linear portion gives a 1:1 change in decibels, the relationship between input and output back-off is $[BO]_o = [BO]_i - 5$ dB. For example, with an input back-off of $[BO]_i = 11$ dB, the corresponding output back-off is $[BO]_o = 11 - 5 = 6$ dB.

If the satellite EIRP for saturation conditions is specified as $[EIRP_S]_D$, then $[EIRP]_D = [EIRP_S]_D - [BO]_o$ and Eq. (10.53) becomes

$$\left[\frac{C}{N_0}\right]_D = [EIRP_S]_D - [BO]_o + \left[\frac{G}{T}\right]_D - [LOSSES]_D - [k] \qquad (10.55)$$

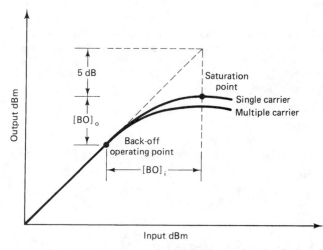

Figure 10.6 Input and output back-off relationship for the satellite traveling-wave-tube amplifier; $[BO]_i = [BO]_0 + 5$ dB.

Example 10.14

The specified parameters for a downlink are satellite saturation value of EIRP, 25 dBW; output back-off, 6 dB; free-space loss, 196 dB; allowance for other downlink losses, 1.5 dB; and earth station G/T, 41 dBK^{-1}. Calculate the carrier-to-noise density ratio at the earth station.

solution

As with the uplink budget calculations the work is best set out in tabular form with the minus signs in Eq.(10.55) attached to the tabulated values.

Quantity	Decilogs
Satellite saturation [EIRP]	25.0
Free-space loss	−196.0
Other losses	−1.5
Output back-off	−6.0
Earth station [G/T]	41.0
−[k]	228.6
Total	91.1

The total gives the carrier-to-noise density ratio at the earth station in dBHz, as calculated from Eq. (10.55).

For the up link, the saturation flux density at the satellite receiver is a specified quantity. For the down link, there is no need to know the saturation flux density at the earth station receiver, since this is a terminal point and the signal is not used to saturate a power amplifier.

10.8.2 Satellite TWTA Output

The satellite power amplifier, which usually is a traveling wave tube amplifier, has to supply the radiated power, plus the transmit feeder losses. These losses include the waveguide, filter, and coupler losses between the TWTA output and the satellite's transmit antenna. Referring back to Eq. (10.3), the power output of the TWTA is given by

$$[P_{\mathrm{TWTA}}] = [\mathrm{EIRP}]_D - [G_T]_D + [\mathrm{TFL}]_D \qquad (10.56)$$

Once $[P_{\mathrm{TWTA}}]$ is found, the saturated power output rating of the TWTA is given by

$$[P_{\mathrm{TWTA}}]_S = [P_{\mathrm{TWTA}}] + [\mathrm{BO}]_o \qquad (10.57)$$

Example 10.15

A satellite is operated at an EIRP of 56 dBW with an output back-off of 6 dB. The transmitter feeder losses amount to 2 dB, and the antenna gain is 50 dB.

Calculate the power output of the TWTA, assuming it may be required to provide the full saturated EIRP.

solution

Equation (10.56) gives

$$[P_{\text{TWTA}}] = [\text{EIRP}]_D - [G_T]_D + [\text{TFL}]_D$$
$$= 56 - 50 + 2$$
$$= 8 \text{ dBW}$$

Equation (10.57) gives

$$[P_{\text{TWTA}}]_S = 8 + 6 = 14 \text{ dBW (or 25 W)}$$

10.8.3 Downlink Rain-Fade Margin

The results given by Eqs. (10.53) and (10.54) are for clear sky conditions. Rainfall introduces attenuation by absorption and scattering of signal energy, and the absorptive attenuation introduces noise as discussed in Sec. 10.5.5. Let $[A]$ dB represent the rain attenuation caused by absorption. The corresponding power loss ratio is $A = 10^{[A]/10}$, and substituting this for L in Eq. (10.29) gives the effective noise temperature of the rain as

$$T_{\text{RAIN}} = T_a\left(1 - \frac{1}{A}\right) \tag{10.58}$$

Here, T_a, which takes the place of T_x in Eq. (10.29), is known as the *apparent absorber temperature*. It is a measured parameter which is a function of many factors including the physical temperature of the rain and the scattering effect of the rain cell on the thermal noise incident upon it (Hogg and Chu, 1975). The value of the apparent absorber temperature lies between 270 and 290 K, with measured values for North America lying close to or just below freezing (273 K). For example, the measured value given by Webber et al., 1986, is 272 K.

The total sky-noise temperature is the clear sky temperature T_{CS} plus the rain temperature:

$$T_{\text{sky}} = T_{\text{CS}} + T_{\text{RAIN}} \tag{10.59}$$

Rainfall therefore degrades the received $[C/N_o]$ in two ways: by attenuating the carrier wave and by increasing the sky-noise temperature.

Example 10.16

Under clear sky conditions the downlink $[C/N]$ is 20 dB, the effective noise temperature of the receiving system being 400 K. If rain attenuation exceeds 1.9 dB for 0.1 percent of the time, calculate the value below which [C/N] falls for 0.1 percent of the time. Assume $T_a = 280$ K.

solution

1.9 dB attenuation is equivalent to a 1.55:1 power loss. The equivalent noise temperature of the rain is therefore

$$T_{RAIN} = 280(1 - 1/1.55) = 99.2 \text{ K}$$

The new system noise temperature is $400 + 99.2 = 499.2$ K. The decibel increase in noise power is therefore [499.2] − [400] = 0.96 dB. At the same time, the carrier is reduced by 1.9 dB and therefore the [C/N] with 1.9-dB rain attenuation drops to $20 - 1.9 - 0.96 = 17.14$ dB. This is the value below which [C/N] drops for 0.1 percent of the time.

It is left as an exercise for the student to show that where the rain attenuation is entirely absorptive, the downlink C/N power ratio is related to the clear sky value by

$$\left(\frac{N}{C}\right)_{RAIN} = \left(\frac{N}{C}\right)_{CS}\left[A + (A - 1)\frac{T_a}{T_{S,CS}}\right] \qquad (10.60)$$

where the subscript CS is used to indicate clear sky conditions and $T_{S,CS}$ is the *system* noise temperature under clear sky conditions.

For low frequencies (6/4 GHz) and low rainfall rates (below about 1 mm/h), the rain attenuation is almost entirely absorptive. At higher rainfall rates, scattering becomes significant, especially at the higher frequencies. When scattering and absorption are both significant, the total attenuation must be used to calculate the reduction in carrier power and the absorptive attenuation to calculate the increase in noise temperature.

As discussed in Chap. 8, a minimum value of [C/N] is required for satisfactory reception. In the case of frequency modulation, the minimum value is set by the threshold level of the FM detector, and a *threshold margin* is normally allowed, as shown in Fig. 8.12. Sufficient margin must be allowed so that rain-induced fades do not take the [C/N] below threshold more than a specified percentage of the time, as shown in Example 10.17.

Example 10.17

In an FM satellite system the clear sky downlink [C/N] ratio is 17.4 dB and the FM detector threshold is 10 dB as shown in Fig. 8.12. (*a*) Calculate the threshold margin at the FM detector, assuming the threshold [C/N] is determined solely by the downlink value. (*b*) Given that $T_a = 272$ K and that $T_{S,CS} = 544$ K, calculate the percentage of time the system stays above threshold. The curve of Fig. 10.7 may be used for the downlink, and it may be assumed that the rain attenuation is entirely absorptive.

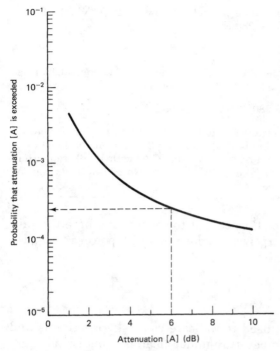

Figure 10.7 Typical rain attenuation curve used in Ex. (10.17).

solution

(a) Since it is assumed that the overall $[C/N]$ ratio is equal to the downlink value, the clear sky input $[C/N]$ to the FM detector is 17.4 dB. The threshold level for the detector is 10 dB, and therefore the rain-fade margin is 17.4 − 10 = 7.4 dB.

(b) The rain attenuation can reduce the $[C/N]$ to the threshold level of 10 dB (that is, it reduces the margin to zero), which is a (C/N) power ratio of 10:1, or a down-link N/C power ratio of 1/10.

For clear sky conditions, $[C/N]_{CS}$ = 17.4 dB, which gives a N/C ratio of 0.0182. Substituting these values in Eq. (10.60) gives

$$0.1 = 0.0182 \times \left[A + \frac{(A - 1) \times 272}{544} \right]$$

Solving this equation for A gives $A = 4$ or approximately 6 dB. From the curve of Fig. 10.7, the probability of exceeding the 6-dB value is 2.5×10^{-4}, and therefore the availability is $1 - 2.5 \times 10^{-4} = 0.99975$ or 99.975 percent.

For digital signals, the required $[C/N_0]$ ratio is determined by the acceptable bit error rate (BER) which must not be exceeded for more than a specified percentage of the time. Figure 9.17 relates the BER to the $[E_b/N_0]$ ratio, and this in turn is related to the $[C/N_0]$ by Eq. (9.20), as discussed in Sec. 9.6.4.

For the downlink, the user does not have control of the satellite [EIRP], and thus the downlink equivalent of uplink power control, described in Sec. 10.7.4, cannot be used. In order to provide the rain-fade margin needed, the gain of the receiving antenna may be increased, by using a larger dish and/or a receiver front end having a lower noise temperature. Both measures increase the receiver $[G/T]$ ratio and thus increase $[C/N_o]$ as shown by Eq. (10.53).

10.9 Combined Uplink and Downlink C/N Ratio

The complete satellite circuit consists of an uplink and a downlink as sketched in Fig. 10.8a. Noise will be introduced on the uplink at the satellite receiver input. Denoting the noise power per unit bandwidth by P_{NU} and the average carrier at the same point by P_{RU}, the carrier-to-noise ratio on the uplink is $(C/N_o)_U = (P_{RU}/P_{NU})$. It is important to note that power levels, and not decibels, are being used here.

The carrier power at the end of the space link is shown as P_R, which of course is also the received carrier power for the downlink. This is equal to γ times the carrier power input at the satellite, where γ is

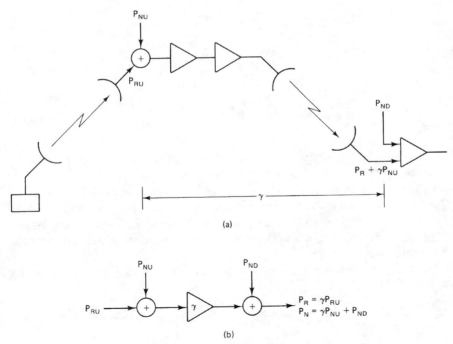

(a)

(b)

Figure 10.8 (a) Combined uplink and downlink; (b) power flow diagram for (a).

the system power gain from satellite input to earth station input, as shown in Fig. 10.8a. It includes the satellite transponder and transmit antenna gains, the downlink losses, and the earth station receive antenna gain and feeder losses.

The noise at the satellite input also appears at the earth station input multiplied by γ, and in addition, the earth station introduces its own noise, denoted by P_{ND}. Thus, the end-of-link noise is $\gamma P_{NU} + P_{ND}$.

The C/N_0 ratio for the downlink alone, not counting the γP_{NU} contribution, is P_R/P_{ND}, and the combined C/N ratio at the ground receiver is $P_R/(\gamma P_{NU} + P_{ND})$. The power flow diagram is shown in Fig. 10.8b. The combined carrier-to-noise ratio can be determined in terms of the individual link values. To show this, it is more convenient to work with the noise-to-carrier ratios rather than the carrier-to-noise ratios, and again, these must be expressed as power ratios, not decibels. Denoting the combined noise-to-carrier ratio value by N_0/C, the uplink value by $(N_0/C)_U$, and the downlink value by $(N_0/C)_D$ then,

$$
\begin{aligned}
\frac{N_0}{C} &= \frac{P_N}{P_R} \\
&= \frac{\gamma P_{NU} + P_{ND}}{P_R} \\
&= \frac{\gamma P_{NU}}{P_R} + \frac{P_{ND}}{P_R} \\
&= \frac{\gamma P_{NU}}{\gamma P_{RU}} + \frac{P_{ND}}{P_R} \\
&= \left(\frac{N_0}{C}\right)_U + \left(\frac{N_0}{C}\right)_D
\end{aligned}
\tag{10.61}
$$

Equation (10.61) shows that, to obtain the combined value of C/N_0, the reciprocals of the individual values must be added to obtain the N_0/C ratio, and then the reciprocal of this taken to get C/N_0. Looked at in another way, the reason for this reciprocal of the sum of the reciprocals method is that a single signal power is being transferred through the system, while the various noise powers which are present are additive.

Similar reasoning applies to the carrier-to-noise spectral density ratio, C/N.

Example 10.18

For a satellite circuit the individual link carrier-to-noise spectral density ratios are: uplink 100 dBHz; downlink 87 dBHz. Calculate the combined C/N_0 ratio.

solution

$$\frac{N_0}{C} = 10^{-10} + 10^{-8.7} = 2.095 \times 10^{-9}$$

Therefore

$$\left[\frac{C}{N_0}\right] = -10 \log 2.095 \times 10^{-9} = 86.79 \text{ dBHz}$$

Example 10.18 illustrates the point that, when one of the link C/N_0 ratios is much less than the other, the combined C/N ratio is approximately equal to the lower (worst) one. The downlink C/N is usually (but not always) less than the uplink C/N, and in many cases it is much less. This is true primarily because of the limited EIRP available from the satellite.

Example 10.19 illustrates how back-off is taken into account in the link budget calculations and how it affects the C/N_0 ratio.

Example 10.19

A multiple carrier satellite circuit operates in the 6/4-GHz band with the following characteristics. *Uplink:* saturation flux density -67.5 dBW/m²; input back-off 11 dB; satellite G/T -11.6 dBK^{-1}. *Downlink:* satellite saturation EIRP 26.6 dBW; output back-off 6 dB; free-space loss 196.7 dB; earth station G/T 40.7 dBK^{-1}. For this example, the other losses may be ignored. Calculate the carrier-to-noise density ratios for both links and the combined value.

solution

As in the previous examples, the data are best presented in tabular form, and values are shown in decilogs. The minus signs in Eq. (10.50) and (10.55) are attached to the tabulated numbers:

	Decilog values
Uplink	
Saturation flux density	-67.5
$[A_0]$ at 6 GHz	-37
Input back-off	-11
Satellite saturation $[G/T]$	-11.6
$-[k]$	228.6
$[C/N_0]$ from Eq. (10.50)	101.5
Downlink	
Satellite [EIRP]	26.6
Output back-off	-6
Free-space loss	-196.7
Earth station $[G/T]$	40.7
$-[k]$	228.6
$[C/N_0]$ from Eq. (10.55)	93.2

Application of Eq. (10.61) provides the combined $[C/N_0]$:

$$\frac{N_0}{C} = 10^{-10.15} + 10^{-9.32} = 5.49 \times 10^{-10}$$

$$\left[\frac{C}{N_0}\right] = -10 \log 5.49 \times 10^{-10} = 92.6 \text{ dBHz}$$

Again, it is seen from Example 10.19 that the combined C/N_0 value is close to the lowest value, which is the downlink value.

So far, only thermal and antenna noise has been taken into account in calculating the combined value of C/N ratio. Another source of noise which must be considered is intermodulation noise, discussed in the next section.

10.10 Intermodulation Noise

Intermodulation occurs where multiple carriers pass through any device with nonlinear characteristics. In satellite communications systems, this most commonly occurs in the traveling wave tube high-power amplifier aboard the satellite, as described in Sec. 10.7.3. Both amplitude and phase nonlinearities give rise to intermodulation products.

As shown in Fig. 6.20, third-order intermodulation products fall on neighboring carrier frequencies, where they result in interference. Where a large number of modulated carriers are present, the intermodulation products are not distinguishable separately but instead appear as a type of noise which is termed *intermodulation noise*.

The carrier-to-intermodulation-noise ratio is usually found experimentally, or in some cases it may be determined by computer methods. Once this ratio is known, it can be combined with the carrier-to-thermal-noise ratio by the addition of the reciprocals in the manner described in Sec. 10.9. By denoting the intermodulation term by $(C/N)_{\text{IM}}$, and bearing in mind that the reciprocals of the C/N power ratios (and not the corresponding dB values) must be added, Eq. (10.61) is extended to

$$\frac{N_0}{C} = \left(\frac{N_0}{C}\right)_U + \left(\frac{N_0}{C}\right)_D + \left(\frac{N_0}{C}\right)_{\text{IM}} \tag{10.62}$$

Example 10.20

For a satellite circuit the carrier-to-noise ratios are: up link 23 dB, down link 20 dB, intermodulation 24 dB. Calculate the overall carrier-to-noise ratio in decibels.

solution

From Eq. (10.62),

Therefore

$$\frac{N}{C} = 10^{-2.4} + 10^{-2.3} + 10^{-2} = 0.0019$$

$$\left[\frac{C}{N}\right] = -10 \log 0.0019 = 17.2 \text{ dB}$$

In order to reduce intermodulation noise, the TWT must be operated in a back-off condition as described previously. Figure 10.9 shows how the $[C/N_0]_{\text{IM}}$ ratio improves as the input back-off is increased, for a typical TWT. At the same time, increasing the back-off decreases both $[C/N_0]_U$ and $[C/N_0]_D$, as shown by Eqs. (10.50) and (10.55). The result is that there is an optimum point where the overall carrier-to-noise ratio is a maximum. The component $[C/N]$ ratios as functions of the TWT input are sketched in Fig. 10.10. The TWT input in dB is $[\psi]_S - [\text{BO}]_i$ and therefore Eq. (10.50) plots as a straight line. Equation (10.55) reflects the curvature in the TWT characteristic through the output back-off, $[\text{BO}]_o$, which is not linearly related to the input back-off, as shown in Fig. 10.7. The intermodulation curve is not easily predictable and only the general trend is shown. The

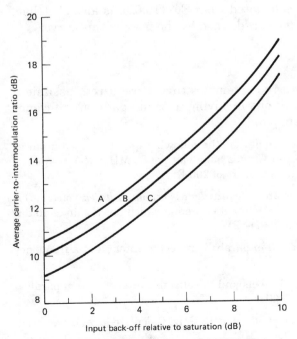

Figure 10.9 Intermodulation in a typical TWT. Curve A, 6 carriers; curve B, 12 carriers; curve C, 500 carriers. (*From CCIR, 1982b. With permission from the International Telecommunications Union.*)

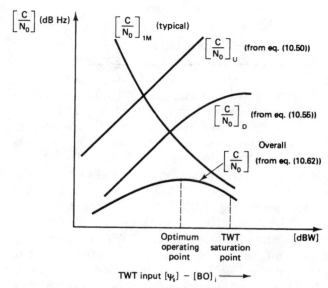

Figure 10.10   Carrier-to-noise density ratios and a function of input back-off.

overall $[C/N_0]$, which is calculated from Eq. (10.62), is also sketched. The optimum operating point is defined by the peak of this curve.

10.11   Problems

Note: In problems where room temperature is required, assume a value of 290 K. In calculations involving antenna gain, an efficiency factor of 0.55 may be assumed.

10.1.   Give the decibel equivalents for the following quantities: (*a*) a power ratio of 30:1; (*b*) a power of 230 W; (*c*) a bandwidth of 36 MHz; (*d*) a frequency ratio of 2 MHz/3 kHz; (*e*) a temperature of 200 K.

10.2.   (*a*) Explain what is meant by *equivalent isotropic radiated power.* (*b*) A transmitter feeds a power of 10 W into an antenna which has a gain of 46 dB. Calculate the EIRP in (*i*) watts; (*ii*) dBW.

10.3.   Calculate the gain of a 3-m parabolic reflector antenna at a frequency of (*a*) 6 GHz; (*b*) 14 GHz.

10.4.   Calculate the gain in decibels and the effective area of a 30-m parabolic antenna at a frequency of 4 GHz.

10.5.   An antenna has a gain of 46 dB at 12 GHz. Calculate its effective area.

10.6.   Calculate the effective area of a 10-ft parabolic reflector antenna at a frequency of (*a*) 4 GHz; (*b*) 12 GHz.

10.7. The EIRP from a satellite is 49.4 dBW. Calculate (a) the power density at a ground station for which the range is 40,000 km and (b) the power delivered to a matched load at the ground station receiver if the antenna gain is 50 dB. The downlink frequency is 4 GHz.

10.8. Calculate the free-space loss as a power ratio and in decibels for transmission at frequencies of (a) 4 GHz, (b) 6 GHz, (c) 12 GHz, and (d) 14 GHz, the range being 42,000 km.

10.9. Repeat the calculation in Prob. 10.7b allowing for a fading margin of 1.0 dB and receiver feeder losses of 0.5 dB.

10.10. Explain what is meant by (a) *antenna noise temperature,* (b) *amplifier noise temperature,* and (c) *system noise temperature* referred to input. A system operates with an antenna noise temperature of 40 K and an input amplifier noise temperature of 120 K. Calculate the available noise power density of the system referred to the amplifier input.

10.11. Two amplifiers are connected in cascade, each having a gain of 10 dB and a noise temperature of 200 K. Calculate (a) the overall gain and (b) the effective noise temperature referred to input.

10.12 Explain what is meant by *noise factor.* For what source temperature is noise factor defined?

10.13. The noise factor of an amplifier is 7:1. Calculate (a) the noise figure and (b) the equivalent noise temperature.

10.14. An attenuator has an attenuation of 6 dB. Calculate (a) its noise figure and (b) its equivalent noise temperature referred to input.

10.15. An amplifier having a noise temperature of 200 K has a 4-dB attenuator connected at its input. Calculate the effective noise temperature referred to the attenuator input.

10.16. A receiving system consists of an antenna having a noise temperature of 60 K, feeding directly into a low-noise amplifier (LNA). The amplifier has a noise temperature of 120 K and a gain of 45 dB. The coaxial feeder between the LNA and the main receiver has a loss of 2 dB, and the main receiver has a noise figure of 9 dB. Calculate the system noise temperature referred to input.

10.17. Explain why the low-noise amplifier of a receiving system is placed at the antenna end of the feeder cable.

10.18. An antenna having a noise temperature of 35 K is connected through a feeder having 0.5-dB loss to an LNA. The LNA has a noise temperature of 90 K. Calculate the system noise temperature referred to (a) the feeder input and (b) the LNA input.

10.19. Explain what is meant by *carrier-to-noise ratio.* At the input to a receiver the received carrier power is 400 pW and the system noise temperature 450 K. Calculate the carrier-to-noise density ratio in dBHz. Given that the bandwidth is 36 MHz, calculate the carrier-to-noise ratio in decibels.

10.20. Explain what is meant by the G/T ratio of a satellite receiving system. A satellite receiving system employs a 5-m parabolic antenna operating at 12 GHz. The antenna noise temperature is 100 K, and the receiver front-end noise temperature is 120 K. Calculate $[G/T]$.

10.21. In a satellite link the propagation loss is 200 dB. Margins and other losses account for another 3 dB. The receiver $[G/T]$ is 11 dB, and the [EIRP] is 45 dBW. Calculate the received $[C/N]$ for a system bandwidth of 36 MHz.

10.22. A carrier-to-noise density ratio of 90 dBHz is required at a receiver having a $[G/T]$ ratio of 12 dB. Given that total losses in the link amount to 196 dB, calculate the [EIRP] required.

10.23. Explain what is meant by *saturation flux density*. The power received by a 1.8-m parabolic antenna at 14 GHz is 250 pW. Calculate the power flux density (a) in W/m^2 and (b) in dBW/m^2 at the antenna.

10.24. An earth station radiates an [EIRP] of 54 dBW at a frequency of 6 GHz. Assuming that total losses amount to 200 dB, calculate the power flux density at the satellite receiver.

10.25. A satellite transponder requires a saturation flux density of -110 dBW/m^2, operating at a frequency of 14 GHz. Calculate the earth station [EIRP] required if total losses amount to 200 dB.

10.26. Explain what is meant by input back-off. An earth station is required to operate at an [EIRP] of 44 dBW in order to produce saturation of the satellite transponder. If the transponder has to be operated in a 10 dB back-off mode, calculate the new value of [EIRP] required.

10.27. Determine the carrier-to-noise density ratio at the satellite input for an uplink which has the following parameters: operating frequency 6 GHz; saturation flux density -95 dBW/m^2; input back-off 11 dB; satellite $[G/T]$ -7 dBK^{-1}; [RFL] are 0.5dB. Tabulate the link budget values as shown in the text.

10.28. For an uplink the required $[C/N]$ ratio is 20 dB. The operating frequency is 30 GHz, and the bandwidth is 72 MHz. The satellite $[G/T]$ is 14.5 dBK^{-1}. Assuming operation with 11 dB input back-off, calculate the saturation flux density. [RFL] are 1dB.

10.29. For the uplink in Prob. 10.28, the total losses amount to 218 dB. Calculate the earth station [EIRP] required.

10.30. An earth station radiates an [EIRP] of 54 dBW at 14 GHz from a 10-m parabolic antenna. The transmit feeder losses between the high-power amplifier (HPA) and the antenna are 2.5 dB. Calculate the output of the HPA.

10.31. The following parameters apply to a satellite downlink: saturation [EIRP] 22.5 dBW; free-space loss 195 dB; other losses and margins 1.5 dB; earth station $[G/T]$ 37.5 dB/K. Calculate the $[C/N_0]$ at the earth station. Assuming an output back-off of 6 dB is applied, what is the new value of $[C/N_0]$?

10.32. The output from a satellite traveling wave tube amplifier is 10 W. This is fed to a 1.2-m parabolic antenna operating at 12 GHz, the feeder loss being 2 dB. Calculate the [EIRP].

10.33. The $[C/N]$ values for a satellite circuit are: uplink 25 dB, downlink 15 dB. Calculate the overall $[C/N]$ value.

10.34. The required $[C/N]$ value at the ground station receiver is 22 dB and the downlink $[C/N]$ is 24 dB. What is the minimum value of $[C/N]$ that the uplink can have in order that the overall value can be achieved?

10.35. A satellite circuit has the following parameters:

	Uplink, decilogs	Downlink, decilogs
[EIRP]	54	34
[G/T]	0	17
[FSL]	200	198
[RFL]	2	2
[AA]	0.5	0.5
[AML]	0.5	0.5

Calculate the overall $[C/N_0]$ value.

10.36. Explain how intermodulation noise originates in a satellite link, and describe how it may be reduced. In a satellite circuit the carrier-to-noise ratios are: uplink 25 dB; downlink 20 dB; intermodulation 13 dB. Calculate the overall carrier-to-noise ratio.

10.37. For the satellite circuit of Prob. 10.36, input back-off is introduced that reduces the carrier-to-noise ratios by the following amounts: uplink 7 dB; downlink 2 dB; and improves the intermodulation by 11 dB. Calculate the new value of overall carrier-to-noise ratio.

Interference

11.1 Introduction

With many telecommunications services using radio transmissions, interference between services can arise in a number of ways. Figure 11.1 shows in a rather general way the possible interference paths between services. It will be seen in Fig. 11.1 that the terms *earth station* and *terrestrial station* are used, and the distinction must be carefully noted. Earth stations are specifically associated with satellite circuits, and terrestrial stations are specifically associated with ground-based microwave line-of-sight circuits. The possible modes of interference shown in Fig. 11.1 are classified by the International Telecommunications Union (ITU, 1985), as follows:

A_1: terrestrial-station transmissions, possibly causing interference to reception by an earth station

A_2: earth-station transmissions, possibly causing interference to reception by a terrestrial station

B_1: space-station transmission of one space system, possibly causing interference to reception by an earth station of another space system

B_2: earth-station transmissions of one space system, possibly causing interference to reception by a space station of another space system

C_1: space-station transmission, possibly causing interference to reception by a terrestrial station

C_2: terrestrial-station transmission, possibly causing interference to reception by a space station

E: space-station transmission of one space system, possibly causing interference to reception by a space station of another space system

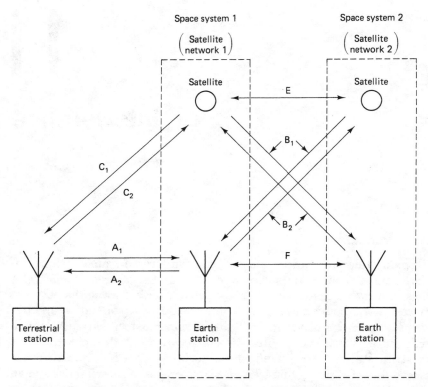

Figure 11.1 Possible interference modes between satellite circuits and a terrestrial station. (*From CCIR Radio Regulations.*)

F: earth-station transmission of one space system, possibly causing interference to reception by an earth station of another space system

A_1, A_2, C_1, and C_2 are possible modes of interference between space and terrestrial services. B_1 and B_2 are possible modes of interference between stations of different space systems using separate uplink and downlink frequency bands, and E and F are extensions to B_1 and B_2 where bidirectional frequency bands are used.

The Radio Regulations (ITU, 1986) specify maximum limits on radiated powers (more strictly, on the distribution of energy spectral density) in an attempt to reduce the potential interference to acceptable levels in most situations. However, interference may still occur in certain cases, and what is termed *coordination* between the telecommunications administrations that are affected is then required. Coordination may require both administrations to change or adjust some of the technical parameters of their systems.

For geostationary satellites, interference modes B_1 and B_2 set a lower limit to the orbital spacing between satellites. To increase the capacity of the geostationary orbit, the Federal Communications Commission (FCC) in the United States has in recent years authorized a reduction in orbital spacing from 4° to 2° in the 6/4-GHz band. Some of the effects that this has on the B_1 and B_2 levels of interference are examined later in the chapter. It may be noted, however, that although the larger, authorized operators will in general be able to meet the costs of technical improvements needed to offset the increased interference resulting from reduced orbital spacing, the same cannot be said for individually owned television receive-only (TVRO) installations (the "home satellite dish"), and these users will have no recourse to regulatory control (Chouinard, 1984).

Interference with individually owned TVRO receivers may also occur from terrestrial station transmissions in the 6/4-GHz band. Although this may be thought of as an A_1 mode of interference, the fact that these home stations are considered by many broadcasting companies to be "pirates" means that regulatory controls to reduce interference are not applicable. Some steps that can be taken to reduce this form of interference are described in a publication by the Microwave Filter Company, 1984.

It has been mentioned that the Radio Regulations places limits on the energy spectral density which may be emitted by an earth station. Energy dispersal is one technique employed to redistribute the transmitted energy more evenly over the transmitted bandwidth. This principle is described in more detail later in this chapter.

Intermodulation interference, briefly mentioned in Sec. 6.7.3, is a type of interference which can occur between two or more carriers using a common transponder in a satellite or a common high-power amplifier in an earth station. For all practical purposes, this type of interference can be treated as noise, as described in Sec. 10.10.

11.2   Interference between Satellite Circuits (B_1 and B_2 Modes)

A satellite circuit may suffer the B_1 and B_2 modes of interference shown in Fig. 11.1 from a number of neighboring satellite circuits, the resultant effect being termed *aggregrate interference*. Because of the difficulties of taking into account the range of variations expected in any practical aggregate, studies of aggregrate interference have been quite limited, with most of the study effort going into what is termed *single-entry interference studies* (see Sharp, 1984a). As the name suggests, single-entry interference refers to the interference produced by a single interfering circuit upon a neighboring circuit.

Interference may be considered as a form of noise, and as with noise, system performance is determined by the ratio of wanted to interfering powers, in this case the wanted carrier to the interfering carrier power or *C/I* ratio. The single most important factor controlling interference is the radiation pattern of the earth station antenna. Comparatively large-diameter reflectors can be used with earth station antennas, and hence narrow beamwidths can be achieved. For example, a 10-m antenna at 14 GHz has a −3-dB beamwidth of about 0.15°. This is very much narrower than the 2° to 4° orbital spacing allocated to satellites. To relate the *C/I* ratio to the antenna radiation pattern, it is necessary first to define the geometry involved.

Figure 11.2 shows the angles subtended by two satellites in geostationary orbit. The orbital separation is defined as the angle α subtended at the center of the earth, known as the geocentric angle. However, from an earth station at point *P* the satellites would appear to subtend an angle ß. Angle ß is referred to as the topocentric angle. In all practical situations relating to satellite interference, the topocentric and geocentric angles may be assumed equal, and in fact making this assumption leads to an overestimate of the interference (Sharp, 1983).

Consider now S_1 as the wanted satellite and S_2 as the interfering satellite. An antenna at *P* will have its main beam directed at S_1 and an off-axis component at angle θ directed at S_2. Angle θ is the same as the topocentric angle, which as already shown may be assumed equal to the geocentric or orbital spacing angle. Therefore, when calculating the antenna sidelobe pattern the orbital spacing angle may be used, as described in Sec. 11.2.4. Orbital spacing angles range from 2° to 4° in 0.5° intervals in the C band.

In Fig. 11.3 the satellite circuit being interfered with is that from earth station *A* via satellite S_1 to receiving station *B*. The B_1 mode of interference can occur from satellite S_2 into earth station *B*, and the

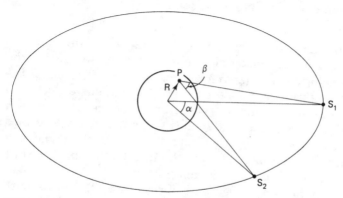

Figure 11.2   Geocentric angle α and the topocentric angle ß.

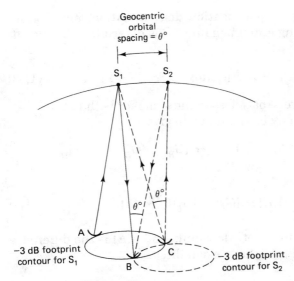

Figure 11.3 Orbital spacing angle θ.

B_2 mode of interference can occur from earth station C into satellite S_1. The total single-entry interference is the combined effect of these two modes. Because the satellites cannot carry very large antenna reflectors, the beamwidth is relatively wide, even for the so-called *spot beams*. For example, a 3.5-m antenna at 12 GHz has a beamwidth of about 0.5°, and the equatorial arc subtended by this angle is about 314 km. In interference calculations, therefore, the earth stations will be assumed to be situated on the −3-dB contours of the satellite footprints, in which case the satellite antennas do not provide any gain discrimination between the wanted and the interfering carriers on either transmit or receive.

11.2.1 Downlink

Equation (10.13) may be used to calculate the wanted and interfering downlink carrier powers received by an earth station. The carrier power [C] in dBW received at station B is

$$[C] = [\text{EIRP}]_1 - 3 + [G_B] - [\text{FSL}] \tag{11.1}$$

Here $[\text{EIRP}]_1$ is the equivalent isotropic radiated power in dBW from satellite 1, the −3 dB accounts for the −3-dB contour of the satellite transmit antenna, G_B is the boresight (on-axis) receiving antenna gain at B, and [FSL] is the free-space loss in decibels. A similar equation may be used for the interfering carrier [I], except an additional term $[Y]_D$ dB, allowing for polarization discrimination, must be included. Also, the receiving antenna gain at B is determined by the off-axis angle θ, giving

$$[I] = [\text{EIRP}]_2 - 3 + [G_B(\theta)] - [\text{FSL}] - [Y]_D \qquad (11.2)$$

It is assumed that the free-space loss is the same for both paths.

These two equations may be combined to give

$$[C] - [I] = [\text{EIRP}]_1 - [\text{EIRP}]_2 + [G_B] - [G_B(\theta)] + [Y]_D$$

or

$$\left[\frac{C}{I}\right]_D = \Delta[E] + [G_B] - [G_B(\theta)] + [Y]_D \qquad (11.3)$$

The subscript D is used to denote downlink, and $\Delta[E]$ is the difference in dB between the [EIRP]s of the two satellites.

Example 11.1

The desired carrier [EIRP] from a satellite is 34 dBW, and the ground station receiving antenna gain is 44 dB in the desired direction and 24.47 dB toward the interfering satellite. The interfering satellite also radiates an [EIRP] of 34 dBW. The polarization discrimination is 4 dB. Determine the carrier-to-interference ratio at the ground receiving antenna.

solution From Eq. (11.3),

$$\left[\frac{C}{I}\right]_D = (34-34) + 44 - 24.47 + 4$$

$$= 23.53 \text{ dB}$$

11.2.2 Uplink

A result similar to Eq. (11.3) can be derived for the uplink. In this situation, however, it is desirable to work with the radiated powers and the antenna transmit gains rather than the EIRPs of the two earth stations. Equation (10.3) may be used to substitute power and gain for EIRP. Also, for the uplink, G_B and $G_B(\theta)$ are replaced by the satellite receive antenna gains, both of which are assumed to be given by the -3-dB contour. Denoting by $\Delta[P]$ the difference in dB between wanted and interfering transmit powers, $[G_A]$ the boresight transmit antenna gain at A, $[G_C(\theta)]$ the off-axis transmit gain at C, it is left as an exercise for the reader to show that Eq. (11.3) is modified to

$$\left[\frac{C}{I}\right]_U = \Delta[P] + [G_A] - [G_C(\theta)] + [Y]_U \qquad (11.4)$$

Example 11.2

Station A transmits at 24 dBW with an antenna gain of 54 dB, and station C transmits at 30 dBW. The off-axis gain in the S_1 direction is 24.47 dB, and the polarization discrimination is 4 dB. Calculate the $[C/I]$ ratio on the uplink.

solution   Equation (11.4) gives

$$\left[\frac{C}{I}\right]_U = (24 - 30) + 54 - 24.47 + 4$$

$$= 27.53 \text{ dB}$$

11.2.3   Combined [C/I] due to Interference on Both Uplink and Downlink

Interference may be considered as a form of noise, and assuming that the interference sources are statistically independent, the interference powers may be added directly to give the total interference at receiver B. The uplink and the downlink ratios are combined in exactly the same manner described in Sec. 10.9, for noise resulting in

$$\left(\frac{I}{C}\right)_{\text{ant}} = \left(\frac{I}{C}\right)_U + \left(\frac{I}{C}\right)_D \qquad (11.5)$$

Here power ratios must be used, not decibels, and the subscript $_{\text{ant}}$ denotes the combined ratio at the output of station B receiving antenna.

Example 11.3

Using the uplink and downlink values of $[C/I]$ determined in Examples 11.1 and 11.2, find the overall ratio $[C/I]_{\text{ant}}$.

solution   For the uplink a $[C/I] = 27.53$ dB gives $(I/C)_U = 0.001766$, and for the downlink, $[C/I] = 23.53$ dB gives $(I/C)_D = 0.004436$. Combining these according to Eq. (11.5) gives

$$\left(\frac{I}{C}\right)_{\text{ant}} = 0.001766 + 0.004436$$

$$= 0.006202$$

Hence

$$\left[\frac{C}{I}\right]_{\text{ant}} = -10 \log 0.006202$$

$$= 20.07 \text{ dB}$$

11.2.4   Antenna Gain Function

The antenna radiation pattern can be divided into three regions: the mainlobe region, the sidelobe region, and the transition region be-

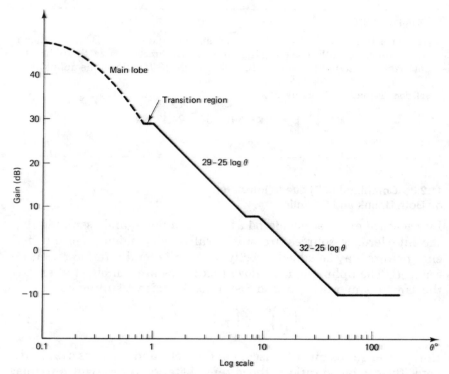

Figure 11.4 Earth-station antenna gain pattern used in FCC/OST R83-2, revised Nov. 30, 1984. (*From Sharp, 1984b.*)

tween the two. For interference calculations the fine detail of the antenna pattern is not required and an envelope curve is used instead.

Figure 11.4 shows a sketch of the envelope pattern used by the FCC. The width of the mainlobe and transition region depend upon the ratio of the antenna diameter to the operating wavelength, and Fig. 11.4 is intended to show only the general shape. The sidelope gain function in decibels is defined for different ranges of θ. Specifying θ in degrees, the sidelobe gain function can be written as follows:

$$[G(\theta)] = \begin{cases} 29 - 25 \log \theta & 1 \le \theta \le 7 \\ +8 & 7 < \theta \le 9.2 \\ 32 - 25 \log \theta & 9.2 < \theta \le 48 \\ -10 & 48 < \theta \le 180 \end{cases} \tag{11.6}$$

For the range of satellite orbital spacings presently in use, it is this sidelobe gain function that determines the interference levels.

Example 11.4

Determine the degradation in the downlink $[C/I]$ ratio when satellite orbital spacing is reduced from 4° to 2°, all other factors remaining unchanged. FCC antenna characteristics may be assumed.

solution The decibel increase in interference will be

$$(29 - 25 \log 2) - (29 - 25 \log 4) = 25 \log 2 = 7.5 \text{ dB}$$

The $[C/I]_D$ will be degraded directly by this amount. Alternatively, from Fig. 11.4,

$$[G(2°)] - [G(4°)] = 21.4 - 13.9 = 7.5 \text{ dB}$$

It should be noted that no simple relationship can be given for calculating the effect of reduced orbital spacing on the overall $[C/I]$. The separate uplink and downlink values must be calculated and combined as described in Sec. 11.2.3. Other telecommunications authorities specify antenna characteristics that differ from the FCC specifications (see, e.g., CCIR Rep. 391–3, 1978).

11.2.5 Passband Interference

In the preceding section, the carrier-to-interference ratio at the receiver input is determined. However, the amount of interference reaching the detector will depend on the amount of frequency overlap between the interfering spectrum and the wanted channel passband.

Two situations can arise as shown in Fig. 11.5. In Fig. 11.5a partial overlap of the interfering signal spectra with the wanted passband is shown. The fractional interference is given as the ratio of the shaded area to the total area under the interference spectrum curve. This is

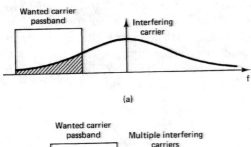

(a)

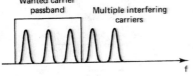

(b)

Figure 11.5 Power spectral density curves for (a) wideband interfering signal and (b) multiple interfering carriers.

denoted by Q (Sharp, 1983), or in decibels as $[Q]$. Where partial over-lap occurs, Q is less than unity, or $[Q] < 0$ dB. Where the interfering spectrum coincides with the wanted passband, $[Q] = 0$ dB. Evaluation of Q usually has to be carried out by computer.

The second situation, illustrated in Fig. 11.5b, is where multiple in-terfering carriers are present within the wanted passband, such as with single carrier per channel (SCPC) operation discussed in Sec. 12.5. Here, Q represents the sum of the interfering carrier powers within the passband, and $[Q] > 0$ dB.

In the FCC Report FCC/OST R83–2 (Sharp, 1983), Q values are computed for a wide range of interfering and wanted carrier combina-tions. Typical $[Q]$ values obtained from the FCC Report are: with the wanted carrier a TV/FM signal, and the interfering carrier a similar TV/FM signal, $[Q] = 0$ dB; with SCPC/PSK interfering carriers, $[Q] = 27.92$ dB; and with the interfering carrier a wideband digital-type sig-nal, $[Q] = -3.36$ dB.

The passband $[C/I]$ ratio is calculated using

$$\left[\frac{C}{I}\right]_{pb} = \left[\frac{C}{I}\right]_{ant} - [Q] \tag{11.7}$$

The positions of these ratios in the receiver chain are illustrated in Fig. 11.6, where it will be seen that both $[C/I]_{ant}$ and $[C/I]_{pb}$ are prede-tection ratios, measured at RF or IF. Interference can also be mea-sured in terms of the postdetector output, shown as $[S/I]$ in Fig. 11.6, and this is discussed in the next section.

11.2.6   Receiver Transfer Characteristic

In some situations a measure of the interference in the postdetection baseband, rather than in the IF or RF passband, is required. Baseband interference is measured in terms of baseband signal-to-in-terference ratio $[S/I]$. To relate $[S/I]$ to $[C/I]_{ant}$ a *receiver transfer char-acteristic* is introduced which takes into account the modulation characteristics of the wanted and interfering signals, and the carrier frequency separation. Denoting the receiver transfer characteristic in decibels by [RTC], the relationship can be written as

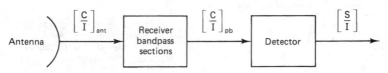

Figure 11.6   Carrier-to-interference ratios and signal-to-interference ratio.

$$\left[\frac{S}{I}\right] = \left[\frac{C}{I}\right]_{\text{ant}} + [\text{RTC}] \qquad (11.8)$$

It will be seen that [RTC] is analogous to the receiver processing gain $[K_R]$ introduced in Sec. 8.6.3. Note that it is the $[C/I]$ at the antenna which is used, not the passband value, the [RTC] taking into account any frequency offset. The [RTC] will always be a positive number of decibels so that the baseband signal-to-interference ratio will be greater than the carrier-to-interference ratio at the antenna.

Calculation of [RTC] for various combinations of wanted and interfering carriers is very complicated and has to be done by computer. As an example, taken from Sharp, 1983, when the wanted carrier is TV/FM with a modulation index of 2.571, and the interfering carrier is TV/FM with a modulating index of 2.560, the carrier frequency separation being zero, the [RTC] is computed to be 31.70 dB. These computations are limited to low levels of interference (see Sec. 11.2.8).

11.2.7 Specified Interference Objectives

Although $[C/I]_{\text{pb}}$ or $[S/I]$ gives a measure of interference, ultimately the effects of interference must be assessed in terms of what is tolerable to the end user. Such assessment usually relies on some form of subjective measurement. For TV, viewing tests are conducted in which a mixed audience of experienced and inexperienced viewers (experienced from the point of view of assessing the effects of interference) assess the effects of interference on picture quality. By gradually increasing the interference level, a quality impairment factor can be established which ranges from 1 to 5. The five grades are defined as (Chouinard, 1984)

5. Imperceptible

4. Perceptible, but not annoying

3. Slightly annoying

2. Annoying

1. Very annoying

Acceptable picture quality requires a quality impairment factor of at least 4.2. Typical values of interference levels which result in acceptable picture quality are: for broadcast TV, $[S/I] = 67$ dB; and for cable TV, $[C/I]_{\text{pb}} = 20$ dB.

For digital circuits the $[C/I]_{\text{pb}}$ is related to the bit error rate (BER) (see, e.g., CCIR Rec. 523, 1978). Values of the required $[C/I]_{\text{pb}}$ used by Sharp, 1983, for different types of digital circuits range from 20 to 32.2 dB.

TABLE 11.1 Summary of Single-Entry Interference Objectives Used in FCC/OST
R83–2, May 1983

[*S/I*] Objectives:

 FDM/FM: 600 pW0p or [*S/I*] = 62.2 dB (62.2 dB). Reference: CCIR Rec. 466–3.

 TV/FM (broadcast quality): [*S/I*] = 67 dB weighted (67 dB; 65.5 dB). References:
 CCIR Rec. 483–1, 354–2, 567, 568.

 CSSB/AM: (54.4 dB; 62.2 dB). No references quoted.

[*C/I*]$_{pb}$ Objectives:

 TV/FM (CATV): [*C/I*]$_{pb}$ = 20 dB (22 dB; 27 dB). Reference: ABC 62 FCC 2d 901
 (1976).

 Digital: [*C/I*]$_{pb}$ = [*C/N*] (at BER = 10^{-6}) + 14 dB (20 to 32.2 dB). Reference: CCIR
 Rec. 523.

 SCPC/PSK: (21.5 dB; 24 dB). No references quoted.

 SCPC/FM: (21.2 dB; 23.2 dB). No references quoted.

 SS/PSK: (11 dB; 0.6 dB). No references quoted.

To give some idea of the numerical values involved, a summary of
the objectives stated in the FCC single-entry interference program is
presented in Table 11.1. In some cases the objective used differed from
the reference objective, and the values used are shown in parentheses.
In some entries in the table, noise is shown measured in units of
"pW0p." Here the "pW" stands for picowatts. The "0" means that the
noise is measured at a "zero-level test point," which is a point in the
circuit where a test-tone signal produces a level of 0 dBm. The final
"p" stands for psophometrically weighted noise, discussed in Sec. 8.6.6.

11.2.8 Protection Ratio

In CCIR Report 634–2, 1982, the International Radio Consultative
Committee (CCIR) specifies the permissible interference level for TV
carriers in terms of a parameter known as the *protection ratio*. The
protection ratio is defined as the minimum carrier-to-interference
ratio at the input to the receiver which results in "just perceptible"
degradation of picture quality. The protection ratio applies only for
wanted and interfering TV carriers at the same frequency, and it is
equivalent to [*C/I*]$_{pb}$ evaluated for this situation. Denoting the quality
impairment factor by Q_{IF} and the protection ratio in decibels by [PR$_0$],
the equation given in CCIR, 1982, is

$$[\mathrm{PR}_0] = 12.5 - 20 \log\left(\frac{\mathrm{Dv}}{12}\right) - Q_{IF} + 1.1 Q_{IF}^{2} \qquad (11.9)$$

Here Dv is the peak-to-peak deviation in megahertz.

Example 11.5

An FM/TV carrier is specified as having a modulation index of 2.571 and a top modulating frequency of 4.2 MHz. Calculate the protection ratio required to give a quality impairment factor of (a) 4.2 and (b) 4.5.

solution The peak-to-peak deviation is $2 \times 2.571 \times 4.2 = 21.6$ MHz. Applying Eq. (11.9) gives the following results.

(a)

$$[PR_0] = 12.5 - 20 \log \frac{21.6}{12} - 4.2 + (1.1)(4.2)^2$$

$$= 22.6 \text{ dB}$$

(b)

$$[PR_0] = 12.5 - 20 \log \frac{21.6}{12} - 4.5 + 1.1 \times 4.5^2$$

$$= 25.2 \text{ dB}$$

It should be noted that the receiver transfer characteristic discussed in Sec. 11.2.6 was developed from the CCIR protection ratio concept (see Jeruchim and Kane, 1970).

11.3   Energy Dispersal

The power in a frequency-modulated signal remains constant, independent of the modulation index. When unmodulated, all the power is at the carrier frequency, and when modulated, the same total power is distributed among the carrier and the sidebands. At low modulation indices the sidebands are grouped close to the carrier, and the power spectral density, or wattage per unit bandwidth, is relatively high in that spectral region. At high modulation indices, the spectrum becomes widely spread, and the power spectral density relatively low.

Use is made of this property in certain situations to keep radiation within CCIR recommended limits. For example, to limit the A_2 mode of interference in the 1- to 15-GHz range for the fixed satellite service, CCIR Radio Regulations state in part that the earth-station EIRP should not exceed 40 dBW in any 4-kHz band for $\theta \leq 0°$, and should not exceed $(40 + 3\theta)$ dBW in any 4-kHz band for $0° < \theta \leq 5°$. The angle θ is the angle of elevation of the horizon viewed from the center of radiation of the earth station antenna. It is positive for angles above the horizontal plane, as illustrated in Fig. 11.7a, and negative for angles below the horizontal plane, as illustrated in Fig. 11.7b.

For space stations transmitting in the frequency range 3400 to 7750 MHz, the limits are specified in terms of power flux density for

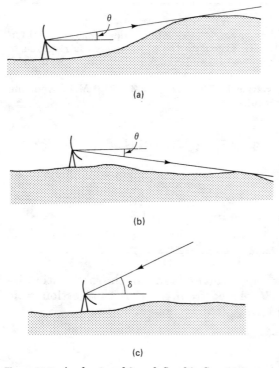

(a)

(b)

(c)

Figure 11.7 Angles θ and δ as defined in Sec. 11.3.

any 4-kHz bandwidth. Denoting the angle of arrival as δ degrees, measured above the horizontal plane as shown in Fig. 11.7c, these limits are

-152 dBW/m^2 in any 4-kHz band for $0° \leq \delta \leq 5°$

$-152 + 0.5\,(\delta - 5)$ dBW/m^2 in any 4-kHz band for $5° < \delta \leq 25°$

-142 dBW/m^2 in any 4-kHz band for $25° < \delta \leq 90°$

Because the specification is in terms of power or flux density in any 4-kHz band, not the total power or the total flux density, a carrier may be within the limits when heavily frequency-modulated, but the same carrier with light frequency modulation may exceed the limits. An energy-dispersal waveform is a low-frequency modulating wave which is inserted below the lowest baseband frequency for the purpose of dispersing the spectral energy when the current value of the modulating index is low. In the Intelsat system for FDM carriers, a symmetrical triangular wave is used, a different fundamental frequency for this triangular wave in the range 20 to 150 Hz being assigned to each FDM carrier. The rms level of the baseband is monitored, and the amplitude of the dispersal waveform automatically adjusted to keep the overall frequency deviation within defined limits. At the receive end the dispersal waveform is removed from the demodulated signal by low-pass filtering.

With television signals the situation is more complicated. The dispersal waveform, usually a sawtooth waveform, must be synchronized with the field frequency of the video signal to prevent video interference, so that for the 525/60 standard a 30-Hz wave is used, and for the 625/50 standard a 25-Hz wave is used. If the TV signal occupies the full bandwidth of the transponder, known as *full-transponder television,* the dispersal level is kept constant at a peak-to-peak deviation of 1 MHz irrespective of the video level. In what is termed *half-transponder television,* where the TV carrier occupies only one-half of the available transponder bandwidth, the dispersal deviation is maintained at 1 MHz peak to peak when video modulation is present, and is automatically increased to 2 MHz when video modulation is absent. At the receiver, video clamping is the most commonly used method of removing the dispersal waveform.

Energy dispersal is effective in reducing all modes of interference but particularly that occurring between earth and terrestrial stations (A_2 mode) and between space and terrestrial stations (C_1). It is also effective in reducing intermodulation noise.

11.4   Coordination

When a new satellite network is in the planning stage certain calculations have to be made to ensure that the interference levels will remain within acceptable limits. These calculations include determining the interference that will be caused by the new system, and interference it will receive from other satellite networks.

In Sec. 11.2, procedures are outlined showing how interference may be calculated by taking into account modulation parameters and carrier frequencies of wanted and interfering systems. These calculations are very complex, and the CCIR uses a simplified method to determine whether *coordination* is necessary. As mentioned previously, where the potential for interference exists, the telecommunication administrations are required to coordinate the steps to be taken to reduce interference, a process referred to as *coordination.*

To determine whether or not coordination is necessary, the interference level is calculated assuming maximum spectrum density levels of the interfering signals, and converted to an equivalent increase in noise temperature. The method is specified in detail in CCIR Report 454–3, 1982, for a number of possible situations. To illustrate the method, one specific situation will be explained here: that where the existing and proposed systems operate on the same uplink and downlink frequencies.

Figure 11.8*a* shows the two networks, R and R'. The method will be described for network R' interfering with the operation of R. Satellite S' can interfere with the earth station E, this being a B_1 mode of interference; and earth station E' can interfere with the satellite S, this being a B_2 mode. Note that the networks need not be physically adjacent to one another.

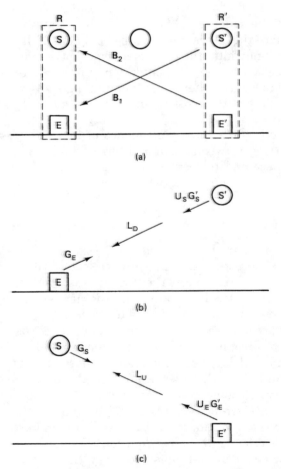

(a)

(b)

(c)

Figure 11.8 (*a*) Interference modes B_1 and B_2 from network R' into network R. (*b*) For the B_1 mode the interfering power in dBW/Hz is $[I_1] = [U_s] + [G'_s] + [G_E] - [L_D]$. (*c*) For the B_2 mode the interfering power in dBW/Hz is $[I_2] = [U_E] + [G'_E] + [G_s] - [L_u]$.

11.4.1 Interference Levels

Consider first the interference B_1. This is illustrated in Fig. 11.8b. Let U_S represent the maximum power density transmitted from satellite S'. The units for U_S are W/Hz or joules (J), and this quantity is explained in more detail shortly. Let the transmit gain of satellite S' in the direction of earth station E be G'_S, and let G_E be the receiving gain of earth station E in the direction of satellite S'. The interfering spectral power density received by the earth station is therefore

$$[I_1] = [U_S] + [G'_S] + [G_E] - [L_D] \tag{11.10}$$

where L_D is the propagation loss for the downlink. The gain and loss factors are power ratios, and the square brackets denote the corresponding decibel values as before. The increase in equivalent noise temperature at the earth-station receiver input can then be defined using Eq. (10.15) as

$$[\Delta T_E] = [I_1] - [k]$$

$$= [I_1] + 228.6 \tag{11.11}$$

Here k is Boltzmann's constant and $[k] = -228.6$ dBJ/K.

A similar argument can be applied to the uplink interference B_2 as illustrated in Fig. 11.8c, giving

$$[I_2] = [U_E] + [G'_E] + [G_S] - [L_U] \tag{11.12}$$

The corresponding increase in the equivalent noise temperature at the satellite receiver input is then

$$[\Delta T_S] = [I_2] + 228.6 \tag{11.13}$$

Here U_E is the maximum power spectral density transmitted by earth station E', G'_E the transmit gain of E' in the direction of S, G_S the receive gain of S in the direction of E', and L_U is the uplink propagation loss.

11.4.2 Transmission Gain

The effect of the equivalent temperature rise ΔT_S must be transferred to the earth station E, and this is done using the transmission gain for system R, which is calculated for the situation shown in Fig. 11.9. Figure 11.9a shows the satellite circuit in block schematic form. U_E represents the maximum power spectral density transmitted by earth station E and G_{TE} the transmit gain of E in direction S. G_{RS} represents the receive gain of S in direction E. The received power spectral density at satellite S is therefore

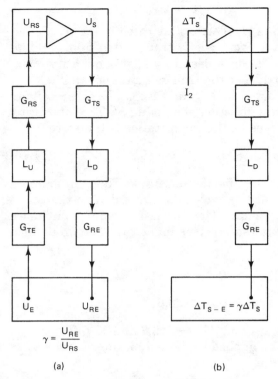

Figure 11.9 (a) Defining the transmission gain γ in Sec. 11.4.2. (b) Use of transmission gain to refer satellite noise temperature to an earth-station.

$$[U_{RS}] = [U_E] + [G_{TE}] + [G_{RS}] - [L_U] \tag{11.14}$$

In a similar way, with satellite S transmitting and earth station E receiving, the received power spectral density at earth station E is

$$[U_{RE}] = [U_S] + [G_{TS}] + [G_{RE}] - [L_D] \tag{11.15}$$

where U_S is the maximum power spectral density transmitted by S, G_{TS} is the transmit gain of S in the direction of E, and G_{RE} is the receive gain of E in the direction of S. It is assumed that the uplink and downlink propagation losses L_U and L_D, are the same as those used for the interference signals.

The transmission gain for network R is then defined as

$$[\gamma] = [U_{RE}] - [U_{RS}] \tag{11.16}$$

Note that this is the same transmission gain shown in Fig. 10.8.

Using the transmission gain, the interference I_2 at the satellite may be referred to the earth-station receiver as γI_2, and hence the noise-temperature rise at the satellite receiver input may be referred to the earth-station receiver input as $\gamma \, \Delta T_S$. This is illustrated in Fig. 11.9b. Expressed in decibel units, the relationship is

$$[\Delta T_{S-E}] = [\gamma] + [\Delta T_S] \tag{11.17}$$

11.4.3 Resultant Noise-Temperature Rise

The overall equivalent rise in noise temperature at earth station E as a result of interference signals B_1 and B_2 is then

$$\Delta T = \Delta T_{S-E} + \Delta T_E \tag{11.18}$$

In this final calculation the dBK values must first be converted to degrees, which are then added to give the resultant equivalent noise-temperature rise at the earth station E receive antenna output.

Example 11.6

Given that $L_U = 200$ dB, $L_D = 196$ dB, $G_E = G'_E = 25$ dB, $G_S = G'_S = 9$ dB, $G_{TE} = G_{RE} = 48$ dB, $G_{RS} = G_{TS} = 19$ dB, $U_S = U'_s = 1$ μJ, and $U_E = U'_E = 10$ μJ, calculate the transmission gain $[\gamma]$, the interference levels $[I_1]$ and $[I_2]$, and the equivalent temperature rise overall.

solution   Using Eq. (11.14) gives

$$[U_{RS}] = -50 + 48 + 19 - 200$$
$$= -183 \text{ dBJ}$$

Using Eq. (11.15) gives

$$[U_{RE}] = -60 + 19 + 48 - 196$$
$$= -189 \text{ dBJ}$$

Therefore,

$$[\gamma] = -189 - (-183)$$
$$= -6 \text{ dB}$$

From Eq. (11.10),

$$[I_1] = -60 + 9 + 25 - 196$$
$$= -222 \text{ dBJ}$$

From Eq. (11.12),

$$[I_2] = -50 + 25 + 9 - 200$$
$$= -216 \text{ dBJ}$$

From Eq. (11.11),

$$[\Delta T_E] = -222 + 228.6$$
$$= 6.6 \text{ dBK} \quad \text{or} \quad \Delta T_E = 4.57 \text{ K}$$

From Eq. (11.13),

$$[\Delta T_S] = -216 + 228.6$$
$$= 12.6 \text{ dBK}$$

From Eq. (11.17),

$$[\Delta T_{S-E}] = -6 + 12.6$$
$$= 6.6 \text{ dBK} \quad \text{or} \quad \Delta T_{S-E} = 4.57 \text{ K}$$

The resultant equivalent noise-temperature rise at the earth station E receive antenna output is $4.57 + 4.57 = 9.14$K.

11.4.4 Coordination Criterion

CCIR Report 454–3, 1982, specifies that the equivalent noise temperature rise should be no more than 4 percent of the equivalent thermal noise temperature of the satellite link. The equivalent thermal noise temperature is defined in the CCIR Radio Regulations, App. 29.

As an example, the CCIR Recommendations for FM Telephony allows up to 10,000 pW0p total noise in a telephone channel. The abbreviation pW0p stands for picowatts at a zero-level test point, psophometrically weighted, as already defined in connection with Table 11.1. The 10,000-pW0p total includes a 1000-pW0p allowance for terrestrial station interference, and 1000 pW0p for interference from other satellite links. Thus the thermal noise allowance is $10,000 - 2000 = 8000$ pW0p. Four percent of this is 320 pW0p. Assuming that this is over a 3.1-kHz bandwidth, the spectrum density is 320/3100 or approximately 0.1 pJ0p (pW0p/Hz). In decibels this is -130 dBJ. This is output noise, and to relate it back to the noise temperature at the antenna, the overall gain of the receiver from antenna to output, including the processing gain discussed in Sec. 8.6.3, must be known. For illustration purposes, assume that the gain is 90 dB, so that the antenna noise is $-130 - 90 = -220$ dBJ. The noise-temperature rise corresponding to this is $-220 + 228.6 = 8.6$ dBK. Converting this to kelvins gives 7.25 K.

11.4.5 Noise Power Spectral Density

The concept of noise power spectral density has been introduced in Sec. 10.5 for a flat frequency spectrum. Where the spectrum is not flat, an average value for the spectral density can be calculated. To il-

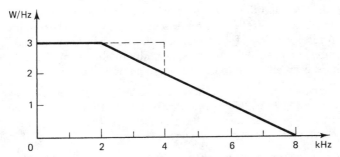

Figure 11.10   Power spectrum density curve (see Sec. 11.4.5).

lustrate this, the very much simplified spectrum curve of Fig. 11.10 will be used. The maximum spectrum density is flat at 3 W/Hz from zero to 2 kHz, and then slopes linearly down to zero over the range from 2 to 8 kHz.

The noise power in any given bandwidth is calculated as the area under the curve, whose width is the value of the bandwidth. Thus, for the first 2 kHz, the noise power is 3 W/Hz $\times$ 2000 = 6000 W. From 2 to 8 kHz, the noise power is 3 $\times$ (8 − 2) $\times$ 1000/2 = 9000 W. The total power is therefore 15,000 W and the average spectral density is 15,000/8000 = 1.875 W/Hz.

The noise power spectral density over the worst 4-kHz bandwidth must include the highest part of the curve and is therefore calculated for the 0- to 4-kHz band. The power over this band is seen to be the area of the rectangle 3 W/Hz $\times$ 4 kHz minus the area of the triangle shown dashed in Fig. 11.10. The power over the 0- to 4-kHz band is therefore (3 $\times$ 4000) − (3 − 2) $\times$ (4 − 2) $\times$ 1000/2 = 11,000 W, and the spectral density is 11,000/4000 = 2.75 W/Hz.

The units for spectral power density are often stated as watts/hertz (W/Hz). Expressed in this manner the units are descriptive of the way in which the power spectral density is arrived at. In terms of fundamental units, watts are equivalent to joules per second, and hertz to cycles per second, or simply seconds^{-1} since cycles are a dimensionless quantity. Thus 1 watt/hertz is equivalent to 1 joule/second ÷ second^{-1}, which is simply joules. The units for power spectral density can therefore be stated as joules. This is verified by Eq. (10.15) for noise power spectral density.

11.5   Problems

11.1.   Describe briefly the modes of interference that can occur in a satellite communications system. Distinguish carefully between satellite and terrestrial modes of interference.

11.2. Define and explain the difference between topocentric angles and geocentric angles as applied to satellite communications. Two geostationary satellites have an orbital spacing of 4°. Calculate the topocentric angle subtended by the satellites, measured (*a*) from the midpoint between the subsatellite points and (*b*) from either of the subsatellite points.

11.3. Westar IV is located at 98.5° W, and Telstar at 96° W. The coordinates for two earth stations are 104° W, 36° N, and 90° W, 32° N. By using the look angle and range formulas given in Sec. 2.11.1, calculate the topocentric angle subtended at each earth station by these two satellites.

11.4. Explain what is meant by *single-entry interference*. Explain why it is the radiation pattern of the earth station antennas, not the satellite antennas, which governs the level of interference.

11.5. A geostationary satellite employs a 3.5-m parabolic antenna at a frequency of 12 GHz. Calculate the −3-dB beamwidth and the spot diameter on the equator.

11.6. Calculate the −3-dB beamwidth for an earth-station antenna operating at 14 GHz. The antenna utilizes a parabolic reflector of 3.5-m diameter. Compare the distance separation of satellites at 2° spacing with the diameter of the beam at the −3-dB points on the geostationary orbit.

11.7. Compare the increase in interference levels expected when satellite orbital spacing is reduced from 4° to 2° for earth station antenna sidelobe patterns of (*a*) $32 - 25 \log \theta$ dB and (*b*) $29 - 25 \log \theta$ dB.

11.8. A satellite circuit operates with an uplink transmit power of 28.3 dBW and an antenna gain of 62.5 dB. A potential interfering circuit operates with an uplink power of 26.3 dBW. Assuming a 4-dB polarization discrimination figure and earth station sidelobe gain function of $32 - 25 \log \theta$ dB, calculate the [*C/I*] ratio at the satellite for 2° satellite spacing.

11.9. The downlink of a satellite circuit operates at a satellite [EIRP] of 35 dBW and a receiving-earth-station antenna gain of 59.5 dB. Interference is produced by a satellite spaced 3°, its [EIRP] also being 35 dBW. Calculate the [*C/I*] ratio at the receiving antenna, assuming 6-dB polarization discrimination. The sidelobe gain function for the earth station antenna is $32 - 25 \log \theta$.

11.10. A satellite circuit operates with an earth-station transmit power of 30 dBW and a satellite [EIRP] of 34 dBW. This circuit causes interference with a neighboring circuit for which the earth-station transmit power is 24 dBW, the transmit antenna gain 54 dB, the satellite [EIRP] is 34 dBW, and the receive-earth-station antenna gain is 44 dB. Calculate the carrier-to-interference ratio at the receive-station antenna. Antenna sidelobe characteristics of $32 - 25 \log \theta$ dB may be assumed, with the satellites spaced at 2° and polarization isolation of 4 dB on uplink and downlink.

11.11. Repeat Prob. 11.10 for a sidelobe pattern of $29 - 25 \log \theta$.

11.12. A satellite TV/FM circuit operates with an uplink power of 30 dBW, an antenna gain of 53.5 dB, and a satellite [EIRP] of 34 dBW. The destination earth station has a receiver antenna gain of 44 dB. An interfering circuit has an uplink power of 11.5 dBW and a satellite [EIRP] of 15.7 dBW. Given that the spectral overlap is [Q] = 3.8 dB, calculate the passband [C/I] ratio. Assume polarization discrimination figures of 4 dB on the uplink and 0 dB on the downlink and an antenna sidelobe pattern of $32 - 25 \log \theta$. The satellite spacing is 2°.

11.13. Repeat Prob. 11.12 for an interfering carrier for which the uplink transmit power is 26 dBW, the satellite [EIRP] is 35.7 dBW, and the spectral overlap figure is −3.36 dB.

11.14. An FDM/FM satellite circuit operates with an uplink transmit power of 11.9 dBW, an antenna gain of 53.5 dB, and a satellite [EIRP] of 19.1 dBW. The destination earth station has an antenna gain of 50.5 dB. A TV/FM interfering circuit operates with an uplink transmit power of 28.3 dBW and a satellite [EIRP] of 35 dBW. Polarization discrimination figures are 6 dB on the uplink and 0 dB on the downlink. Given that the receiver transfer characteristic for wanted and interfering signals is [RTC] = 60.83 dB, calculate the baseband [S/I] ratio for an antenna sidelobe pattern of $29 - 25 \log \theta$. The satellite spacing is 2°.

11.15. Repeat Prob. 11.14 for an interfering circuit operating with an uplink transmit power of 27 dBW, a satellite [EIRP] of 34.2 dBW, and [RTC] = 37.94 dB.

11.16. Explain what is meant by *single-entry interference objectives*. Show that an interference level of 600 pW0p is equivalent to a [S/I] ratio of 62.2 dB.

11.17. For the wanted circuit in Probs. 11.12 and 11.13, the specified interference objective is $[C/I]_{pb} = 22$ dB. Is this objective met?

11.18. For broadcast TV/FM the permissible video-to-noise objective is specified as [S/N] = 53 dB. The CCIR recommendation for interference is that the total interference from all other satellite networks not exceed 10% of the video noise, and that the single-entry interference not exceed 40% of this total. Show that this results in a single-entry objective of [S/I] = 67 dB.

11.19. Explain what is meant by *protection ratio*. A TV/FM carrier operates at a modulation index of 2.619, the top modulating frequency being 4.2 MHz. Calculate the protection ratio required for quality factors of (*a*) 4.2 and (*b*) 4.5. How do these values compare with the specified interference objective of $[C/I]_{pb} = 22$ dB?

11.20. Explain what is meant by *energy dispersal*, and how this may be achieved.

11.21. Explain what is meant by *coordination* in connection with interference assessment in satellite circuits.

Satellite Access

12.1 Introduction

A transponder channel aboard a satellite may be fully loaded by a single transmission from an earth station. This is referred to as a *single-access* mode of operation. It is also possible, and more common, for a transponder to be loaded by a number of carriers. These may originate from a number of earth stations geographically separate, and each earth station may transmit one or more of the carriers. This mode of operation is termed *multiple access*. The need for multiple access arises because more than two earth stations, in general, will be within the service area of a satellite. Even so-called spot beams from satellite antennas cover areas several hundred miles across.

The two most commonly used methods of multiple access are *frequency-division multiple access* (FDMA) and *time-division multiple access* (TDMA). These are analogous to frequency-division multiplexing (FDM) and time-division multiplexing (TDM) described in Chaps. 8 and 9. However, multiple access and multiplexing are different concepts, and as pointed out in CCIR Report 708, 1982, modulation (and hence multiplexing) is essentially a transmission feature, whereas multiple access is essentially a traffic feature.

A third category of multiple access is *code-division multiple access* (CDMA). In this method each signal is associated with a particular code that is used to spread the signal in frequency and or time. All such signals will be received simultaneously at an earth station, but by using the key to the code the station can recover the desired signal by means of correlation. The other signals occupying the transponder channel appear very much like random noise to the correlation decoder. The two subsets of CDMA are *spread-spectrum multiple access* (SSMA) and *pulse-address multiple access* (PAMA).

Multiple access may also be classified by the way in which circuits are assigned to users (*circuits* in this context implies one communication channel through the multiple-access transponder). Circuits may be *preassigned,* which means they are allocated on a fixed or partially fixed basis to certain users. These circuits are therefore not available for general use. Preassignment is simple to implement, but is efficient only for circuits with *continuous heavy* traffic.

An alternative to preassignment is *demand-assigned multiple access* (DAMA). In this method, all circuits are available to all users and are assigned according to the demand. DAMA results in more efficient overall use of the circuits but is more costly and complicated to implement.

Both FDMA and TDMA can be operated as preassigned or demand-assigned systems. CDMA is a random access system, there being no control over the timing of the access or of the frequency slots accessed.

These multiple-access methods refer to the way in which a single *transponder* channel is utilized. A satellite carries a number of transponders, and normally each covers a different frequency channel, as shown in Fig. 6.13. This provides a form of frequency-division multiple access to the whole satellite. It is also possible for transponders to operate at the same frequency but to be connected to different spot beam antennas. These allow the satellite as a whole to be accessed by earth stations widely separated geographically but transmitting on the same frequency. This is termed *frequency reuse.* This method of access is referred to as *space-division multiple access* (SDMA). It should be kept in mind that each spot beam may itself be carrying signals in one of the other multiple-access formats.

12.2   Single Access

With single access, a single modulated carrier occupies the whole of the available bandwidth of a transponder. Single-access operation is used on heavy traffic routes, and requires large earth station antennas such as the class A antenna shown in Fig. 7.7. As an example, Telesat Canada provides heavy route message facilities, with each transponder channel being capable of carrying 960 one-way voice circuits on an FDM/FM carrier as illustrated in Fig. 12.1. The earth station employs a 30-m-diameter antenna and a parametric amplifier, which together provide a minimum $[G/T]$ of 37.5 dB/K.

12.3   Preassigned FDMA

Frequency slots may be preassigned to analog and digital signals, and to illustrate the method, analog signals in the FDM/FM/FDMA format will be considered first. As the acronyms indicate, the signals are fre-

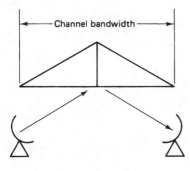

Figure 12.1 Heavy route message (frequency modulation—single access). (*From Telsat Canada, 1983.*)

quency-division multiplexed, frequency modulated (FM), with frequency-division multiple access to the satellite. In Chap. 8, FDM/FM signals are discussed. It will be recalled that the voice-frequency (telephone) signals are first SSBSC amplitude modulated onto voice carriers in order to generate the single sidebands needed for the frequency-division multiplexing. For the purpose of illustration, each earth station will be assumed to transmit a 60-channel supergroup. Each 60-channel supergroup is then frequency modulated onto a carrier which is then upconverted to a frequency in the satellite uplink band.

Figure 12.2 shows the situation for three earth stations: one in Ottawa, one in New York, and one in London. All three earth stations access a single satellite transponder channel simultaneously, and each communicates with both of the others. Thus it is assumed that the satellite receive and transmit antenna beams are *global,* encompassing all three earth stations. Each earth station transmits one uplink carrier modulated with a 60-channel supergroup and receives two similar downlink carriers.

The earth station at New York is shown in more detail. One transmit chain is used and this carries telephone traffic for both Ottawa and London. On the receive side, two receive chains must be provided, one for the Ottawa-originated carrier and one for the London-originated carrier. Each of these carriers will have a mixture of traffic, and in the demultiplexing unit, only those telephone channels intended for New York are passed through. These are remultiplexed into an FDM/FM format which is transmitted out along the terrestrial line to the New York switching office. This earth-station arrangement should be compared with that shown in Fig. 7.6.

Figure 12.3 shows a hypothetical frequency assignment scheme for the hypothetical network of Fig. 12.2. Uplink carrier frequencies of 6253, 6273, and 6278 MHz are shown for illustration purposes. For the satellite transponder arrangement of Fig. 6.13, these carriers would be translated down to frequencies of 4028, 4048, and 4053 MHz

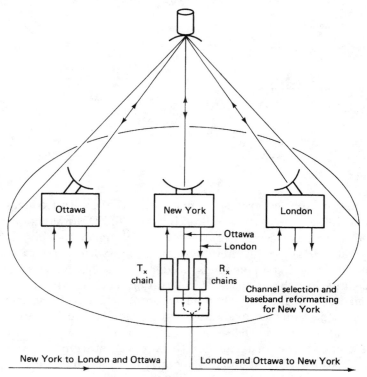

Figure 12.2   Three earth stations transmitting and receiving simultaneously through the same satellite transponder, using fixed-assignment FDMA.

(i.e., the corresponding 4-GHz-band downlink frequencies) and sent to transponder 9 of the satellite. Typically, a 60-channel FDM/FM carrier occupies 5 MHz of transponder bandwidth, including guardbands. A total frequency allowance of 15 MHz is therefore required for the three stations, and each station receives all the traffic. The remainder of the transponder bandwidth may be unused, or it may be occupied by other carriers, which are not shown.

As an example of preassignment, suppose that each station can transmit up to 60 voice circuits, and that 40 of these are preassigned to the New York–London route. If these 40 circuits are fully loaded, additional calls on the New York–London route will be blocked even though there may be idle circuits on the other preassigned routes.

Telesat Canada operates medium-route message facilities utilizing FDM/FM/FDMA. Figure 12.4 shows how five carriers may be used to support 168 voice channels. The earth station that carries the full load has a $[G/T]$ of 37.5 dB/K, and the other four have $[G/T]$'s of 28 dB/K.

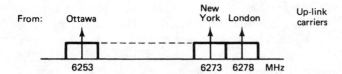

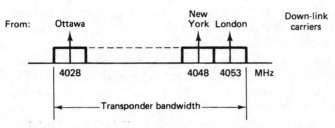

Figure 12.3 Transponder channel assignments for the earth stations shown in Fig. 12.2.

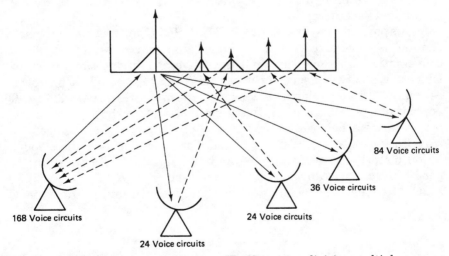

Figure 12.4 Medium route message traffic (frequency-division multiple access, FM/FDMA). (*From Telesat Canada, 1983.*)

Preassignment may also be made on the basis of a single channel per carrier (SCPC). This refers to a single voice (or data) channel per carrier, not a transponder channel, which may in fact carry some hundreds of voice channels by this method. The carriers may be frequency modulated or phase-shift modulated, and an earth station may be capable of transmitting one or more SCPC signals simultaneously.

Figure 12.5 shows the Intelsat SCPC channeling scheme for a 36-MHz transponder. The transponder bandwidth is subdivided into 800 channels each 45 kHz wide. The 45 kHz, which includes a guardband, is required for each digitized voice channel, which utilizes QPSK modulation. The channel information signal may be digital data or PCM voice signals (see Chap. 9). A pilot frequency is transmitted for the purpose of frequency control, and the adjacent channel slots on either side of the pilot are left vacant to avoid interference. The scheme therefore provides a total of 798 one-way channels or up to 399 full-duplex voice circuits. In duplex operation the frequency pairs are separated by 18.045 MHz, as shown in Fig. 12.5.

The frequency tolerance relative to the assigned values is within ± 1 kHz for the received SCPC carrier, and must be within ± 250 Hz for the transmitted SCPC carrier (Miya, 1981). The pilot frequency is transmitted by one of the earth stations designated as a primary station. This provides a reference for automatic frequency control (AFC) (usually through the use of phase-locked loops) of the transmitter frequency synthesizers and receiver local oscillators. In the event of failure of the primary station, the pilot frequency is transmitted from a designated backup station.

An important feature of the Intelsat SCPC system is that each channel is voice-activated. This means that on a two-way telephone conversation, only one carrier is operative at any one time. Also, in long pauses between speech the carriers are switched off. It has been estimated that for telephone calls, the one-way utilization time is 40 percent of the call duration. Using voice activation the average number of carriers being amplified at any one time by the transponder traveling wave tube (TWT) is reduced. For a given level of intermodulation distortion (see Secs. 6.7.3 and 10.10), the TWT power output per FDMA carrier can therefore be increased.

SCPC systems are widely used on lightly loaded routes, this type of

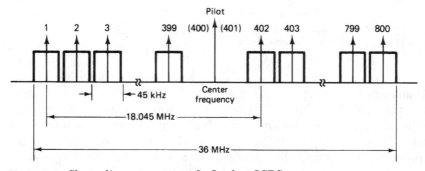

Figure 12.5 Channeling arrangement for Intelsat SCPC system.

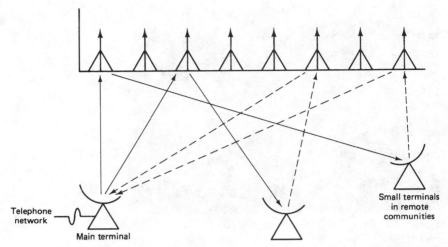

Figure 12.6   Thin route message traffic (single channel per carrier, SCPC/FDMA). (*From Telesat Canada, 1983.*)

service being referred to as a *thin route service*. It enables remote earth stations in sparsely populated areas to connect into the national telephone network in a reasonably economical way. A main earth station is used to make the connection to the telephone network as illustrated in Fig. 12.6. The Telesat Canada Thin Route Message Facilities provides up to 360 two-way circuits using PSK/SCPC (PSK = phase-shift keying). The remote terminals operate with 4.6-m-diameter antennas with [G/T] values of 19.5 or 21 dB/K. Transportable terminals are also available, one of these being shown in Fig. 12.7. This is a single-channel station that uses a 3.6-m antenna and comes complete with a desktop electronics package which can be installed on the customers' premises.

12.4   Demand-Assigned FDMA

In the demand-assigned mode of operation, the transponder frequency bandwidth is subdivided into a number of channels. A channel is assigned to each carrier in use, giving rise to the single channel per carrier mode of operation discussed in the preceding section. As in the preassigned access mode, carriers may be frequency modulated with analog information signals, these being designated FM/SCPC, or they may be phase modulated with digital information signals, these being designated as PSK/SCPC.

Demand assignment may be carried out in a number of ways. In the polling method, a master earth station continuously polls all the

Figure 12.7 Transportable message station. (*From Telesat Canada, 1983.*)

earth stations in sequence, and if a *call request* is encountered, frequency slots are assigned from the pool of available frequencies. The polling delay with such a system tends to become excessive as the number of participating earth stations increases.

Instead of using a polling sequence, earth stations may request calls through the master earth station as the need arises. This is referred to as centrally controlled random access. The requests go over a digital orderwire, which is a narrowband digital radio link or a circuit through a satellite transponder reserved for this purpose. Frequencies are assigned, if available, by the master station, and when the call is completed, the frequencies are returned to the pool. If no frequencies are available, the blocked call requests may be placed in a queue, or a second call attempt initiated by the requesting station.

As an alternative to centrally controlled random access, control may be exercised at each earth station, this being known as distributed-control random access. A good illustration of such a system is provided by the Spade system operated by Intelsat on some of its satellites. This is described in the next section.

12.5 Spade System

The word Spade is a loose acronym for single-channel-per-carrier pulse-code-modulated multiple-access demand-assignment equipment. Spade was developed by Comsat for use on the Intelsat satellites (see, e.g., Martin, 1978), and is compatible with the Intelsat SCPC preassigned system described in Sec. 12.3. However, the distributed-demand assignment facility requires a common signaling channel (CSC). This is shown in Fig. 12.8. The CSC bandwidth is 160

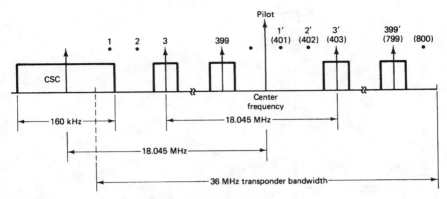

Figure 12.8   Channeling scheme for the Spade system.

kHz, and its center frequency is 18.045 MHz below the pilot frequency, as shown in Fig. 12.8. To avoid interference with the CSC, voice channels 1 and 2 are left vacant, and to maintain duplex matching the corresponding channels 1' and 2' are also left vacant. Recalling from Fig. 12.5 that channel 400 must also be left vacant, this requires that channel 800 be left vacant for duplex matching. Thus six channels are removed from the total of 800, leaving a total of 794 one-way or 397 full-duplex voice circuits, the frequencies in any pair being separated by 18.045 MHz, as shown in Fig. 12.8. (An alternative arrangement is shown in Freeman, 1981).

All the earth stations are permanently connected through the common signaling channel (CSC). This is shown diagrammatically in Fig. 12.9 for six earth stations, *A, B, C, D, E,* and *F.* Each earth station has the facility for generating any one of the 794 carrier frequencies using frequency synthesizers. Furthermore, each earth station has a memory containing a list of the frequencies currently available, and this list is continuously updated through the CSC. To illustrate the procedure, suppose that a call to station *F* is initiated from station *C* in Fig. 12.9. Station *C* will first select a frequency pair at random from those currently available on the list, and signal this information to station *F* through the CSC. Station *F* must acknowledge, through the CSC, that it can complete the circuit. Once the circuit is established, the other earth stations are instructed, through the CSC, to remove this frequency pair from the list.

The round-trip time between station *C* initiating the call and station *F* acknowledging it is about 600 ms. During this time, the two frequencies chosen at station *C* may be assigned to another circuit. In this event, station *C* will receive the information on the CSC update and will immediately choose another pair at random, even before hearing back from station *F.*

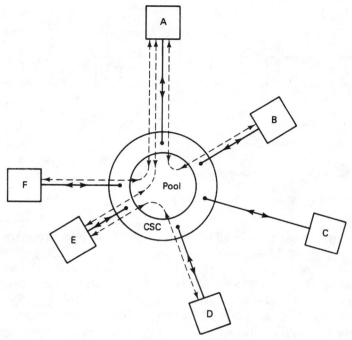

Figure 12.9 Diagrammatic representation of a Spade communications system.

Once a call has been completed and the circuit disconnected, the two frequencies are returned to the pool, the information again being transmitted through the CSC to all the earth stations.

As well as establishing the connection through the satellite, the CSC passes signaling information from the calling station to the destination station, in the example above from station C to station F. Signaling information in the Spade system is routed through the CSC rather than being sent over a voice channel. Each earth station has equipment called the demand assignment signaling and switching (DASS) unit which performs the functions required by the CSC.

Some type of multiple access to the CSC must be provided for all the earth stations using the Spade system. This is quite separate from the SCPC multiple access of the network's voice circuits. Time-division multiple access, described in Sec. 12.7.8, is used for this purpose, allowing up to 49 earth stations to access the common signaling channel.

12.6 Bandwidth-Limited and Power-Limited TWT Amplifier Operation

A transponder will have a total bandwidth B_{TR}, and it is apparent that this can impose a limitation on the number of carriers which can access the transponder in an FDMA mode. For example, if there are K carriers each of bandwidth B, then the best that can be achieved is $K = B_{TR}/B$. Any increase in the transponder EIRP will not improve on this, and the system is said to be *bandwidth-limited*. Likewise, for digital systems, the bit rate is determined by the bandwidth which again will be limited to some maximum value by B_{TR}.

Power limitation occurs where the EIRP is insufficient to meet the [C/N] requirements as shown by Eq. (10.34). The signal bandwidth will be approximately equal to the noise bandwidth, and if the EIRP is below a certain level, the bandwidth will have to be correspondingly reduced to maintain the [C/N] at the required value. These limitations are discussed in more detail in the next two sections.

12.6.1 FDMA downlink analysis

To see the effects of intermodulation noise which results with FDMA operation, consider the overall carrier-to-noise ratio as given by Eq. (10.62). In terms of noise power rather than noise power density, Eq. (10.62) states

$$\left(\frac{N}{C}\right) = \left(\frac{N}{C}\right)_U + \left(\frac{N}{C}\right)_D + \left(\frac{N}{C}\right)_{IM} \qquad (12.1)$$

A certain value of carrier-to-noise ratio will be needed, as specified in the system design, and this will be denoted by the subscript REQ. The overall C/N must be at least as great as the required value, a condition which can therefore be stated as

$$\left(\frac{N}{C}\right)_{REQ} \geq \left(\frac{N}{C}\right) \qquad (12.2)$$

Note that, because the noise-to-carrier ratio rather than the carrier to noise ratio is involved, the actual value is equal to or less than the required value. Using Eq. (12.1) the condition can be rewritten as

$$\left(\frac{N}{C}\right)_{REQ} \geq \left(\frac{N}{C}\right)_U + \left(\frac{N}{C}\right)_D + \left(\frac{N}{C}\right)_{IM} \qquad (12.3)$$

The right-hand side of Eq. (12.3) is usually dominated by the downlink ratio. With FDMA, back-off is utilized to reduce the intermodulation noise to an acceptable level, and as shown in Sec. 10.9 the uplink noise contribution is usually negligible. Thus, the expression can be approximated by

$$\left(\frac{N}{C}\right)_{REQ} \geq \left(\frac{N}{C}\right)_D$$

or

$$\left(\frac{C}{N}\right)_{REQ} \leq \left(\frac{C}{N}\right)_D \qquad (12.4)$$

Consider the situation where each carrier of the FDMA system occupies a bandwidth B and has a downlink power denoted by $[EIRP]_D$. Equation (10.54) gives

$$\left[\frac{C}{N}\right]_D = [EIRP]_D + \left[\frac{G}{T}\right]_D - [LOSSES] - [k] - [B] \qquad (12.5)$$

where it is assumed that $B_N \cong B$. This can be written in terms of the required carrier-to-noise ratio as

$$\left[\frac{C}{N}\right]_{REQ} \leq [EIRP]_D + \left[\frac{G}{T}\right]_D - [LOSSES] - [k] - [B] \qquad (12.6)$$

To set up a reference level, consider first single carrier operation. The satellite will have a saturation value of EIRP, and a transponder bandwidth B_{TR}, both of which are assumed fixed. With single-carrier access, no back-off is needed and Eq. (12.6) becomes

$$\left[\frac{C}{N}\right]_{REQ} \leq [EIRP_S] + \left[\frac{G}{T}\right]_D - [LOSSES] - [k] - [B_{TR}] \qquad (12.7)$$

or

$$\left[\frac{C}{N}\right]_{REQ} - [EIRP_S] - \left[\frac{G}{T}\right]_D + [LOSSES] + [k] + [B_{TR}] \leq 0 \qquad (12.8)$$

If the system is designed for single-carrier operation then the equality sign applies and the reference condition is

$$\left[\frac{C}{N}\right]_{REQ} - [EIRP_S] - \left[\frac{G}{T}\right]_D + [LOSSES] + [k] + [B_{TR}] = 0 \qquad (12.9)$$

Consider now the effect of power limitation imposed by the need for back-off. Suppose the FDMA access provides for K carriers which share the output power equally, and each requires a bandwidth B. The output power can be written as

$$[EIRP]_D = [EIRP_S] - [BO]_O - [K] \qquad (12.10)$$

The transponder bandwidth B_{TR} will be shared between the carriers, but not all of B_{TR} can be utilized because of the power limitation. Let α represent the fraction of the total bandwidth actually occupied, such that $KB = \alpha B_{TR}$, or in terms of decilogs

$$[B] = [\alpha] + [B_{TR}] - [K] \qquad (12.11)$$

Substituting these relationships in Eq. (12.6) gives

$$\left[\frac{C}{N}\right]_{REQ} \le [EIRP_S] - [BO]_O + \left[\frac{G}{T}\right]_D - [LOSSES] - [k] - [B_{TR}] - [\alpha] \tag{12.12}$$

It will be noted that the [K] term cancels out. The expression can be rearranged as

$$\left[\frac{C}{N}\right]_{REQ} - [EIRP_S] - \left[\frac{G}{T}\right]_D + [LOSSES] + [k] + [B_{TR}] \le - [BO]_O - [\alpha] \tag{12.13}$$

But as shown by Eq. (12.9), the left-hand side is equal to zero if the single carrier access is used as reference, and hence

$$0 \le - [BO]_O - [\alpha]$$

or

$$[\alpha] \le - [BO]_O \tag{12.14}$$

The best that can be achieved is to make $[\alpha] = -[BO]_O$, and since the back-off is a positive number of decibels, $[\alpha]$ must be negative, or equivalently, α is fractional. The following example illustrates the limitation imposed by back-off.

Example 12.1

A satellite transponder has a bandwidth of 36 MHz and a saturation EIRP of 27 dBW. The earth station receiver has a G/T ratio of 30 dB/K, and the total link losses are 196 dB. The transponder is accessed by FDMA carriers each of 3-MHz bandwidth, and 6-dB output back-off is employed. Calculate the downlink carrier-to-noise ratio for single-carrier operation, and the number of carriers which can be accommodated in the FDMA system. Compare this with the number which could be accommodated if no back-off were needed. The carrier-to-noise ratio determined for single-carrier operation may be taken as the reference value, and it may be assumed that the uplink noise and intermodulation noise are negligible.

solution

Note: for convenience in the Mathcad solution, decibel or decilog values will be indicated by dB. For example, the output back-off in decibels is shown as $BOdB_O$.

Transponder bandwidth:

$$B_{TR} := 36 \cdot MHz \qquad BdB_{TR} := 10 \cdot log\left(\frac{B_{TR}}{Hz}\right)$$

Carrier bandwidth:

$$B := 3 \cdot MHz \qquad BdB := 10 \cdot log\left(\frac{B}{Hz}\right)$$

Saturation eirp:

$$eirpdBW_S := 27$$

Output back-off:

$$BOdB_O := 6$$

Total losses:

$$LOSSESdB := 196$$

Ground station G/T:

$$GTRdB := 30$$

$$CNRdB_D := eirpdBW_S + GTRdB - LOSSESdB + 228.6 - BdB_{TR} \qquad eq. \ (10.54)$$

$$CNRdB_D = 14$$
$$========$$

$$\alpha dB := -BOdB_O \qquad eq. \ (12.14)$$

$$KdB := \alpha dB + BdB_{TR} - BdB \qquad eq. \ (12.11)$$

$$K := 10^{\frac{KdB}{10}} \qquad K = 3$$
$$======$$

If back-off was not required, the number of carriers which could be accommodated would be:

$$\frac{B_{TR}}{B} = 12$$

12.7 TDMA

With time-division multiple access, only one carrier uses the transponder at any one time, and therefore intermodulation products, which result from the nonlinear amplification of multiple carriers, are absent. This leads to one of the most significant advantages of TDMA, which is that the transponder traveling wave tube (TWT) can be operated at maximum power output or saturation level.

Because the signal information is transmitted in bursts, TDMA is only suited to digital signals. Digital data can be assembled into burst format for transmission and reassembled from the received bursts, through the use of digital buffer memories.

Figure 12.10 illustrates the basic TDMA concept, in which the stations transmit bursts in sequence. Burst synchronization is required, and in the system illustrated in Fig. 12.10, one station is assigned solely for the purpose of transmitting *reference bursts* to which the others can be synchronized. The time interval from the start of one reference burst to the next is termed a *frame*. A frame contains the reference burst *R,* and the bursts from the other earth stations, these being shown as *A, B,* and *C* in Fig. 12.10.

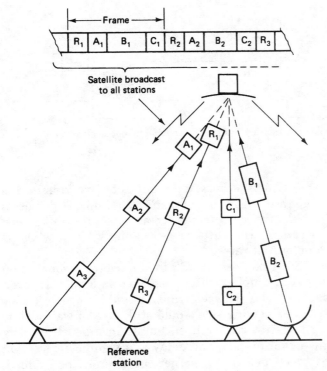

Figure 12.10 Time-division multiple access (TDMA) using a reference station for burst synchronization.

Figure 12.11 illustrates the basic principles of burst transmission for a single channel. Overall, the transmission appears continuous because the input and output bit rates are continuous and equal. However, within the transmission channel, input bits are temporarily stored and transmitted in bursts. Since the time interval between bursts is the frame time T_F, the required buffer capacity is

$$M = R_b T_F \tag{12.15}$$

The buffer memory fills up at the input bit rate R_b during the frame time interval. These M bits are transmitted as a burst in the next frame, without any break in continuity of the input. The M bits are transmitted in the burst time T_B, and the *transmission rate,* which is equal to the burst bit rate, is

$$R_T = \frac{M}{T_B} \tag{12.16}$$

$$= R_b \frac{T_F}{T_B}$$

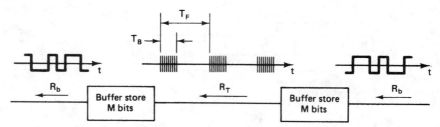

Figure 12.11   Burst-mode transmission linking two continuous-mode streams.

This is also referred to as the *burst rate,* but note that this means the instantaneous bit rate within a burst (not the number of bursts per second, which is simply equal to the frame rate). It will be seen that the *average* bit rate for the burst mode is simply M/T_F, which is equal to the input and output rates.

The frame time T_F will be seen to add to the overall propagation delay. For example, in the simple system illustrated in Fig. 12.11, even if the actual propagation delay between transmit and receive buffers is assumed to be zero, the receiving side would still have to wait a time T_F before receiving the first transmitted burst. In a geostationary satellite system the actual propagation delay is a significant fraction of a second, and excessive delays from other causes must be avoided. This sets an upper limit to the frame time, although with current technology, other factors restrict the frame time to well below this limit. The frame period is usually chosen to be a multiple of 125 µs, which is the standard sampling period used in pulse-code modulation (PCM) telephony systems, as this ensures that the PCM samples can be distributed across successive frames at the PCM sampling rate.

Figure 12.12 shows some of the basic units in a TDMA ground station, which for discussion purposes is labeled earth station *A.* Terrestrial links coming into earth station A carry digital traffic addressed to destination stations, labeled *B, C, X.* It is assumed that the bit rate is the same for the digital traffic on each terrestrial link. In the units labeled *terrestrial interface modules* (TIMs) the incoming continuous bit-rate signals are converted into the intermittent burst rate mode. These individual burst-mode signals are *time-division multiplexed* in the time-division multiplexer (MUX), so that the traffic for each destination station appears in its assigned time slot within a burst.

Certain time slots at the beginning of each burst are used to carry timing and synchronizing information. These time slots collectively are referred to as the *preamble.* The complete burst containing the preamble and the traffic data is used to phase-modulate the radio-frequency (RF) carrier. Thus the composite burst which is transmitted at

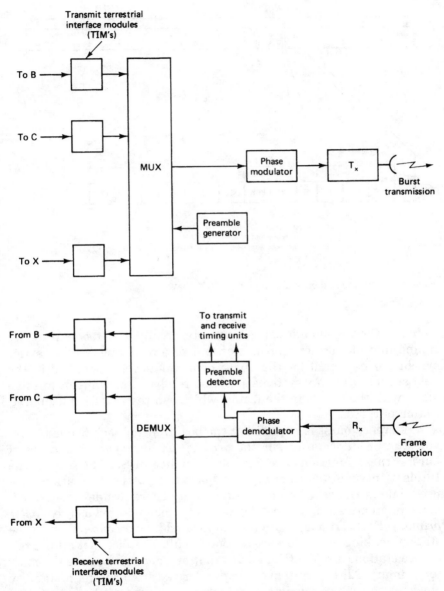

Figure 12.12 Some of the basic equipment blocks in a TDMA system.

RF consists of a number of time slots as shown in Fig. 12.13. These will be described in more detail shortly.

The received signal at an earth station consists of bursts from all transmitting stations arranged in the frame format shown in Fig. 12.13. The RF carrier is converted to intermediate frequency (IF),

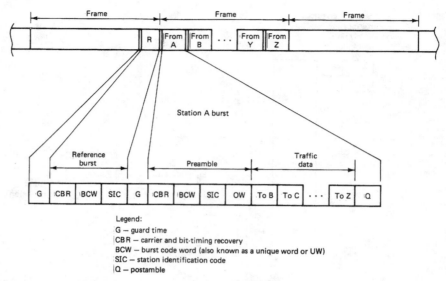

Legend:
G — guard time
CBR — carrier and bit-timing recovery
BCW — burst code word (also known as a unique word or UW)
SIC — station identification code
Q — postamble

Figure 12.13 Frame and burst formats for a TDMA system.

which is then demodulated. A separate preamble detector provides timing information for transmitter and receiver along with a carrier synchronizing signal for the phase demodulator, as described in the next section. In many systems a station receives its own transmission along with the others in the frame, which can then be used for burst-timing purposes.

A reference burst is required at the beginning of each frame to provide timing information for the *acquisition* and *synchronization* of bursts (these functions are described further in Sec. 12.7.4). In the Intelsat international network at least two reference stations are used, one in the east and one in the west. These are designated *primary* reference stations, one of which is further selected as the *master primary*. Each primary station is duplicated by a *secondary* reference station, making four reference stations in all. The fact that all the reference stations are identical means that any one can become the master primary. All the system timing is derived from the high stability clock in the master primary, which is accurate to 1 part in 10^{11} (Lewis, 1982). A clock on the satellite is locked to the master primary, and this acts as the clock for the other participating earth stations. The satellite clock will provide a constant frame time, but the participating earth stations must make corrections for variations in the satellite range, since the transmitted bursts from all the participating earth stations must reach the satellite in synchronism. Details of the timing requirements will be found in Spilker, 1977.

In the Intelsat system two reference bursts are transmitted in each frame. The first reference burst, which marks the beginning of a frame, is transmitted by a master primary (or a primary) reference station, and contains the timing information needed for the acquisition and synchronization of bursts. The second reference burst, which is transmitted by a secondary reference station, provides synchronization but not acquisition information. The secondary reference burst is ignored by the receiving earth stations unless the primary or master primary station fails.

12.7.1 Reference burst

The reference burst that marks the beginning of a frame is subdivided into time slots or channels used for various functions. These will differ in detail for different networks, but Fig. 12.13 shows some of the basic channels that are usually provided. These can be summarized as follows:

Guard time (*G*). A guard time is necessary between bursts to prevent the bursts from overlapping. The guard time will vary from burst to burst depending on the accuracy with which the various bursts can be positioned within each frame.

Carrier and bit-timing recovery (*CBR*). To perform coherent demodulation of the phase-modulated carrier as described in Secs. 9.6.1 and 9.6.2, a coherent carrier signal must first be recovered from the burst. An unmodulated carrier wave is provided during the first part of the CBR time slot. This is used as a synchronizing signal for a local oscillator at the detector, which then produces an output coherent with the carrier wave. The carrier in the subsequent part of the CBR time slot is modulated by a known phase-change sequence which enables the bit timing to be recovered. Accurate bit timing is needed for the operation of the sample-and-hold function in the detector circuit (see Figs. 9.13 and 9.23). Carrier recovery is described in more detail in Sec. 12.7.3.

Burst code word (*BCW*). (Also known as a *unique word.*) This is a binary word, a copy of which is stored at each earth station. By comparing the incoming bits in a burst with the stored version of the BCW, the receiver can detect when a group of received bits matches the BCW, and this in turn provides an accurate time reference for the burst position in the frame. A known bit sequence is also carried in the BCW, which enables the phase ambiguity associated with coherent detection (see Sec. 9.6.1) to be resolved.

Station identification code (*SIC*). This identifies the transmitting station.

Figure 12.14 shows the makeup of the reference bursts used in certain of the Intelsat networks. The numbers of symbols and the corresponding time intervals allocated to the various functions are shown. In addition to the channels already described, a *coordination and delay channel* (sometimes referred to as the control and delay channel) is provided. This channel carries the identification number of the earth station being addressed, and various codes used in connection with the acquisition and synchronization of bursts at the addressed earth station. It is also necessary for an earth station to know the propagation time delay to the satellite, to implement burst acquisition and synchronization. In the Intelsat system the propagation delay is computed from measurements made at the reference station and transmitted to the earth station in question through the coordination and delay channel.

The other channels in the Intelsat reference burst are the following:

TTY: telegraph order-wire channel, used to provide telegraph communications between earth stations.

SC: service channel which carries various network protocol and alarm messages.

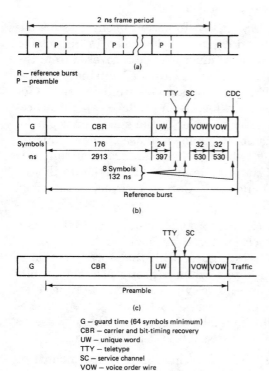

Figure 12.14 (*a*) Intelsat 2-ms frame; (*b*) composition of the reference burst *R*; (*c*) composition of the preamble *P*. (QPSK modulation is used, giving 2 bits per symbol. Approximate time intervals are shown.)

VOW: voice-order-wire channel used to provide voice communications between earth stations. Two VOW channels are provided.

12.7.2   Preamble and postamble

The *preamble* is the initial portion of a traffic burst which carries information similar to that carried in the reference burst. In some systems the channel allocations in the reference bursts and the preambles are identical. No traffic is carried in the preamble. In Fig. 12.13, the only difference between the preamble and the reference burst is that the preamble provides an order-wire (OW) channel.

For the Intelsat format shown in Fig. 12.14, the preamble differs from the reference burst in that it does not provide a coordination and delay channel (CDC). Otherwise, the two are identical.

As with the reference bursts, the preamble provides a carrier and bit-timing recovery channel and also a burst-code-word channel for burst-timing purposes. The burst code word in the preamble of a traffic burst is different from the burst code word in the reference bursts, which enables the two types of bursts to be identified.

In certain phase detection systems the phase detector must be allowed time to recover from one burst before the next burst is received by it. This is termed *decoder quenching,* and a time slot, referred to as a *postamble,* is allowed for this function. The postamble is shown as *Q* in Fig. 12.13. Many systems are designed to operate without a postamble.

12.7.3   Carrier recovery

A factor which must be taken into account with TDMA is that the various bursts in a frame lack coherence so that carrier recovery must be repeated for each burst. This applies to the traffic as well as the reference bursts. Where the carrier recovery circuit employs a phase-locked loop such as shown in Fig. 9.23, a problem known as *hang-up* can occur. This arises when the loop moves to an unstable region of its operating characteristic. The loop operation is such that it eventually returns to a stable operating point, but the time required to do this may be unacceptably long for burst-type signals.

One alternative method utilizes a narrowband tuned circuit filter to recover the carrier. An example of such a circuit for quadrature phase-shift keying (QPSK), taken from Miya, 1981, is shown in Fig. 12.15. The QPSK signal, which has been downconverted to a standard IF of 140 MHz, is quadrupled in frequency to remove the modulation, as described in Sec. 9.7. The input frequency must be maintained at the resonant frequency of the tuned circuit, which requires some form of automatic frequency control. Because of the diffi-

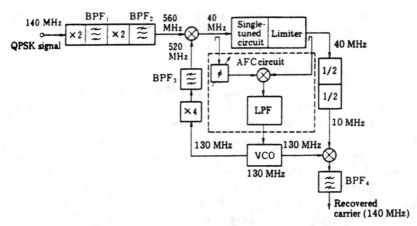

Figure 12.15 An example of carrier recovery circuit with a single-tuned circuit and AFC. (*From Miya, 1981.*)

culties inherent in working with high frequencies, the output frequency of the quadrupler is downconverted from 560 MHz to 40 MHz, and the AFC is applied to the voltage-controlled oscillator (VCO) used to make the frequency conversion. The AFC circuit is a form of phase-locked loop (PLL) in which the phase difference between input and output of the single-tuned circuit is held at zero, which ensures that the 40 MHz input remains at the center of the tuned circuit response curve. Any deviation of the phase difference from zero generates a control voltage which is applied to the VCO in such a way as to bring the frequency back to the required value.

Interburst interference may be a problem with the tuned-circuit method, because of the energy stored in the tuned circuit for any given burst. Avoidance of interburst interference requires careful design of the tuned circuit (Miya, 1981), and possibly the use of a postamble as mentioned in the previous section.

Other methods of carrier recovery are discussed in Gagliardi, 1991.

12.7.4   Network synchronization

Network synchronization is required to ensure that all bursts arrive at the satellite in their correct time slots. As previously mentioned, timing markers are provided by the reference bursts, which are tied to a highly stable clock at the reference station and transmitted through the satellite link to the traffic stations. At any given traffic station, detection of the unique word (or burst code word) in the reference burst signals the *start of receiving frame* (SORF), the marker coinciding with the last bit in the unique word.

It would be desirable to have the highly stable clock located aboard the satellite as this would eliminate the variations in propagation delay arising from the uplink for the reference station, but this is not practical because of weight and space limitations. However, the reference bursts retransmitted from the satellite can be treated, for timing purposes, as if they originated from the satellite (Spilker, 1977).

The network operates what is termed a *burst time plan,* a copy of which is stored at each earth station. The burst time plan shows each earth station where the receive bursts intended for it are relative to the SORF marker. This is illustrated in Fig. 12.16. At earth station A the SORF marker is received after some propagation delay t_A, and the burst time plan tells station A that a burst intended for it follows at time T_A after the SORF marker received by it. Likewise for station B the propagation delay is t_B, and the received bursts start at T_B after the SORF markers received at station B. The propagation delays for each station will differ, but typically they are in the region of 120 ms each.

The burst time plan also shows a station when it must transmit its bursts in order to reach the satellite in the correct time slots. A major advantage of the TDMA mode of operation is that the burst time plan is essentially under software control, so that changes in traffic patterns can be accommodated much more readily than is the case with FDMA, where modifications to hardware are required. Against this, implementation of the synchronization is a complicated process. Corrections must be included for changes in propagation delay which result from the slowly varying position of the satellite (see Sec. 6.4). In general the procedure for transmit timing control has two stages. First there is the need for a station just entering, or reentering after a

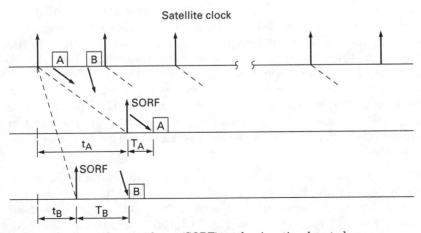

Figure 12.16 Start of receive frame (SORF) marker in a time burst plane.

long delay, to acquire its correct slot position, this being referred to as *burst position acquisition.* Once the time slot has been acquired, the traffic station must maintain the correct position, this being known as *burst position synchronization.*

Open-loop timing control. This is the simplest method of transmit timing. A station transmits at a fixed interval following reception of the timing markers, according to the burst time plan, and sufficient guard time is allowed to absorb the variations in propagation delay. The burst position error can be large with this method and longer guard times are necessary, which reduces frame efficiency (see Sec. 12.7.7). However, for frame times longer than about 45 ms, the loss of efficiency is less than 10 percent. In a modified version of the open-loop method known as *adaptive open-loop timing,* the range is computed at the traffic station from orbital data or from measurements, and the traffic earth station makes its own corrections in timing to allow for the variations in the range. It should be noted that with open-loop timing, no special acquisition procedure is required.

Loopback timing control. Loopback refers to the fact that an earth station receives its own transmission from which it can determine range. It follows that the loopback method can only be used where the satellite transmits a global or regional beam encompassing all the earth stations in the network. A number of methods are available for the acquisition process (see, for example, Gagliardi, 1991), but basically these all require some form of ranging to be carried out so that a close estimate of the slot position can be acquired. In one method, the traffic station transmits a low-level burst consisting of the preamble only. The power level is 20 to 25 dB below the normal operating level (Ha, 1990) to prevent interference with other bursts, and the short burst is swept through the frame until it is observed to fall within the assigned time slot for the station. The short burst is then increased to full power, and fine adjustments in timing are made to bring it to the beginning of the time slot. Acquisition can take up to about 3 s in some cases. Following acquisition the traffic data can be added, and synchronization maintained by continuously monitoring the position of the loopback transmission with reference to the SORF marker. The timing positions are reckoned from the last bit of the unique word in the preamble (as is also the case for the reference burst). The loopback method is also known as *direct closed-loop feedback.*

Feedback timing control. Where a traffic station lies outside the satellite beam containing its own transmission, loopback of the transmission does not of course occur, and some other method must be used for the station to receive ranging information. Where the synchronization information is transmitted back to an earth station from a distant station, this is termed *feedback closed-loop control.* The distant station may be a reference station, as in the Intelsat network, or it may be another traffic station which is a designated "partner." During the acquisition stage, the distant station can feed back information to guide the positioning of the short burst, and once the correct time slot is acquired, the necessary synchronizing information can be fed back on a continuous basis.

Figure 12.17 illustrates the feedback closed-loop control method for two earth stations A and B. The SORF marker is used as a reference point for the burst transmissions. However, the reference point which denotes the start of transmit frame (SOTF) has to be delayed by a certain amount, shown as D_A for earth station A and D_B for earth station B. This is necessary so that the SOTF reference points for each earth station coincide at the satellite transponder, and the traffic bursts, which are transmitted at their designated times after the SOTF, arrive in their correct relative positions at the transponder as shown in Fig. 12.17. The total time delay between any given satellite clock pulse and the corresponding SOTF is a constant, shown as C in Fig. 12.17. C is equal to $2t_A + D_A$ for station A and $2t_B + D_B$ for station B. In general, for earth station i, the delay D_i is determined by

$$2t_i + D_i = C \tag{12.17}$$

In the Intelsat network, $C = 288$ ms.

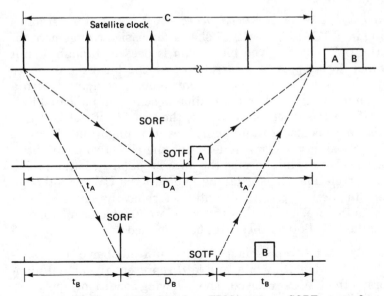

Figure 12.17 Timing relationships in a TDMA system. SORF, start of receive frame; SOTF, start of transmit frame.

For a truly geostationary satellite, the propagation delay t_i would be constant. However, as shown in Sec. 6.4, station-keeping maneuvers are required to keep a geostationary satellite at its assigned orbital position, and hence this position can be held only within certain tolerances. For example, in the Intelsat network the variation in satellite position can lead to a variation of up to $\pm$ 0.55 ms in the propagation delay (Intelsat, 1980). In order to minimize the guard time needed between bursts, this variation in propagation delay must be taken into account in determining the delay D_i required at each traffic station. In the Intelsat network, the D_i numbers are updated every 512 frames, which is a period of 1.024 s, based on measurements and calculations of the propagation delay times made at the reference station. The D_i numbers are transmitted to the earth stations through the CDC channel in the reference bursts. (It should be noted that the open-loop synchronization described previously amounts to using a constant D_i value).

The use of traffic-burst preambles along with reference bursts to achieve synchronization is the most common method, but at least one other method, not requiring preambles, has been proposed by Nuspl and de Buda, 1974. It should also be noted that there are certain types of "packet satellite networks," for example the basic Aloha system (Rosner, 1982), which are closely related to TDMA, in which synchronization is not used.

12.7.5 Unique word detection

The *unique word* (UW) or *burst code word* (BCW) is used to establish burst timing in TDMA. Figure 12.18 shows the basic arrangement for detecting the UW. The received bit stream is passed through a shift register which forms part of a correlator. As the bit stream moves through the register the sequence is continuously compared with a stored version of the UW. When correlation is achieved, indicated by a high output from the threshold detector, the last bit of the UW provides the reference point for timing purposes. It is important therefore to know the probability of error in detecting the UW. Two possibilities have to be considered. One, termed the *miss probability,* is the probability of the correlation detector failing to detect the UW even though it is present in the bit stream. The other, termed the *probability of false alarm,* is the probability that the correlation detector misreads a sequence as the UW. Both of these will be examined in turn.

Miss Probability. Let E represent the maximum number of errors allowed in the UW of length N bits, and let I represent the actual number of errors in the UW as received. The following conditions apply:

When $I \le E$ the detected sequence is declared to be the UW.

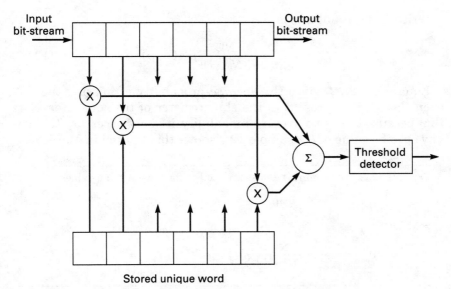

Figure 12.18 Basic arrangement for detection of the unique word (UW).

When $I > E$ the detected sequence N is declared not to be the UW; that is, the unique word is missed.

Let p represent the average probability of error in transmission (the BER). The probability of receiving a sequence N containing I errors in any one particular arrangement is

$$p_I = p^I(1 - p)^{N-I} \tag{12.18}$$

The number of combinations of N things taken I at a time, usually written as $_NC_P$ is given by

$$_NC_I = \frac{N!}{I!(N - I)!} \tag{12.19}$$

The probability of receiving a sequence of N bits containing I errors is therefore

$$P_I = {_NC_I}p_I \tag{12.20}$$

Now since the UW is just such a sequence, Eq. (12.20) gives the probability of a UW containing I errors. The condition for a miss occurring is that $I > E$ and therefore the miss probability is

$$P_{\text{miss}} = \sum_{I=E+1}^{N} P_I \tag{12.21}$$

Written out in full this is

$$P_{\text{miss}} = \sum_{I=E+1}^{N} \frac{N!}{I!(N-I)!} p^I (1-p)^{N-I} \qquad (12.22)$$

Equation (12.22) gives the *average* probability of missing the UW even though it is present in the shift register of the correlator. Note that because this is an average probability, it is not necessary to know any specific value of I. Example 12.2 shows this worked in Mathcad.

Example 12.2.   Determine the miss probability for the following values:

$$N := 40 \qquad E := 5 \qquad p := 10^{-3}$$

solution

$$P_{\text{miss}} := \sum_{I=E+1}^{N} \frac{N!}{I! \cdot (N-I)!} \cdot p^I \cdot (1-p)^{N-I} \qquad P_{\text{miss}} = 3.7 \cdot 10^{-12}$$

False detection probability.   Consider now a sequence of N which is not the UW, but which would be interpreted as the UW even if it differs from it in some number of bit positions E, and let I represent the number of bit positions by which the random sequence actually does differ from the UW. Thus E represents the number of acceptable "bit errors" considered from the point of view of the UW, although they may not be errors in the message they represent. Likewise I represents the actual number of "bit errors" considered from the point of view of the UW although they may not be errors in the message they represent. As before, the number of combinations of N things taken I at a time is given by Eq. (12.19) and hence the number of words acceptable as the UW is

$$W = \sum_{I=0}^{E} {}_N C_I \qquad (12.23)$$

The number of words which can be formed from a random sequence of N bits is 2^N, and on the assumption that all such words are equiprobable, the probability of receiving any one particular word is 2^{-N}. Hence the probability of a false detection is

$$P_F = 2^{-N} W \qquad (12.24)$$

Written out in full this is

$$P_F = 2^{-N} \sum_{I=0}^{E} \frac{N!}{I!(N-I)!} \qquad (12.25)$$

Again it will be noticed that because this is an average probability it is not necessary to know a specific value of *I*. Also, in this case, the BER does not enter into the calculation. A Mathcad calculation is given in Example 12.3.

Example 12.3. Determine the probability of false detection for the following values:

$$N := 40 \qquad E := 5$$

solution

$$P_F := 2^{-N} \cdot \sum_{I=0}^{E} \frac{N!}{I! \cdot (N-I)!} \qquad P_F = 6.9 \cdot 10^{-7}$$

========

From Examples 12.2 and 12.3 it is seen that the probability of a false detection is much higher than the probability of a miss, and this is true in general. In practice, once frame synchronization has been established, a time window can be formed around the expected time of arrival for the UW, such that the correlation detector is only in operation for the window period. This greatly reduces the probability of false detection.

12.7.6 Traffic data

The traffic data immediately follow the preamble in a burst. As shown in Fig. 12.13, the traffic data subburst is further subdivided into time slots addressed to the individual destination stations. Any given destination station selects only the data in the time slots intended for that station. As with FDMA networks, TDMA networks can be operated with both preassigned and demand assigned channels, and examples of both types will be given shortly.

The greater the fraction of frame time that is given over to traffic, the higher the efficiency. The concept of *frame efficiency* is discussed in the next section.

12.7.7 Frame Efficiency and Channel Capacity

The frame efficiency is a measure of the fraction of frame time used for the transmission of traffic. Frame efficiency may be defined as

$$\text{Frame efficiency} = \eta_F = \frac{\text{traffic bits}}{\text{total bits}} \qquad (12.26)$$

Alternatively, this can be written as

$$\eta_F = 1 - \frac{\text{overhead bits}}{\text{total bits}} \qquad (12.27)$$

In these equations, bits per frame are implied. The overhead bits consist of the sum of the preamble, the postamble, the guard intervals, and the reference-burst bits per frame. The equations may be stated in terms of symbols rather than bits, or the actual times may be used.

For a fixed overhead, Eq. (12.27) shows that a longer frame, or greater number of total bits, results in higher efficiency. However, longer frames require larger buffer memories and also add to the propagation delay. Synchronization may also be made more difficult, keeping in mind that the satellite position is varying with time. It is clear that a lower overhead also leads to higher efficiency, but again, reducing synchronizing and guard times may result in more complex equipment being required.

Example 12.4

Calculate the frame efficiency for an Intelsat frame given the following information:

> Total frame length = 120,832 symbols
>
> Traffic bursts per frame = 14
>
> Reference bursts per frame = 2
>
> Guard interval = 103 symbols

solution   From Fig. 12.14, the preamble symbols add up to

$$P = 176 + 24 + 8 + 8 + 32 + 32$$

$$= 280$$

With the addition of the CDC channel the reference channel symbols add up to

$$R = 280 + 8$$

$$= 288$$

Therefore, the overhead symbols are

$$OH = 2 \times (103 + 288) + 14 \times (103 + 280)$$

$$= 6144 \text{ symbols}$$

Therefore, from Eq. (12.27),

$$\eta_F = 1 - \frac{6144}{120,832} = 0.949$$

The voice-channel capacity of a frame, which is also the voice-channel capacity of the transponder being accessed by the frame, can be found from a knowledge of the frame efficiency and the bit rates. Let R_b be the bit rate of a voice channel, and let there be a total of n voice channels shared between all the earth stations accessing the transponder. The total incoming *traffic* bit rate to a frame is nR_b. The traffic bit rate of the frame is $\eta_F R_T$ and therefore

$$nR_b = \eta_F R_T$$

or

$$n = \frac{\eta_F R_T}{R_b} \tag{12.28}$$

Example 12.5

Calculate the voice-channel capacity for the Intelsat frame in Example 12.2, given that the voice-channel bit rate is 64 kb/s, and that QPSK modulation is used. The frame period is 2 ms.

solution The number of symbols per frame is 120,832, and the frame period is 2 ms. Therefore, the symbol rate is 120,832/2 ms = 60.416 megasymbols/second. QPSK modulation utilizes 2 bits per symbol, and therefore the transmission rate is $R_T = 60.416 \times 2 = 120.832$ Mb/s.

Using Eq. (12.28) and the efficiency as calculated in Example 12.4,

$$n = 0.949 \times 120.832 \times \frac{10^3}{64} = 1792$$

12.7.8 Preassigned TDMA

An example of a preassigned TDMA network is the common signaling channel (CSC) for the Spade network described in Sec. 12.5. The frame and burst formats are shown in Fig. 12.19. The CSC can accommodate up to 49 earth stations in the network plus one reference station, making a maximum of 50 bursts in a frame.

All the bursts are of equal length. Each burst contains 128 bits and occupies a 1-ms time slot. Thus the bit rate is 128 kb/s. As discussed in Sec. 12.5, the frequency bandwidth required for the CSC is 160 kHz.

The *signaling unit* (SU) shown in Fig. 12.19 is that section of the data burst which is used to update the other stations on the status of the frequencies available for the SCPC calls. It also carries the signaling information as described in Sec. 12.5.

Another example of a preassigned TDMA frame format is the intelsat frame shown in simplified form in Fig. 12.20. In the Intelsat system, preassigned and demand-assigned voice channels are carried together, but for clarity, only a preassigned traffic burst is shown. The traffic burst is subdivided into time slots, termed *satellite channels* in

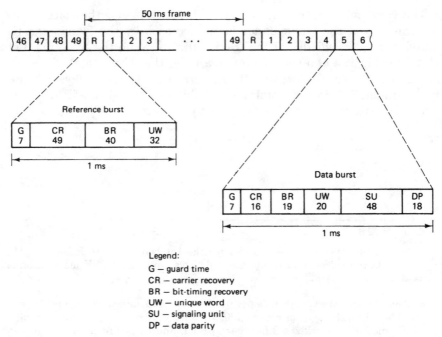

Figure 12.19 Frame and bit formats for the common signaling channel (CSC) used with the Spade system. (*Data from Miya, 1981.*)

the Intelsat terminology, and there can be up to 128 of these in a traffic burst. Each satellite channel is further subdivided into 16 time slots termed *terrestrial channels,* each terrestrial channel carrying one PCM sample of an analog telephone signal. QPSK modulation is used and therefore there are two bits per symbol as shown. Thus each terrestrial channel carries 4 symbols (or 8 bits). Each satellite channel carries $4 \times 16 = 64$ symbols, and at its maximum of 128 satellite channels, the traffic burst carries 8192 symbols.

As discussed in Sec. 9.2, the PCM sampling rate is 8 kHz and with 8 bits per sample, the PCM bit rate is 64 kb/s. Each satellite channel can accommodate this bit rate. Where input data at a higher rate must be transmitted, multiple satellite channels are used. The maximum input data rate which can be handled is $128(SC) \times 64$ kb/s = 8.192 Mb/s.

The Intelsat frame is 120,832 symbols or 241,664 bits long. The frame period is 2 ms, and therefore the burst bit rate is 120.832 Mb/s.

As mentioned previously, preassigned and demand-assigned voice channels can be accommodated together in the Intelsat frame format. The demand-assigned channels utilize a technique known as digital speech interpolation (DSI), which is described in the next section. The

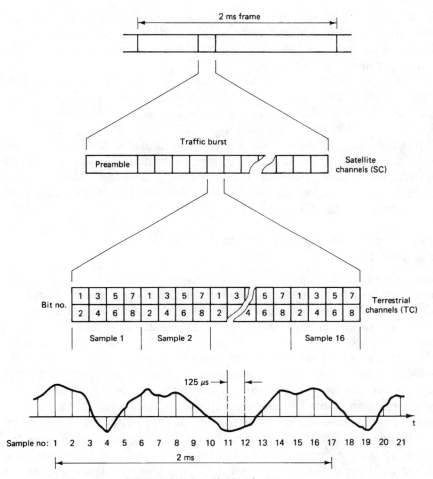

Figure 12.20 Preassigned TDMA frame in the Intelsat system.

preassigned channels are referred to as digital noninterpolated (DNI) channels.

12.7.9 Demand-assigned TDMA

With TDMA, the burst and subburst assignments are under software control, compared to hardware control of the carrier frequency assignments in FDMA. Consequently, compared to FDMA networks, TDMA networks have more flexibility in reassigning channels, and the changes can be made more quickly and easily.

A number of methods are available for providing traffic flexibility with TDMA. The burst length assigned to a station may be varied as the traffic demand varies. A central control station may be employed by the network to control the assignment of burst lengths to each participating station. Alternatively, each station may determine its own burst-length requirements and assign these in accordance with a prearranged network discipline.

As an alternative to burst-length variation, the burst length may be kept constant and the number of bursts per frame used by a given station varied as demand requires. In one proposed system (CCIR Report 708, 1982), the frame length is fixed at 13.5 ms. The basic burst time slot is 62.5 μs, and stations in the network transmit information bursts varying in discrete steps over the range 0.5 ms (8 basic bursts) to 4.5 ms (72 basic bursts) per frame. Demand assignment for speech channels takes advantage of the intermittent nature of speech as described in the next section.

12.7.10 Speech interpolation and prediction

Because of the intermittent nature of speech, a speech transmission channel lies inactive for a considerable fraction of the time it is in use. A number of factors contribute to this. The talk/listen nature of a normal two-way telephone conversation means that transmission in any one direction occurs only about 50 percent of the time. In addition, the pauses between words and phrases may further decrease this to about 33 percent. If further allowance is made for "end-party" delays such as the time required for a party to answer a call, the average fraction of the total connect time may drop to as low as 25 percent. The fraction of time a transmission channel is active is known as the *telephone load activity factor,* and for system design studies the value of 0.25 is recommended by Comité Consutatif Internationale Télégraphique et Téléphonique (CCITT), although higher values are also used (Pratt and Bostian, 1986). The point is that for a significant fraction of the time the channel is available for other transmissions, and advantage is taken of this in a form of demand assignment known as *digital speech interpolation.*

Digital speech interpolation may be implemented in one of two ways, these being digital *time assignment speech interpolation* (digital TASI) and *speech predictive encoded communications* (SPEC).

Digital TASI. The traffic-burst format for an Intelsat burst carrying demand-assigned channels and preassigned channels is shown in Fig. 12.21. As mentioned previously, the demand-assigned channels utilize digital TASI, or what is referred to in the Intelsat nomenclature as DSI for *digital speech interpolation.* These are shown by the block labeled "interpolated" in Fig. 12.21. The first satellite channel (channel 0) in this block is an assignment channel, labeled DSI-AC. No traffic

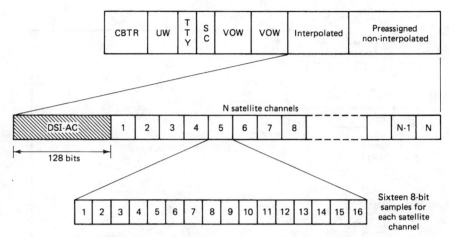

Figure 12.21 Intelsat traffic burst structure. (*From Intelsat, 1983. With permission.*)

is carried in the assignment channel; it is used to transmit channel assignment information as will be described shortly.

Figure 12.22 shows in outline the DSI system. Basically, the system allows N terrestrial channels to be carried by M satellite channels, where $N > M$. For example, in the Intelsat arrangement, $N = 240$ and $M = 127$.

On each incoming terrestrial channel, a speech detector senses when speech is present, the intermittent speech signals being referred to as *speech spurts*. A speech spurt lasts on average about 1.5 seconds (Miya, 1981). A control signal from the speech detector is sent to the channel assignment unit, which searches for an empty TDMA buffer. Assuming that one is found, the terrestrial channel is assigned to this satellite channel, and the speech spurt is stored in the buffer, ready for transmission in the DSI subburst. A delay is inserted in the speech circuit as shown in Fig. 12.22, to allow some time for the assignment process to be completed. However, this delay cannot exactly compensate for the assignment delay, and the initial part of the speech spurt may be lost. This is termed a *connect clip*.

In the Intelsat system an intermediate step occurs where the terrestrial channels are renamed *international channels* before being assigned to a satellite channel (Pratt and Bostian, 1986). For clarity this step is not shown in Fig. 12.22.

At the same time as an assignment is made, an assignment message is stored in the assignment channel buffer which informs the receive stations which terrestrial channel is assigned to which satellite channel. Once an assignment is made, it is not interrupted, even during pauses between spurts, unless the pause times are required for

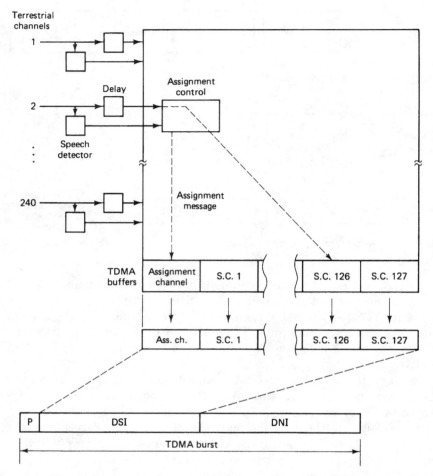

Figure 12.22 Digital speech interpolation. DSI = digital speech interpolation; DNI = digital noninterpolation.

another DSI channel. This reduces the amount of information needed to be transmitted over the assignment channel.

At the receive side, the traffic messages are stored in their respective satellite-channel buffers. The assignment information ensures that the correct buffer is read out to the corresponding terrestrial channel during its sampling time slot. During speech pauses when the channel has been reassigned, a low-level noise signal is introduced at the receiver to simulate a continuous connection.

It has been assumed that a free satellite channel will be found for any incoming speech spurt, but of course there is a finite probability that all channels will be occupied and the speech spurt lost. Losing a speech spurt in this manner is referred to as *freeze-out,* and the freeze-out fraction is the ratio of the time the speech is lost to the average spurt duration. It is found that a design objective of 0.5 percent for a freeze-out fraction is satisfactory in practice. This means that the probability of a freeze-out occurring is 0.005.

Another source of signal mutilation is the *connect clip* mentioned earlier. Again, it is found in practice that clips longer than about 50 ms are very annoying to the listener. An acceptable design objective is to limit the fraction of clips which are equal to or greater than 50 ms, to a maximum of 2 percent of the total clips. In other words, the probability of encountering a clip that exceeds 50 ms is 0.02.

The *DSI gain* is the ratio of the number of terrestrial channels to number of satellite channels, or *N/M*. The DSI gain depends on the number of satellite channels provided as well as the design objectives stated above. Typically, DSI gains somewhat greater than 2 can be achieved in practice.

Speech predictive encoded communications (SPEC). The block diagram for the SPEC system is shown in Fig. 12.23 (Sciulli and Campanella, 1973). In this method the incoming speech signals are converted to a PCM multiplexed signal using 8 bits per sample quantization. With 64 input lines and sampling at 125 μs, the output bit rate from the multiplexer is $8 \times 64/125 = 4.096$ Mb/s.

The digital voice switch following the PCM multiplexer is time-shared between the input signals. It is voice-activated to prevent

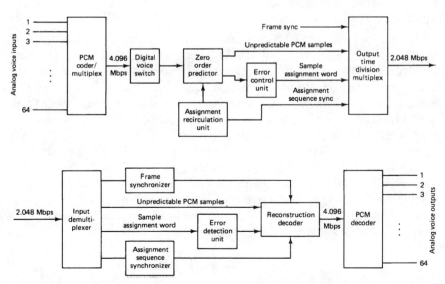

Figure 12.23 (a) SPEC transmitter, (b) SPEC receiver. (*From Sciulli and Campanella, 1973. © 1973—IEEE.*)

transmission of noise during silent intervals. When the zero-order predictor receives a new sample, it compares it with the previous sample for that voice channel, which it has stored, and transmits the new sample only if it differs from the preceding one by a predetermined amount. These new samples are labeled *unpredictable PCM samples* in Fig. 12.23a.

For the 64 channels a 64-bit assignment word is also sent. A logic 1 in the channel for the assignment word means that a new sample was sent for that channel, and a logic 0 means that the sample was unchanged. At the receiver, the sample assignment word either directs the new (unpredictable) sample into the correct channel slot, or it results in the previous sample being regenerated in the reconstruction decoder. The output from this is a 4.096-Mb/s PCM multiplexed signal which is demultiplexed in the PCM decoder.

By removing the redundant speech samples and silent periods from the transmission link, a doubling in channel capacity is achieved. As shown in Fig. 12.23, the transmission is at 2.048 Mb/s for an input–output rate of 4.096 Mb/s.

An advantage of the SPEC method over the DSI method is that freeze-out does not occur during overload conditions. During overload, sample values which should change may not. This effectively leads to a coarser quantization and therefore an increase in quantization noise. This is subjectively more tolerable than freeze-out.

12.7.11 Downlink analysis for digital transmission

As mentioned in Sec. 12.6, the transponder power output and bandwidth both impose limits on the system transmission capacity. With TDMA, the TWT back-off is not generally required, which allows the transponder to operate at saturation. One drawback arising from this is that the uplink station must be capable of saturating the transponder, which means that even a low-traffic-capacity station requires comparatively large power output compared to what would be required for FDMA. This point is considered further in Sec. 12.7.12.

As with the FDM/FDMA system analysis, it will be assumed that the overall carrier-to-noise ratio is essentially equal to the downlink carrier-to-noise ratio. With a power-limited system this C/N ratio is one of the factors that determines the maximum digital rate, as shown by Eq. (9.24). Equation (9.24) can be rewritten as

$$[R_b] = \left[\frac{C}{N_0} \right] - \left[\frac{E_b}{N_0} \right] \tag{12.29}$$

The $[E_b/N_0]$ ratio is determined by the required bit-error rate as shown in Fig. 9.17 and described in Sec. 9.6.4. For example, for a BER of 10^{-5} an $[E_b/N_0]$ of 9.6 dB is required. If the rate R_b is specified, then the $[C/N_0]$ ratio is determined, as shown by Eq. (12.29), and this value is used in the link-budget calculations as required by Eq. (10.57). Alternatively, if the $[C/N_0]$ ratio is fixed by the link-budget parameters as given by Eq. (10.57), the bit rate is then determined by Eq. (12.29).

The bit rate is also constrained by the IF bandwidth. As shown in Sec. 9.5, the ratio of bit rate to IF bandwidth is given by

$$\frac{R_b}{B_{IF}} = \frac{m}{1+\rho}$$

where $m = 1$ for BPSK and $m = 2$ for QPSK and ρ is the roll-off factor. The value of 0.2 is commonly used for the roll-off factor, and therefore the bit rate for a given bandwidth becomes

$$R_b = \frac{mB_{IF}}{1.2} \tag{12.30}$$

Example 12.6

Using Eq. (10.57), a downlink $[C/N_0]$ of 87.3 dBHz is calculated for a TDMA circuit that uses QPSK modulation. A BER of 10^{-5} is required. Calculate the maximum transmission rate. Calculate also the IF bandwidth required assuming a roll-off factor of 0.2.

solution Figure 9.17 is applicable for QPSK and BPSK. From this figure, $[E_b/N_0] = 9.5$ dB for a BER of 10^{-5}. Hence

$$[R_b] = 87.3 - 9.5 = 77.8 \text{ dB b/s}$$

This is equal to 60.25 Mb/s.

For QPSK $m = 2$ and using Eq. (12.30), we have

$$B_{IF} = 60.25 \times \frac{1.2}{2} = 36.15 \text{ MHz}$$

From Example 12.6 it will be seen that if the satellite transponder has a bandwidth of 36 MHz, and an EIRP that results in a $[C/N_0]$ of 87.3 dBHz at the receiving ground station, the system is optimum in that the power and the bandwidth limits are reached together.

12.7.12 Comparison of uplink power requirements for FDMA and TDMA

With frequency-division multiple access, the modulated carriers at the input to the satellite are retransmitted from the satellite as a combined frequency-division-multiplexed signal. Each carrier retains its modulation, which may be analog or digital. For this comparison, digital modulation will be assumed. The modulation bit rate for each carrier is equal to the input bit rate [adjusted as necessary for forward error correction (FEC)]. The situation is illustrated in Fig. 12.24a, where for simplicity, the input bit rate R_b is assumed to be the same for each earth station. The [EIRP] is also assumed to be the same for each earth station.

With time-division multiple access, the uplink bursts which are displaced in time from one another are retransmitted from the satellite as a combined time-division-multiplexed signal. The uplink bit rate is

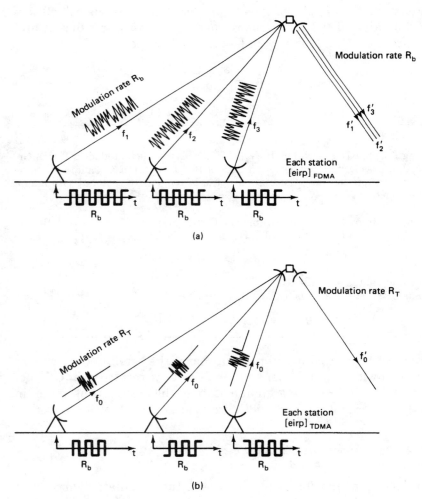

Figure 12.24 (*a*) FDMA network; (*b*) TDMA network.

equal to the downlink bit rate in this case, as illustrated in Fig. 12.24b. As described in Sec. 12.7, compression buffers are needed in order to convert the input bit rate R_b to the transmitted bit rate R_T.

Because the TDMA earth stations have to transmit at a higher bit rate compared to FDMA, a higher [EIRP] is required, as can be deduced from Eq. (9.24). Equation (9.24) states that

$$\left[\frac{C}{N_0}\right] = \left[\frac{E_b}{N_0}\right] + [R]$$

where $[R]$ is equal to $[R_b]$ for an FDMA uplink and $[R_T]$ for a TDMA uplink.

For a given bit error rate (BER) the $[E_b/N_0]$ ratio is fixed as shown by Fig. 9.17. Hence, assuming that $[E_b/N_0]$ is the same for the TDMA and the FDMA uplinks, an increase in $[R]$ requires a corresponding increase in $[C/N_0]$. Assuming that the TDMA and FDMA uplinks operate with the same [LOSSES] and satellite $[G/T]$, Eq. (10.39) shows that the increase in $[C/N_0]$ can be achieved only through an increase in the earth station [EIRP] and therefore

$$[\text{EIRP}]_{\text{TDMA}} - [\text{EIRP}]_{\text{FDMA}} = [R_T] - [R_b] \qquad (12.31)$$

For the same earth station antenna gain in each case, the decibel increase in earth station transmit power for TDMA compared to FDMA is

$$[P]_{\text{TDMA}} - [P]_{\text{FDMA}} = [R_T] - [R_b] \qquad (12.32)$$

Example 12.7

A 14-GHz uplink operates with transmission losses and margins totaling 212 dB and a satellite $[G/T] = 10$ dB/K. The required uplink $[E_b/N_0]$ is 12 dB. (a) Assuming FDMA operation and an earth station uplink antenna gain of 46 dB, calculate the earth station transmitter power needed for transmission of a T1 baseband signal. (b) If the downlink transmission rate is fixed at 74 dBb/s, calculate the uplink power increase required for TDMA operation.

solution (a) From Sec. 9.4 the T1 bit rate is 1.544 Mb/s or $[R] = 62$ dBb/s. Using the $[E_b/N_0] = 12$-dB value specified, Eq. (9.24) gives

$$\left[\frac{C}{N_0}\right] = 12 + 62 = 74 \text{ dBHz}$$

From Eq. (10.39),

$$[\text{EIRP}] = \left[\frac{C}{N_0}\right] - \left[\frac{G}{T}\right] + [\text{LOSSES}] - 228.6$$

$$= 74 - 10 + 212 - 228.6$$

$$= 47.4 \text{ dBW}$$

Hence the transmitter power required is

$$[P] = 47.4 - 46 = 1.4 \text{ dBW or } 1.38 \text{ W}$$

(*b*) With TDMA operation the rate increase is $74 - 62 = 12$ dB. All other factors being equal, the earth station [EIRP] must be increased by this amount, and hence

$$[P] = 1.4 + 12 = 13.4 \text{ dBW or } 21.9 \text{ W}$$

For small satellite business systems it is desirable to be able to operate with relatively small earth stations, which suggests that FDMA should be the mode of operation. On the other hand, TDMA permits more efficient use of the satellite transponder by eliminating the need for back-off. This suggests that it might be worthwhile to operate a hybrid system in which FDMA is the uplink mode of operation, with the individual signals converted to a time-division-multiplexed format in the transponder before being amplified by the TWTA. This would allow the transponder to be operated at saturation as in TDMA. Such a hybrid mode of operation would require the use of a signal-processing transponder as discussed in the next section.

12.8   On-Board Signal Processing for FDMA/TDM Operation

As seen in the preceding section, for small earth stations carrying digital signals at relatively low data rates, there is an advantage to be gained in terms of earth-station power requirements, by using FDMA. On the other hand, TDMA signals make more efficient use of the transponder because back-off is not required.

Market studies show that what is termed *customer premises services* (CPS) will make up a significant portion of the satellite demand over the decade 1990–2000 (Stevenson et al., 1984). Multiplexed digital transmission will be used, most likely at the T1 rate. This bit rate provides for most of the popular services, such as voice, data, and videoconferencing, but specifically excludes standard television signals. Customer premises services is an ideal candidate for the FDMA/TDM mode of operation mentioned in the preceding section. To operate in this mode requires the use of *signal processing transponders*, in which the FDMA uplink signals are converted to the TDM format for retransmission on the downlink. It should also be noted that the use of signal-processing transponders "decouples" the uplink from the downlink. This is important, because it allows the performance of each link to be optimized independently of the other.

A number of signal-processing methods have been proposed. One conventional approach is illustrated in the simplified block schematic of Fig. 12.25a. Here the individual uplink carriers at the satellite are selected by frequency filters and detected in the normal manner. The baseband signals are then combined in the baseband processor, where they are converted to a time-division-multiplexed format for remodulation onto a downlink carrier. More than one downlink carrier may be provided, but only one is shown for simplicity. The disadvantages of the conventional approach are those of excessive size, weight, and power consumption, since the circuitry must be duplicated for each input carrier.

The disadvantages associated with processing each carrier separately can be avoided by means of *group processing,* in which the

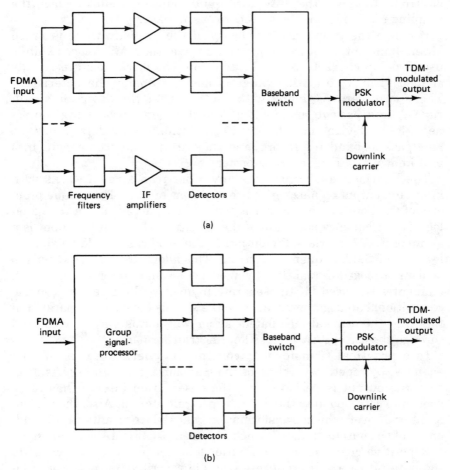

Figure 12.25 On-board signal processing for FDMA/TDM operation; (a) conventional approach; (b) group signal processing.

input FDMA signals are demultiplexed as a group in a single processing circuit, illustrated in Fig. 12.25b. Feasibility studies are being conducted into the use of digital-type group processors, although it would appear that these may require very high speed integrated circuits (VHSICs) not presently available. A different approach to the problem of group processing has been proposed, which makes use of an analog device known as a surface acoustic wave (SAW) Fourier transformer (Atzeni et al., 1975; Hays et al., 1975; Hays and Hartmann, 1976; Maines and Paige, 1976; Nud and Otto, 1975).

In its basic form, the surface acoustic wave device consists of two electrodes deposited on the surface of a piezoelectric dielectric. An electrical signal applied to the input electrode sets up a surface acoustic wave which induces a corresponding signal in the output electrode. In effect, the SAW device is a coupled circuit in which the coupling mechanism is the surface acoustic wave.

Because the propagation velocity of the acoustic wave is much lower than that of an electromagnetic wave, the SAW device exhibits useful delay characteristics. In addition, the electrodes are readily shaped to provide a wide range of useful transfer characteristics. These two features, along with the fact that the device is small, rugged, and passive, makes it a powerful signal-processing component. SAW devices may be used conventionally as delay lines, as bandpass or bandstop filters, and they are the key component in a unit known as a Fourier transformer.

The Fourier transformer, like any other transformer, works with input and output signals which are functions of time. The unique property of the Fourier transformer is that the output signal is a time analog of the frequency spectrum of the input signal. When the input is a group of FDMA carriers, the output in the ideal case would be an analog of the FDMA frequency spectrum. This allows the FDMA signals to be demultiplexed in real time by means of a commutator switch, which eliminates the need for the separate frequency filters required in the conventional analog approach. Once the signals have been separated in this way, the original modulated-carrier waveforms may be recovered through the use of SAW inverse Fourier transformers.

In a practical transformer, continuous operation can only be achieved by repetitive cycling of the transformation process. As a result, the output is periodic, and the observation interval has to be chosen to correspond to the desired spectral interval. Also, the periodic interval over which transformation takes place results in a broadening of the output "pulses" which represent the FDMA spectra.

Repetitive operation of the Fourier transformer at a rate equal to the data bit rate will produce a suitable repetitive output. The relative positions of the output pulses will remain unchanged, fixed by

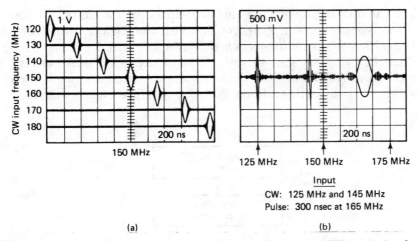

Figure 12.26 Prototype chirp transform of (a) seven successive CW input signals and (b) three simultaneous input signals, including CW and pulsed RF. 200 ns/div; 31.5 MHz./μs chirp rate. (*From Hays and Hartmann, 1976.* © *1976—IEEE.*)

the frequencies of the FDMA carriers. The PSK modulation on the individual FDMA carriers appears in the phase of the carriers within each output pulse. Thus the FDMA carriers have been converted to a pulsed TDM signal. Further signal processing is required before this can be retransmitted as a TDM signal.

Figure 12.26 shows the output obtained from a practical Fourier transformer for various input signals. For Fig. 12.26a the input was seven continuous-wave (CW) signals applied in succession. The output is seen to be pulses corresponding to the line spectra for these waves. The broadening of the lines is a result of the finite time gate over which the Fourier transformer operates. It is important to note that the horizontal axis in Fig. 12.26 is a time axis on which the equivalent frequency points are indicated.

Figure 12.26b shows the output obtained with three simultaneous inputs, two CW waves and one pulsed carrier wave. Again, the output contains two pulses corresponding to the CW signals, and a time function which has the shape of the spectrum for the pulsed wave (Hays and Hartmann, 1976). A detailed account of SAW devices will be found in Morgan, 1985, and in the *IEEE Proceedings,* 1976.

12.9   Satellite-Switched TDMA

More efficient utilization of satellites in the geostationary orbit can be achieved through the use of antenna spot beams, as discussed in Sec. 5.8. The use of spot beams is also referred to as *space-division multi-*

plexing. Further improvements can be realized by switching the antenna interconnections in synchronism with the TDMA frame rate, this being known as *satellite-switched TDMA* (SS/TDMA).

Figure 12.27*a* shows in simplified form the SS/TDMA concept (Scarcella and Abbott, 1983). Three antenna beams are used, each beam serving two earth stations. A 3 × 3 satellite switch matrix is

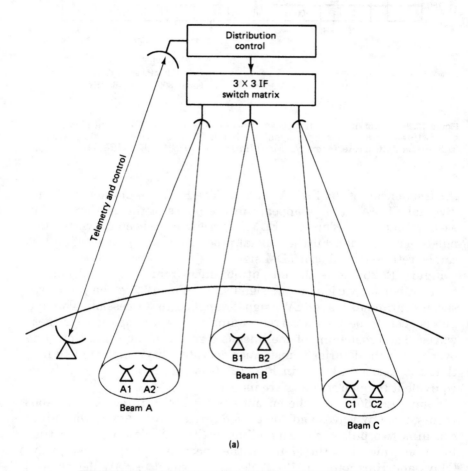

(a)

Input	Output					
	Mode 1	Mode 2	Mode 3	Mode 4	Mode 5	Mode 6
A	A	A	B	C	B	C
B	B	C	A	A	C	B
C	C	B	C	B	A	A

Figure 12.27 (*a*) Satellite switching of three spot beams; (*b*) connectivities or modes.

shown. This is the key component that permits the antenna interconnections to be made on a switched basis. A *switch mode* is a connectivity arrangement. With three beams, six modes would be required for full interconnectivity as shown in Fig. 12.27b, and in general with N beams, $N!$ modes are required for full interconnectivity. Full interconnectivity means that the signals carried in each beam are transferred to each of the other beams at some time in the switching sequence. This includes the loopback connection, where signals are returned along the same beam, enabling intercommunications between stations within a beam. Of course, the uplink and downlink microwave frequencies are different.

Because of beam isolation, one frequency can be used for all uplinks, and a different frequency for all downlinks (e.g., 14 GHz and 12 GHz in the Ku band). To simplify the satellite switch design, the switching is carried out at the intermediate frequency that is common to uplinks and downlinks. The basic block schematic for the 3×3 system is shown in Fig. 12.28.

A *mode pattern* is a repetitive sequence of satellite switch modes, also referred to as *SS/TDMA frames*. Successive SS/TDMA frames need not be identical, as there is some redundancy between modes. For example, in Fig. 12.27b beam A interconnects with beam B in modes 3 and 5, and thus not all modes need be transmitted during each SS/TDMA frame. However, for full interconnectivity the *mode pattern* must contain all modes.

All stations within a beam receive all the TDM frames transmitted in the downlink beam. Each frame is a normal TDMA frame consisting of bursts, addressed to different stations in general. As mentioned, successive frames may originate from different transmitting stations and therefore have different burst formats. The receiving station in a beam recovers the bursts addressed to it in each frame.

The two basic types of switch matrix are the *crossbar matrix* and the

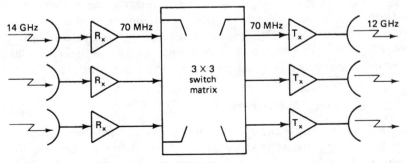

Figure 12.28 Switch matrix in the R.F. link.

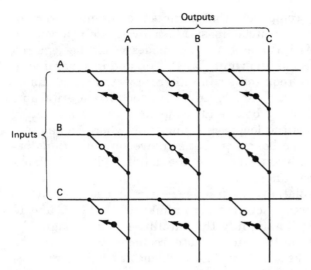

Figure 12.29 3×3 crossbar matrix switch, showing input B connected in the broadcast mode.

rearrangeable network. The crossbar matrix is easily configured for the *broadcast mode,* in which one station transmits to all stations. The broadcast mode with the rearrangeable network type switch is more complex, and this can be a deciding factor in favor of the crossbar matrix (Watt, 1986). The schematic for a 3×3 crossbar matrix is shown in Fig. 12.29, which also shows input beam B connected in the broadcast mode.

The switching elements may be ferrites, diodes, or transistors. The dual-gate FET appears to offer significant advantages over the other types and is considered by some to be the most promising technology (Scarcella and Abbott, 1983).

The schematic for a 4×4 matrix switch as used on the European Olympus satellite is shown in Fig. 12.30 (Watt, 1986). This arrangement is derived from the crossbar matrix. It permits broadcast mode operation, but does not allow more than one input to be connected to one output. Diodes are used as switching elements, and as shown, diode quads are used which provide redundancy against diode failure. It is clear that satellite-switched TDMA adds to the complexity of the on-board equipment and to synchronization requirements.

12.10   Code-Division Multiple Access

In code-division multiple access, signals access the satellite simultaneously. Each signal uses the same carrier frequency and occupies the full transponder bandwidth, but the individual carriers are identified by means of coding. At the earth station transmitter the carrier is code-modulated in a particular way. The earth station receiver at the other end of the link has the code key that enables it to respond to the correct carrier.

The code modulation used for this purpose is distinct from the traffic coding and modulation (although the modulation processes may be

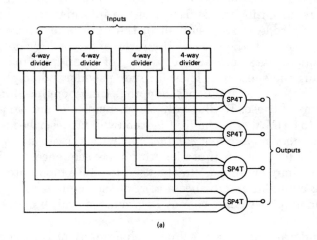

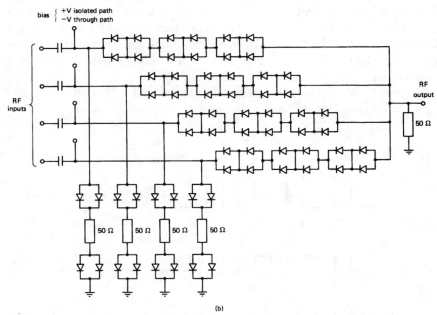

Figure 12.30   (*a*) 4 × 4 switch matrix; (*b*) circuit diagram of redundant SP4T switch element. (*From Watt, 1986; reprinted with permission of IEE, London.*)

combined). The code rate used for the CDMA purpose has to be much greater than the traffic rate, and it is purposely chosen to spread the signal spectrum over the available bandwidth. The method is therefore also referred to as *spread-spectrum multiple access* (SSMA). The abbreviation CDMA will be used here to avoid possible confusion with SS/TDMA, discussed in the preceding section.

Pickholtz et al., 1982, define spread spectrum as a means of transmission in which the signal occupies a bandwidth in excess of the minimum required for transmission of the information and in which the spreading code is independent of the data; and where synchronized reception with the code at the receiver is used for despreading and subsequent data recovery. A number of techniques may be used for spreading the spectrum. These include pseudorandom phase modulation of the carrier, frequency hopping, time hopping, and hybrid combinations of these. To illustrate the principles of spread spectrum as applied to CDMA, only the first method will be described.

Pseudorandom means *like random* but with certain nonrandom or deterministic features. A finite-length binary sequence, in which the bits are randomly arranged, may be used to phase modulate a carrier. By making available in advance a copy of the sequence at the receiver, the random phase changes may be predicted and thus the modulation is "pseudorandom."

The pseudorandom binary sequence has many of the properties of wideband noise, and such a sequence is known generally as a pseudonoise, or *PN sequence*. The sequence is clocked in a regular fashion, as shown in Fig. 12.31. What would be known as *bits* for a binary signal are referred to as *chips* for the PN sequence. The clock rate is measured in chips per second (c/s).

Figure 12.32 shows how the spectrum may be spread. Data are modulated onto the carrier in the normal manner. For clarity it is as-

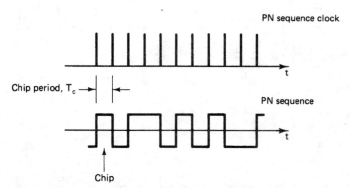

Figure 12.31 PN binary sequence. One element is known as a chip.

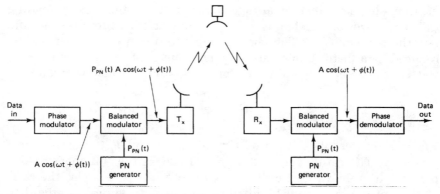

Figure 12.32 Spread spectrum achieved using a PN sequence.

sumed that the data modulator stage is separate from the code modulator stage. The data-modulated carrier is represented by $A \cos [\omega t + \phi(t)]$, where $\phi(t)$ is the phase modulation. This signal is passed on to the balanced modulator used for code modulation, where it is multiplied by the PN sequence. Representing the PN sequence by $P_{PN}(t)$, the output from the transmitter-balanced modulator is

$$e_T = P_{PN}(t)A \cos [\omega t + \phi(t)] \qquad (12.33)$$

At the receiver this signal is multiplied by the stored replica of $P_{PN}(t)$. Assuming that proper synchronization is achieved between transmit and receive codes, the output from the receiver-balanced modulator is

$$e_R = P_{PN}^2(t)A \cos [\omega t + \phi(t)] \qquad (12.34)$$

But because $P_{PN}(t)$ is a binary signal, $P_{PN}^2(t) = 1$, and hence the output from the receiver balanced modulator is

$$e_R = A \cos [\omega t + \phi t)] \qquad (12.35)$$

The data modulation can now be recovered from this in the normal manner.

It will be seen that if the receiver and transmitter codes do not match, either because they are not properly synchronized, or because the wrong code is used at the receiver, the $P_{PN}^2(t) = 1$ condition will not occur and random phase modulation results which corrupts the data.

The need to maintain highly accurate synchronization between transmit and receive coding, and the problems associated with the initial acquisition of sync, does not make the use of CDMA attractive for commercial satellite systems. However, spread-spectrum communications have distinct advantages for military systems. By spreading the signal over a wide bandwidth it becomes difficult to intercept and jam the transmission. The jamming, or interference signal energy, is reduced by a factor known as the *processing gain* (analogous in some ways to the processing gain used for FM systems), which is given by

$$G_P = \frac{T_b}{T_c} \tag{12.36}$$

where T_b represents the period for a data bit and T_c the period for a code chip. The ratio expressed by Eq. (12.36) is the "number of chips per bit." It is the ratio of coding rate to data rate and is also the ratio of the spread-spectrum bandwidth to the data signal bandwidth. Typically, the number of chips per bit may range from about 50 to 500.

If E_J represents the jamming signal energy at the input to the despreader, the jamming energy at the output will be E_J/G_P or $[E_J] - [G_P]$ dB. Thus the output signal-to-interference ratio will be improved by this amount. It should be noted, however, that spread-spectrum techniques do not improve performance as regards wideband noise such as thermal noise. A detailed account of spread-spectrum systems will be found in *IEEE Transactions on Communications,* 1982, and Dixon, 1984.

12.11   Problems

12.1.   Explain what is meant by a *single access* in relation to a satellite communications network. Give an example of the type of traffic route where single access would be used.

12.2.   Distinguish between *preassigned* and *demand-assigned traffic* in relation to a satellite communications network.

12.3.   Explain what is meant by *frequency-division multiple access,* and show how this differs from frequency-division multiplexing.

12.4.   Explain what the acronym SCPC stands for. Explain in detail the operation of a preassigned SCPC network.

12.5.   Explain what is meant by *thin route service.* What type of satellite access is most suited for this type of service?

12.6.   Briefly describe the ways in which demand assignment may be carried out in an FDMA network.

12.7.   Explain in detail the operation of the Spade system of demand assignment. What is the function of the common signaling channel?

12.8.   Explain what is meant by *power limited* and *bandwidth limited* operation as applied to an FDMA network. In an FDMA scheme the carriers utilize equal powers and equal bandwidths, the bandwidth in each case being 5 MHz. The transponder bandwidth is 36 MHz. The saturation EIRP for the downlink is 34 dBW, and an output backoff of 6 dB is employed. The downlink losses are 201 dB, and the destination earth station has a G/T ratio of 35 dBK^{-1}. Determine the $[C/N]$ value assuming this is set by single carrier operation. Determine also the number of carriers which can access the system, and state, with reasons, whether the system is power limited or bandwidth limited.

12.9. Distinguish between *bandwidth-limited* and *power-limited operation* as applied to an FDMA network.

12.10 In some situations it is convenient to work in terms of the carrier-to-noise temperature. Show that $[C/T] = [C/N_0] + [k]$. The downlink losses for a satellite circuit are 196 dB. The earth station $[G/T]$ ratio is 35 dB/K and the received $[C/T]$ ratio is -138 dBW/K. Calculate the satellite [EIRP].

12.11. The earth station receiver in a satellite downlink has an FM detector threshold level of 10 dB and operates with a 3-dB threshold margin. The emphasis improvement figure is 4 dB, and the noise-weighting improvement figure is 2.5 dB. The required $[S/N]$ ratio at the receiver output is 46 dB. Calculate the receiver processing gain. Explain how the processing gain determines the IF bandwidth.

12.12. A 252-channel FM/FDM telephony carrier is transmitted on the downlink specified in Prob. 12.10. The peak/rms ratio factor 2[g] is 10 dB, and the baseband bandwidth extends from 12 to 1052 kHz. The voice-channel bandwidth is 3.1 kHz. Calculate the peak deviation, and hence, using Carson's rule, calculate the IF bandwidth.

12.13. Given that the IF bandwidth for a 252-channel FM/FDM telephony carrier is 7.52 MHz and that the required $[C/N]$ ratio at the earth station receiver is 13 dB, calculate (a) the $[C/T]$ ratio and (b) the satellite [EIRP] required if the total losses amount to 200 dB, and the earth station $[G/T]$ ratio is 37.5 dB/K.

12.14. Determine how many carriers can access an 80-MHz transponder in the FDMA mode, given that each carrier requires a bandwidth of 6 MHz, allowing for 6.5-dB output back-off. Compare this number with the number of carriers possible without back-off.

12.15. (a) Television transmissions may be classified as *full-transponder* or *half-transponder transmissions*. State what this means in terms of transponder access. (b) A composite TV signal (video plus audio) has a top baseband frequency of 6.8 MHz. Determine for a 36-MHz transponder the peak frequency deviation limit set by (1) half-transponder and (2) full-transponder transmission.

12.16. Describe the general operating principles of a *time-division multiple access network*. Show how the transmission bit rate is related to the input bit rate.

12.17. Explain the need for a reference burst in a TDMA system.

12.18. Explain the function of the preamble in a TDMA traffic burst. Describe and compare the channels carried in a preamble with those carried in a reference burst.

12.19. What is the function of (*a*) the burst-code word and (*b*) the carrier and bit-timing recovery channel in a TDMA burst?

12.20. Explain what is meant by (*a*) *initial acquisition* and (*b*) *burst synchronization* in a TDMA network. (*c*) The nominal range to a geostationary satellite is 42,000 km. Using the station-keeping tolerances stated in Sec. 6.4 in connection with Fig. 6.10, determine the variation expected in the propagation delay.

12.21. (*a*) Define and explain what is meant by *frame efficiency* in relation to TDMA operation. (*b*) In a TDMA network the reference burst and the preamble each requires 560 bits, and the nominal guard interval between bursts is equivalent to 120 bits. Given that there are eight traffic bursts and one reference burst per frame, and the total frame length is equivalent to 40,800 bits, calculate the frame efficiency.

12.22. Given that the frame period is 2 ms and the voice-channel bit rate is 64 kb/s, calculate the equivalent number of voice channels that can be carried by the TDMA network specified in Prob. 12.21.

12.23. Calculate the frame efficiency for the CSC shown in Fig. 12.19.

12.24. (*a*) Explain why the frame period in a TDMA system is normally chosen to be an integer multiple of 125 µs. (*b*) Referring to Fig. 12.20 for the Intelsat preassigned frame format, show that there is no break in the timing interval for sample 18 when this is transferred to a burst.

12.25. Show that, all other factors being equal, the ratio of uplink power to bit rate is the same for FDMA and TDMA. In a TDMA system the preamble consists of the following slots, assigned in terms of number of bits: bit timing recovery 304; unique word 48; station identification channel 8; order wire 64. The guard slot is 120 bits, the frame reference burst is identical to the preamble, and the burst traffic is 8192 bits. Given that the frame accommodates 8 traffic bursts, calculate the frame efficiency. The traffic is preassigned PCM voice channels for which the bit rate is 64 kb/s, and the satellite transmission rate is nominally 60 Mb/s. Calculate the number of voice channels which can be carried.

12.26. In comparing design proposals for multiple access, the two following possibilities were considered: (1) uplink FDMA with downlink TDM and (2) uplink TDMA with downlink TDM. The incoming baseband signal is at 1.544 Mb/s in each case and the following table shows values in decilogs:

	Uplink	Downlink
$[E_b/N_0]$	12	12
$[G/T]$	10	19.5
[Losses]	212	210
[EIRP]	—	48
Transmit antenna gain $[G_T]$	45.8	—

Determine (*a*) the downlink TDM bit rate, and (*b*) the transmit power required at the uplink earth station for each proposal.

12.27. A TDMA network utilizes QPSK modulation and has the following symbol allocations: guard slot 32; carrier and bit timing recovery 180; burst code word (unique word) 24; station identification channel 8; order wire 32; management channel (reference bursts only) 12; service channel (traffic bursts only) 8. The total number of traffic symbols per frame is 115010, and a frame consists of two reference bursts and 14 traffic bursts. The frame period is 2 ms. The input consists of PCM channels each with a bit rate of 64 kb/s. Calculate the frame efficiency and the number of voice channels that can be accommodated.

12.28. For the network specified in problem 12.27 the BER must be at most 10^{-5}. Given that the receiving earth station $[G/T]$ value is 30 decilogs, and total losses are 200 dB, calculate the satellite [EIRP] required.

12.29. Discuss briefly how demand assignment may be implemented in a TDMA network. What is the advantage of TDMA over FDMA in this respect?

12.30. Define and explain what is meant by the terms *telephone load activity factor* and *digital speech interpolation*. How is advantage taken of the load activity factor in implementing digital speech interpolation?

12.31. Define and explain the terms *connect clip* and *freeze-out* used in connection with digital speech interpolation.

12.32. Describe the principles of operation of a speech predictive encoded communications (SPEC) system, and state how this compares with digital speech interpolation.

12.33. Determine the bit rate that can be transmitted through a 36-MHz transponder, assuming a roll-off factor of 0.2 and QPSK modulation.

12.34. On a satellite downlink, the $[C/N_0]$ ratio is 86 dBHz, and an $[E_b/N_o]$ of 12 dB is required at the earth station. Calculate the maximum bit rate that can be transmitted.

12.35. FDMA is used for uplink access in a satellite digital network, with each earth station transmitting at the T1 bit rate of 1.544 Mb/s. Calculate (*a*) the uplink $[C/N_0]$ ratio required to provide a $[E_b/N_0] = 14$ dB ratio at the satellite and (*b*) the earth-station [EIRP] needed to realize the $[C/N_0]$ value. The satellite $[G/T]$ value is 8 dB/K, and total uplink losses amount to 210 dB.

12.36. In the satellite network of Prob. 12.35, the downlink bit rate is limited to a maximum of 74.1 dBb/s, with the satellite TWT operating at saturation. A 5-dB output back-off is required to reduce intermodulation products to an acceptable level. Calculate the number of earth stations that can access the satellite on the uplink.

12.37. The [EIRP] of each earth station in an FDMA network is 47 dBW, and the input data are at the T1 bit rate with 7/8 FEC added. The downlink bit rate is limited to a maximum of 60 Mb/s with 6-dB output back-off applied. Compare the [EIRP] needed for the earth stations in a TDMA network utilizing the same transponder.

12.38. (*a*) Describe the general features of an on-board signal processing transponder that would allow a network to operate with FDMA uplinks and a

TDM downlink. (*b*) In such a network, the overall BER must not exceed 10^{-5}. Calculate the maximum permissible BER of each link, assuming that each link contributes equally to the overall value.

12.39. Explain what is meant by *full interconnectivity* in connection with satellite switched TDMA. With four beams, how many switch modes would be required for full interconnectivity?

12.40. Identify all the redundant modes in Fig. 12.27.

12.41. Describe the principles of operation of code-division multiple access. What effect do the unwanted signals have on the wanted signal?

Satellite Services

13.1 Introduction

The idea that three geostationary satellites could provide communications coverage for the whole of the earth, apart from relatively small regions at the north and south poles, is generally credited to Arthur C. Clarke, 1945. The basic idea was sound, but of course the practicalities led to the development of a much more complex undertaking than perhaps was envisaged originally. Technological solutions were found to the many problems that were encountered, and as a result, satellite services expanded into many new areas. Geostationary satellites are still the most numerous, and well in excess of three! If an average of 2° spacing is assumed, the geostationary orbit could hold 180 such satellites. Of course the satellites are not deployed evenly around the orbit but are clustered over regions where services are most in demand. Direct-to-home broadcasting, referred to as *direct broadcast satellite* (DBS) service in the United States, represents one major development in the field of geostationary satellites. Another is the use of *very small aperture terminals* (VSATs) for business applications. A third geostationary development is *mobile satellite service* (MSAT) which extends geostationary satellites services into mobile communications for vehicles, ships, and aircraft.

Rapid development has also been taking place in services using nongeostationary satellites. *Radarsat* is a large polar-orbiting satellite designed to provide environmental monitoring services. Possibly the most notable development in the area of nongeostationary satellites is the *Global Positioning Satellite* (GPS) system which has come into everyday use for surveying and position location generally.

Although the developments in satellites have generally led to a need for larger satellites, a great deal has been happening at the other end of the size spectrum, in what is referred to as *microsats* (see for example Sweeting, 1992, and Williamson, 1994). Two major communications systems using small, nongeostationary satellites are *Orbcomm,* a packet data system being developed by Orbital Sciences Corporation, and *Iridium,* a personal communications system being developed by Motorola.

In this final chapter, brief descriptions are given of some of these services to illustrate the range of applications now found for satellites. At the time of writing (1994) some of the services are in the proposal stage, and others are going through an evolutionary period while still in the planning stage. For example, the launch of MSAT, originally planned for 1992, was postponed to take place in 1994, with full service being offered in 1995, the pretrial phase of the MSAT project having been discontinued. Likewise, Iridium is still considered a controversial, and possibly impractical, project.

13.2   Direct Broadcast Satellite Services

Satellites provide *broadcast* transmissions in the fullest sense of the word, since the antenna footprints can be made to cover large areas of the earth. The idea of using satellites to provide direct transmissions into the home has been around for many years. A comprehensive overview covering the state of DBS in Europe, the United States, and other countries is given by Pritchard and Ogata, 1990. Some of the regulatory and commercial aspects of European DBS are given by Chaplin, 1992, and the U.S. market is discussed by Reinhart, 1990. Reinhart defines three categories of United States DBS systems which are summarized in Table 1.4, and repeated here in Table 13.1 for convenience.

The frequency bands used for DBS television vary from region to region throughout the world, but for the present discussion, which is concerned mainly with downlink analysis, a typical frequency of 12 GHz will be assumed. A typical uplink frequency is 14 GHz, these being in the Ku band as described in Chap. 1. Satellites operating at these frequencies carry the frequency designation 14/12 GHz. Also as mentioned in Chap. 1, the C band is designated as 6/4 GHz, and it is not uncommon to find dual-band satellites which provide services in both bands simultaneously.

There are a number of reasons for the use of the higher frequency Ku band (rather than the C band) for DBS services. One major reason is the availability of frequencies at the higher frequencies. Also the Ku band is not a designated band for terrestrial microwave systems, and so the interference problems associated with the C band, as described in Chap. 11, do not arise. A feature readily observed with Ku-band reception is the comparatively small antennas which can be

TABLE 13.1   Defining Characteristics of Three Categories of United States DBS systems

	High power	Medium power	Low power
Band	Ku	Ku	C
Downlink frequency allocation, GHz	12.2–12.7	11.7–12.2	3.7–4.2
Uplink frequency allocation, GHz	17.3–17.8	14–14.5	5.925–6.425
Space service*	BSS	FSS	FSS
Primary intended use	DBS	Point to point	Point to point
Allowed additional use	Point to point	DBS	DBS
Terrestrial interference possible?	No	No	Yes
Satellite spacing, degrees	9	2	2–3
Satellite spacing determined by†	ITU	FCC	FCC
Adjacent satellite interference possible?	No	Yes	Yes
Satellite EIRP range, dBW	51–60	40–48	33–37

SOURCE: Reinhart, 1990.

*BSS = broadcasting satellite service; FSS = fixed satellite service.

†ITU = International Telecommunications Union; FCC = Federal Communications Commission.

used. The antennas commonly utilize parabolic reflectors (or "dishes"), as described in Sec. 5.13. Ku-band dishes are around 60 cm in diameter, compared to about 3 m for C-band antennas. The smaller antenna size at the receiver is a direct result of the higher EIRP available from the DBS satellites operating at Ku-band frequencies, compared to the C-band satellites. As shown in Sec. 10.3.1 the received power is independent of frequency for a given EIRP, and therefore reducing the receive antenna size results in a reduction in received power, all other factors remaining constant.

One antenna advantage does result from the higher-frequency operation. As shown by Eq. (5.30) the antenna −3-dB beamwidth in degrees is given by

$$\theta_{-3dB} = 70\frac{\lambda}{D} \tag{13.1}$$

A 3-m dish receiving at 4 GHz has a beamwidth of 1.75°. At 12 GHz, since the frequency is increased by a factor of 3, the dish size can be decreased by a factor of 3 for the same beamwidth, and thus a 1-m dish would suffice. In practice, the beamwidth has to be sufficiently narrow to prevent interference from adjacent satellites, as described in Chap. 11, and a 60-cm dish proves to be more than adequate for the satellite spacings used in the DBS service.

Another factor which has to be considered is the surface irregularities which occur in manufacture of antennas. Irregularities cause a reduction in gain by scattering the signal. The reduction can be expressed as a function of the root-mean-square (rms) deviation of the surface, with reference to the ideal parabolic surface. This is a manufacturing tolerance which is more easily held the smaller the antenna. The reduction in gain is given by (see, for example, Baylin and Gale, 1991)

$$\eta_{rms} = e^{-8.8\sigma/\lambda} \tag{13.2}$$

where σ is the rms tolerance in the same units as wavelength λ. For example, at 12 GHz (wavelength 2.5 cm) and for an rms tolerance of 1 mm, the gain is reduced by the factor

$$\eta_{rms} = e^{-8.8\,(0.1/2.5)}$$

$$= 0.7 \tag{13.3}$$

This is a reduction of about 1.5 dB.

At the 1983 Regional Administrative Radio Conference the regulations for DBS were set out, and among these was the recommendation that circular polarization be used for DBS (Pritchard and Ogata, 1990, p. 1132). However, in the United States this applies only to high-power DBS. The advantages of circular polarization are discussed in Chap. 4. Table 13.2 summarizes the DBS position in the United States.

TABLE 13.2   Summary Comparison of United States DBS Systems

Parameter	High power	Medium power	Low power
Downlink band, GHz	12.2–12.7	11.7–12.2	3.7–4.2
Satellite coverage*	F and H	F and H	F
Transponder power, W	100–260	15–45	5–10
Transponder EIRP, dBW	51–58	40–48	33–37
Polarization†	RHC and LHC	H and V	H and V
Typical number of transponders per satellite	8–16	10–16	24
Transponder bandwidth, MHz	24	43–72	36
Home antenna diameter, m	0.3–0.8	1–1.6‡	2.5–4.8
Receiver noise temperature, K	100–200	100–200	35–80
TVRO cost installed, US$	400–800	600–1200	2000–4000

SOURCE: Reinhart, 1990.

*F = full CONUS; H = half CONUS. (CONUS = contiguous United States.)

†RHC = right-hand circular; LHC = left-hand circular.

‡Estimate for future dedicated system with very large number of TV receive-only (TVRO) stations.

The video programming used in the United States varies, as encryption is used for some channels. Encryption methods include General Instrument's *video-cipher* and Scientific Atlanta's *BMAC* system. Most other channels are transmitted using frequency-modulated NTSC 525-line standard M (Reinhart, 1990).

The acronym *MAC* stands for *multiplexed analog components* and is a system of modulation introduced first in Europe, to eliminate certain forms of distortion present in the conventional modulation systems (PAL, SECAM, NTSC) and which arise from interference between the luminance and chroma signals (see Sec. 8.5). In the basic MAC system the luminance and chroma signals are time-division-multiplexed during the horizontal line scan. A number of variants of the MAC system have been developed although the European Broadcasting Union has recommended the adoption of C-MAC for European DBS TV. The main characteristics of the competing MAC systems are shown in Table 13.3.

This review has been mainly concerned with some of the technical aspects of DBS. Commercial factors are not so readily analyzed, as competition with cable companies, and the installation of fiber optics, along with the prospect of telephone companies offering video services

TABLE 13.3   MAC Systems

Attribute	C-MAC	D2-MAC	B-MAC (625)	B-MAC (525)
Lines/frame rate	625/25	625/25	625/25	525/29.97
Line frequency, Hz	15,625	15,625	15,625	15,625
Time increments per line	1296	1296	1365	1365
Active lines	574	574	574	483
Aspect ratio	4:3	4:3	4:3	4:3
Luminance compression	3:2	3:2	3:2	3:2
Chrominance compression	3:1	3:1	3:1	3:1
Sound/data	QPSK/BPSK	Duobinary	Quaternary/binary	Quaternary/binary
Bit rate, Mb/s	3.08	1.54	1.59	1.6
Multiplexing	Separate RF carrier	Packet multiplexer at baseband	Baseband	Baseband
Baseband, MHz	5.6	5.6	5	4.2
RF bandwidth, MHz	27	27	27	24
Used in	Scandinavia	TV satellite TDF-1, Europe	Australia	U.S./Canada

SOURCE: Pritchard and Ogata, 1990.

which might eventually include TV programming all are factors which have to be considered. Satellite direct broadcast might also be used for services other than television, for example, audio programming and data services. It is an active field of satellite development in many countries and its future is intimately linked with developments in other fields, such as high-definition television. According to Chaplin, 1992, the number of transponders available for the European markets is likely to increase from 130 to nearly 300 by the mid-1990s. After allowing for projected use by business, and additional TV programming, the number of idle or occasional-use transponders could amount to about 65, or 23 percent of the total available.

13.3 MSAT

The Mobile Satellite (MSAT) system is intended to complement existing mobile communications systems such as those provided by cellular radio operators, radio common carriers, and telephone companies. Existing mobile services do not adequately cover many rural and remote areas, this being especially true in Canada where the MSAT concept was first proposed. The MSAT system is expected to be in service by 1995, and initially will consist of one Canadian and one American satellite. The Canadian satellite is operated by TMI Communications; the original company, Telesat Mobile Inc. or TMI, filed for bankruptcy protection in April of 1993, and the new company was formed by BCE Inc., Canada's largest telecommunications company (MSAT, 1994). The American satellite is operated by American Mobile Satellite Consortium (AMSC).

The satellite-to-mobile links will use L-band frequencies, the downlink frequencies being in the range 1550 to 1559 MHz, and the uplink frequencies in the range 1626.5 to 1660.5 MHz. These bands are divided into subbands as allocated at the World Administrative Radio Conference (WARC), in 1987, some of which are reserved for special services. Communications between the satellites and the fixed stations will use frequencies in the Ku band. Studies have shown that there will be little call for mobile-to-mobile communications, and as a result this link-up will be made as a double hop through the network control center. Because of the limited bandwidth available in the L band, 5-kHz channels have been allocated, which require the use of bandwidth reduction schemes such as amplitude-companded single sideband (ACSSB) for analog channels and linear predictive coding (LPC) for digital channels.

In the baseline design presented in MSAT, 1988, each satellite will employ two large (at least 5-m-diameter) L-band reflector antennas which will generate nine beams covering Canada and the United

States, with the possibility of adding two more beams to provide mobile satellite services to Mexico. Frequency reuse in the multiple-beam system, and frequency sharing and coordination with other users, will be required. At least two antenna types are available for use with the mobile units, an omnidirectional antenna with a gain of 4 dBi, and a higher gain steerable antenna with a gain of 10 dBi. Circular polarization will be used for the L-band satellite-to-mobile links, and linear polarization for the Ku-band links. Typical MSAT service applications are shown in Table 13.4.

The services offered by MSAT can be classified as:

Mobile radio trunking service (MRTS): This is a mobile dispatch service similar to the terrestrial radio services presently available, but of course it offers wide-area coverage not possible with the terrestrial systems. The service can be private or shared. In the private network, the base station will be owned and operated by the customer's organization, while in the shared network, the base station will be shared among several customers, with the service provider operating the station. Access to the public telephone switched network (PSTN) will be available only on a very restricted basis.

Interconnected mobile radio service (IMRS): This is a radio service which connects users to the PSTN where access is not otherwise available. Calls from mobile units will be forwarded through the MSAT to a gateway earth station, which connects up to the PSTN. In addition to mobile units, fixed stations can also use this service. Here, *fixed* refers to transportable earth stations which typically might be installed in remote communities.

TABLE 13.4   Typical MSAT Applications

Public safety	Police, ambulance, search and rescue, fire fighting, and emergency relief operations
Aeronautical	Operational communications to commercial aircraft, public correspondence, air traffic control and safety
Marine	Operational communications to domestic coastal fishing vessels, Coast Guard operations, research vessels, oceanography data acquisition, and electronic data broadcast to marine vessels
Land applications	Construction projects in remote areas, resource development; forestry and oil and mineral explorations, environmental monitoring, pipeline and oil well operations, and hydroelectric generation; transportation: shipment of hazardous cargo, just-in-time operations, wide-area vehicle monitoring and vehicle positioning

SOURCE: MSAT, 1988.

Mobile data service (MDS): This service is to provide two-way data links between mobile and fixed terminals and should be of particular interest to the transportation industry. Some suggested services (MSAT, 1988) are interactive two-way digital messaging; vehicle dispatch and position location; data acquisition and control; data broadcast; wide-area paging and page alert; automatic wide-area vehicle monitoring; one-way emergency signaling; and electronic data interexchange. It is foreseen that periodic communication of vehicle information and precoded messages would provide position location, cargo monitoring, data collection, and two-way messaging. The mobile data terminal will be similar to the mobile transreceiver with the addition of an alphanumeric display and a keypad or keyboard.

13.4   VSATs

VSAT stands for *very small aperture terminal* system. This is the distinguishing feature of a VSAT system, the earth station antennas being typically less than 2.4 m in diameter (Rana et al., 1990). The trend is toward even smaller dishes, not more than 1.5 m in diameter (Hughes et al., 1993). In this sense, the small TVRO terminals described in Sec. 13.2 for direct broadcast satellites could be labeled as VSATs, but the appellation is usually reserved for private networks, mostly providing two-way communications facilities. Typical user groups include banking and financial institutions, airline and hotel booking agencies, and large retail stores with geographically dispersed outlets.

The basic structure of a VSAT network consists of a hub station which provides a broadcast facility to all the VSATs in the network, and the VSATs themselves which access the satellite in some form of multiple-access mode. The hub station is operated by the service provider, and it may be shared among a number of users, but of course each user organization has exclusive access to its own VSAT network. Time-division multiplex is the normal downlink mode of transmission from hub to the VSATs, and the transmission can be broadcast for reception by all the VSATs in a network, or address coding can be used to direct messages to selected VSATs.

Access the other way, from the VSATs to the hub, is more complicated, and a number of different methods are in use, many of them proprietary. A comprehensive summary of methods is given in Rana et al., 1990. The most popular access method is frequency-division multiple access (FDMA), which allows the use of comparatively low-power VSAT terminals (see Sec. 12.7.12). Time-division multiple access (TDMA) can also be used, but is not efficient for low-density up-link traffic from the VSAT. The traffic in a VSAT network is mostly data transfer of a bursty nature, examples being inventory control, credit verification, and reservation requests, occurring at random and possibly infrequent intervals, so that allocation of time slots in the normal TDMA mode can lead to low channel occupancy. A form of demand assigned multiple access (DAMA) is employed in some systems in which channel capacity is assigned in response to the fluctuating demands of the VSATs in the network. DAMA can be used with FDMA as well as TDMA, but the disadvantage of the method is that a *reserve channel* must be instituted through which the VSATs can make requests for channel allocation. As pointed out by Abramson, 1990, the problem of access then shifts to how the users may access the reserve channel in an efficient and equitable manner. Abramson presents a method of code-division multiple access (CDMA) using spread-spectrum techniques, coupled with the Aloha protocol. The basic Aloha method is a random-access method in which packets are transmitted at random in defined time slots. The system is used where the packet time is small compared to the slot time, and provision is made for dealing with packet collisions which can occur with packets sent up from different VSATs. Abramson calls the method *spread Aloha,* and presents theoretical results which show that the method provides the highest throughput for small earth stations.

VSAT systems operate in a star configuration which means that the connection of one VSAT to another must be made through the hub. This requires a double-hop circuit with a consequent increase in propagation delay, and twice the necessary satellite capacity is required compared to a single-hop circuit (Hughes et al., 1993). In Hughes, a proposal is presented for a VSAT system which provides for *mesh connection,* where the VSATs can connect with one another through the satellite in a single hop.

Most VSAT systems operate in the Ku band, although there are some C-band systems in existence (Rana et al., 1990). For a fixed area coverage by the satellite beam, the system performance is essentially independent of the carrier frequency. For fixed area coverage, the beamwidth and hence the ratio λ/D is a constant [see Eq. (5.30)]. The satellite antenna gain is therefore constant [see Eq. (5.29)], and for a given high-power amplifier output, the satellite EIRP remains constant. As shown in Sec. 10.3.1, for a given size of antenna at the earth station and a fixed EIRP from the satellite, the received power at the earth station is independent of frequency. This ignores the propagation margins needed to combat atmospheric and rain attenuation. As shown in Hughes et al., 1993, the necessary fade margins are not excessive for a Ka-band VSAT system, and the performance otherwise is comparable to a Ku-band system. (From Table 1.1, the K band covers 18 to 24 GHz and Ka band covers 24 to 40 GHz. In Hughes, 1993, results are presented for frequencies of 18.7 and 28.5 GHz).

As summarized in Rana et al., 1990, the major shortcomings of present-day VSAT systems are the high initial costs, the tendency toward optimizing systems for large networks (typically more than 500 VSATs), and the lack of direct VSAT-to-VSAT links. Technological improvements, especially in the areas of microwave technology and digital signal processing (Hughes et al., 1993) will result in VSAT systems in which most if not all of these shortcomings will be overcome.

13.5 Radarsat

Radarsat is an earth-resources remote-sensing satellite which is part of the Canadian space program. The objectives of the Radarsat program, as stated by the Canadian Space Agency are to:

Provide applications benefits for resource management and maritime safety

Develop, launch, and operate an earth observation satellite with synthetic aperture radar (SAR)

Establish a Canadian mission control facility

Market Radarsat data globally through a commercial distributor

Make SAR data available for research

Map the whole world with stereo radar

Map Antarctica in two seasons

The applications seen for Radarsat are

- Shipping and fisheries
- Ocean feature mapping
- Oil pollution monitoring
- Sea ice mapping (including dynamics)
- Iceberg detection
- Crop monitoring
- Forest management
- Geological mapping (including stereo SAR)
- Topographic mapping
- Land use mapping

Radarsat is planned to fly in a low-earth near-circular orbit. The orbital details are given in Table 13.5.

TABLE 13.5   Radarsat Orbital Parameters

Geometry	Circular, sun-synchronous (dawn–dusk)
Altitude (local)	798 km
Inclination	98.6°
Period	100.7 min
Repeat cycle	24 days

It will be seen from the orbital parameters that Radarsat flies in an orbit similar to the NOAA satellites described in Chap. 1; in particular, it is sun-synchronous. There are fundamental differences, however. Radarsat carries only C-band radar as the sensing mechanism whereas the NOAA satellites carry a wide variety of instruments as described in Secs. 1.5 and 6.11. Even though it is known that C-band radar is not the optimum sensing mechanism for all the applications listed, the rationale for selecting it is that it does penetrate cloud cover, smoke, and haze, and it does operate in darkness. Much of the sensing is required at high latitudes where solar illumination of the earth can be poor, and where there can be persistent cloud cover.

It will also be seen that the orbit is described as *dawn to dusk*. What this means is that the satellite is in view of the sun for the ascending and descending passages. With the radar sensor it is not necessary to have the earth illuminated under the satellite; in other words, the sun's rays reach the orbital plane in a broadside fashion. The main operational advantage, suggested in Raney et al., 1991, is that the radar becomes fully dependent on solar power rather than battery power for both the ascending and descending passes. Since there is no operational need to distinguish between the ascending and descending passes, nearly twice as many observations can be made than otherwise would be possible. Also, as Raney et al. point out, the downlink periods for data transmission from Radarsat will take place at times well-removed from those used by other remote-sensing satellites. Further advantages stated by the Canadian Space Agency are that the solar arrays do not have to rotate; the arrangement leads to a more stable thermal design for the spacecraft; the spacecraft design is simpler; and it provides for better power-raising capabilities. With this particular dawn-to-dusk orbit the satellite will be eclipsed by the earth in the southern hemisphere, from May 15 to July 30. The eclipse period changes gradually from zero to a maximum of about 15 minutes and back again to zero as shown in Fig. 13.1. The battery backup consists of three 50 A·h nickel-cadmium batteries.

Radarsat, shown in Fig. 13.2, is a comparatively large spacecraft, the total mass in orbit being about 3100 kg. The radar works at a car-

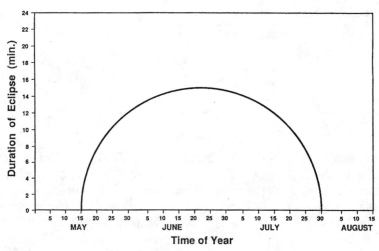

Figure 13.1   Duration of eclipse vs. time of year, dawn-dusk orbit.

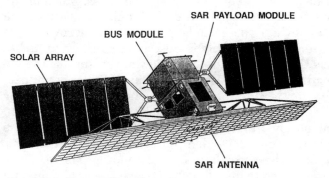

Figure 13.2   Spacecraft configuration.

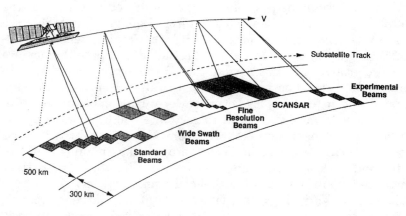

Figure 13.3   SAR operating modes.

TABLE 13.6   SAR Modes

	Swath, km	Resolution	Incidence angle, degrees
Operational	1	28 m × 30 m (4 looks)	20–50
High resolution	50	10 m × 10 m (1 look)	30–50
Experimental	100	28 m × 30 m (4 looks)	50–60
Scan SAR	500	100 m × 100 m (6 looks)	20–50

rier frequency of 5.3 GHz, which can be modulated with three different pulse widths, depending on resolution requirements. The SAR operating modes are illustrated in Fig. 13.3. The swath illuminated by the radar lies 20° to the east, and parallel to the subsatellite path. As shown, different beam configurations can be achieved, giving different resolutions. The mode characteristics are summarized in Table 13.6.

The satellite completes 14 + 7/24 revolutions per day. The separation between equatorial crossings is 116.8 km. According to Raney et al., 1991, the scanning SAR is the first implementation of a special radar technique. In summary, Radarsat is intended as a rapid response system providing earth imagery for a range of operational applications and is intended to complement other earth resources satellites.

13.6   Global Positioning Satellite System

In the Global Positioning Satellite (GPS) system, a constellation of 24 satellites circles the earth in near-circular inclined orbits. By receiving signals from at least four of these satellites, the receiver position (latitude, longitude, and altitude) can be accurately determined. In effect, the satellites substitute for the geodetic position markers used in terrestrial surveying. In terrestrial surveying, it is only necessary to have three such markers to determine the three unknowns of latitude, longitude, and altitude by means of triangulation. With the GPS system a time marker is also required, which necessitates getting simultaneous measurements from four satellites.

The GPS system uses one-way transmissions, from satellites to users, so that the user does not require a transmitter, only a GPS receiver. The only quantity the receiver has to be able to measure is time, from which propagation delay, and hence the range to each satellite, can be determined. Each satellite broadcasts its ephemeris (which is a table of the orbital elements as described in Chap. 2), from which its position can be calculated. Knowing the range to three of the satellites, and their positions, it is possible to compute the position of the observer (user). The geocentric-equatorial coordinate system (see Sec. 2.9.6) is used with the GPS system, where it is called the *earth-centered, earth-fixed* (ECEF) coordinate system.

As mentioned above, if the positions of three points relative to the coordinate system are known, and the distance from an observer to each of the points can be measured, then the position of the observer relative to the coordinate system can be calculated. In the GPS system the three points are provided by three satellites. Of course the satellites are moving, so their positions must be accurately tracked. The satellite orbits can be predicted from the orbital parameters (as described in Chap. 2). These parameters are continually updated by a master control station which transmits them up to the satellites where they are broadcast as part of the navigational message from each satellite.

Just as in a land-based system, better accuracy is obtained by using reference points well-separated in space. For example, the range measurements made to three reference points clustered together will yield nearly equal values. Position calculations involve range differences, and where the ranges are nearly equal, any error is greatly magnified in the difference. This effect, brought about as a result of the satellite geometry, is known as *dilution of precision* (DOP). This means that range errors which occur from other causes, such as timing errors, are magnified by the geometrical effect. With the GPS system, dilution of position is taken into account through a factor known as the *position dilution of precision* (PDOP) *factor*. This is the factor by which the range errors are multiplied to get the position error. The GPS system has been designed to keep the PDOP factor less than 6 most of the time (Langley, 1991c).

The GPS constellation consists of 24 satellites in six near-circular orbits, at an altitude of approximately 20,000 km (Daly, 1993). The ascending nodes of the orbits are separated by 60°, and the inclination of each orbit is 55°. The four satellites in each orbit are irregularly spaced to keep the PDOP factor within the limits referred to above.

It was stated earlier that three satellites are needed to fix position. In the GPS system, a minimum of four satellites must be observed, for reasons which will be explained shortly. Where more than four satellites are in view, the additional data are used to minimize errors by using the method of least squares.

Each satellite broadcasts its *ephemeris* which contains the orbital elements needed to calculate its position, and as previously mentioned the ephemerides are updated and corrected continuously from an earth control station. It should be mentioned that the GPS system is first and foremost intended for military use. However, civilian applications have now become quite extensive and are an accepted part of the GPS program.

Time enters into the position determination in two ways. First, the ephemerides must be associated with a particular time or epoch (as described in Chap. 2). The standard timekeeper is an atomic standard, maintained at the U.S. Naval Observatory, and the resultant time is known as *GPS time*. Each satellite carries its own atomic clock. The time broadcasts from the satellites are monitored by the control station which transmits back to the satellites any errors detected in timing, relative to GPS time. No attempt is made to correct the clocks aboard the satellites; rather, the error information is rebroadcast to the user stations, where corrections can be implemented in the calculations.

Second, time markers are needed to show when transmissions leave the satellites, so that, by measuring propagation times and knowing the speed of propagation, the ranges can be calculated. Therein lies a problem, as the user stations have no direct way of telling when a transmission from a satellite commenced. The problem is overcome by having the satellite transmit a continuous-wave carrier which is modulated by a clocking signal, both the carrier and the clocking signal being derived from the atomic clock aboard the satellite. At a user station, the receiver generates a replica of the modulated signal from its own atomic clock. The satellite signal and its replica are compared in a correlator at the receiver, and the replica is shifted in time until exact correlation is achieved. If the receiver clock kept exactly the same time as the satellite clock, the time shift as measured by the correlator would give the propagation delay. However, the receiver clock in general will be offset from the satellite clock (which is synchronized to GPS time) by an unknown amount. This offset will be the same for the signals received from the four satellites, and hence by obtaining four range measurements, four equations can be set up in terms of the x,y,z position vectors for the user and the time offset. The four equations can then be solved for these four unknowns. All of this requires quite sophisticated microprocessing in the receiver. Also the composition of the GPS signal is much more complex than indicated here, utilizing spread-spectrum techniques. The reader is referred to the following for details: Langley, 1990*a*, 1991*b*, 1991*c*; Kleusberg and Langley, 1990; Mattos, 1992, 1993*a*, 1993*b*, 1993*c*, 1993*d*, 1993*e*.

13.7   Orbcomm

The Orbital Communications Corporation (Orbcomm) system is a low-earth-orbiting (LEO) satellite system intended to provide two-way message and data communications services and position determination. In the Orbcomm submission to the FCC (Orbcomm, 1993), the planned launch dates extend from about mid-1994 to early 1998, with near-full availability expected by December 1995, contingent on licensing factors.

There are to be four main orbital planes, each containing eight satellites spaced from each other by 45° ± 5°. The inclination of each of the main planes is 45°, and the altitude is 775 km (424 nautical miles, or 482 statute miles). The main plane orbits are circular (eccentricity zero). Two supplemental orbits at 70° inclination, each having two satellites, will complete the constellation. The latter two orbits are intended to provide enhanced polar coverage. The satellites and their orbital parameters are listed in Table 13.7.

The ground segment consists of the subscribers (mobile or stationary), gateway earth stations (GES) which provide access to terrestrial systems such as the public switched telephone network and other mobile systems, a network control center (NCC), and a satellite control center (SCC). The network control center and the satellite control center are collocated at the Dulles, Virginia, Orbcomm facility, and the four gateway earth stations for the U.S. services are located near the corners of the contiguous United States (CONUS) in Washington state, Arizona, Georgia, and New York state. Figure 13.4 illustrates the system.

The satellites are small compared to the geostationary satellites in use, as shown in Fig. 13.5. The VHF/UHF antennas are seen to extend in a lengthwise manner, with the solar panels opening like lids top and bottom. Before launch the satellites are in the shape of a disk, and the launch vehicle, a Pegasus XL space booster [developed by Orbital Sciences Corporation (OSC), the parent company of Orbcomm] can deploy eight satellites at a time into the same orbital plane. For launch the satellites are stacked like a roll of coins, in what the company refers to as "an eight-pack."

Attitude control is required to keep the antennas pointing downward and at the same time to keep the solar panels in sunlight (battery backup is provided for eclipse periods). A three-axis magnetic control system, which makes use of the earth's magnetic field, and gravity gradient stabilization are employed. A small weight is added at the end of the antenna extension to assist in the gravity stabilization. Thus the satellite antennas hang down as depicted in Fig. 13.4. At launch the initial separation velocity is provided by springs used to separate the satellites, and a braking maneuver is used when the satellites reach their specified 45° in-plane separation. Intraplane spacing is maintained by a proprietary station-keeping technique which, it is claimed, has no cost in terms of fuel usage (Orbcomm, 1993). Because no on-board fuel is required to maintain the intraplane spacing between satellites, the satellites have a design lifetime

TABLE 13.7   Constellation Orbital Parameters

Sat. no.	Alt, km	Inc, deg	Ecc, deg	ArgP, deg	Ra, deg	M, deg	S, deg
1	775	45	0	0	0	0	360
2	775	45	0	0	0	45	360
3	775	45	0	0	0	90	360
4	775	45	0	0	0	135	360
5	775	45	0	0	0	180	360
6	775	45	0	0	0	225	360
7	775	45	0	0	0	270	360
8	775	45	0	0	0	315	360
9	775	45	0	0	135	0	360
10	775	45	0	0	135	45	360
11	775	45	0	0	135	90	360
12	775	45	0	0	135	135	360
13	775	45	0	0	135	180	360
14	775	45	0	0	135	225	360
15	775	45	0	0	135	270	360
16	775	45	0	0	135	315	360
17	775	45	0	0	270	0	360
18	775	45	0	0	270	45	360
19	775	45	0	0	270	90	360
20	775	45	0	0	270	135	360
21	775	45	0	0	270	180	360
22	775	45	0	0	270	225	360
23	775	45	0	0	270	270	360
24	775	45	0	0	270	315	360
25	775	45	0	0	405	0	360
26	775	45	0	0	405	45	360
27	775	45	0	0	405	90	360
28	775	45	0	0	405	135	360
29	775	45	0	0	405	180	360
30	775	45	0	0	405	225	360
31	775	45	0	0	405	270	360
32	775	45	0	0	405	315	360
33	775	70	0	0	0	0	360
34	775	70	0	0	0	180	360
35	775	70	0	0	180	90	360
36	775	70	0	0	180	270	360

Abbreviations: Alt = altitude, Ecc = eccentricity, M = mean anomaly, S = service arc, Inc = inclination, ArgP = argument of perigee, Ra = right ascension of the ascending node.

SOURCE: Orbcomm 1993.

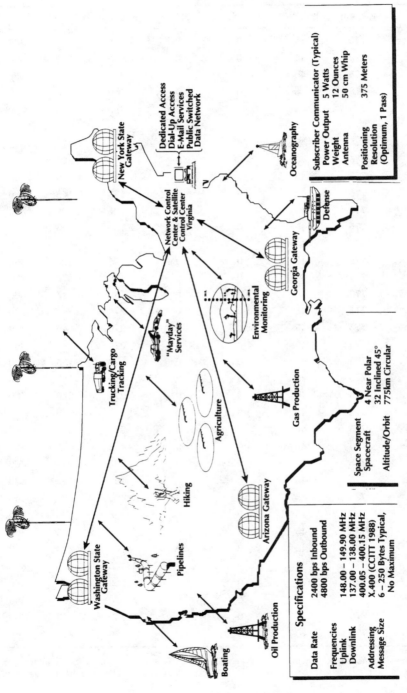

Figure 13.4 The Orbcomm system. (*Courtesy Orbital Communications Corporation.*)

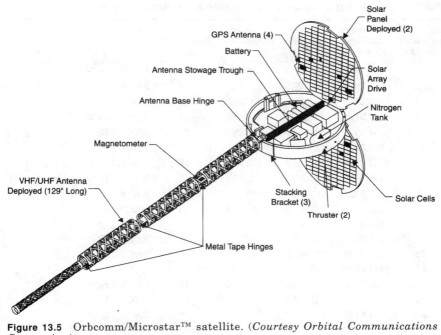

Figure 13.5 Orbcomm/Microstar™ satellite. (*Courtesy Orbital Communications Corporation.*)

of 4 years, which is based on the projected degradation of the power subsystem (solar panels and batteries).

The messaging and data channels are located in the VHF band, the satellites receiving in the 148- to 149.9-MHz band and transmitting in the 137- to 138-MHz band. Circular polarization is used, and a summary of the frequencies and polarization plans is given in Table 13.8. In planning the frequency assignments, great care has been taken to avoid interference to, and from, other services in the VHF bands; the reader is referred to Orbcomm, 1993, for details. In particular, the subscriber-to-satellite uplink channels utilize what is termed a *dynamic channel activity assignment system* (DCAAS), in which a scanning receiver aboard the satellite measures the interference received in small bandwidths, scanning the entire band every 5 s or less. The satellite receiver can then prepare a list of available channels (out of a total of 760 at present) and prioritize these according to interference levels expected. A summary of the link budget calculations for the uplinks and downlinks is given in Tables 13.9 to 13.14.

The Orbcomm system is capable of providing subscribers with a basic position determination service through the use of Doppler positioning, which fixes position to within a few hundred meters. The beacon signal at 400.1 MHz is used to correct for errors in timing

TABLE 13.8   137-MHz Downlink Channelization and Polarization Plan

Subscriber channel number	Lower band edge, MHz	Upper band edge, MHz	Center frequency, MHz	Bandwidth, kHz	Polarization*	Comment
1	137.1927	137.2075	137.200	15	RHCP, LHCP	
GES	137.2075	137.2575	137.235	50	RHCP	GES Downlink
2	137.2575	137.2725	137.265	15	RHCP, LHCP	
3	137.2725	137.2875	137.280	15	RHCP, LHCP	
4	137.2875	137.3025	137.295	15	RHCP, LHCP	
5	137.3025	137.3175	137.310	15	RHCP, LHCP	
6	137.3175	137.3325	137.325	15	RHCP, LHCP	
7	137.3675	137.3825	137.375	15	RHCP, LHCP	
8	137.3825	137.975	137.390	15	RHCP, LHCP	
9	137.3975	137.4125	137.405	15	RHCP, LHCP	
10	137.4125	137.4275	137.420	15	RHCP, LHCP	
11	137.4275	137.4425	137.435	15	RHCP, LHCP	
12	137.4425	137.4575	137.450	15	RHCP, LHCP	
13	137.4575	137.4725	137.465	15	RHCP, LHCP	
14	137.5175	137.5325	137.525	15	RHCP, LHCP	
15	137.5325	137.5475	137.540	15	RHCP, LHCP	
16	137.5475	137.5625	137.555	15	RHCP, LHCP	
17	137.5626	137.5775	137.570	15	RHCP, LHCP	
18	137.5775	137.5925	137.585	15	RHCP, LHCP	
UHF	400.075	400.125	400.100	18	RHCP	

*RHCP = right hand circular polarization, LHCP = left hand circular polarization.
SOURCE: Orbcomm 1993.

TABLE 13.9   148-MHz Uplink Channelization and Polarization Plan

Channel number	Lower band edge, MHz	Upper band edge, MHz	Center frequency, MHz	Bandwidth, kHz	Polarization	Comment
1	149.585	149.635	149.610	50	LHCP	GES uplink
2	148.000	149.900	Dynamic*	10	LHCP	760-channel uplink, DCAAS subscriber channels

*The DCAAS system will operate within the 148–149.9-MHz band, autonomously selecting unused channels.
SOURCE: Orbcomm 1993.

measurements introduced by the ionosphere (these errors are also present in the GPS system described in Sec. 13.6, and two frequencies are used in that situation for correction purposes). When used in conjunction with the VHF downlink signal, the beacon signal enables the

TABLE 13.10   Gateway Earth Station-to-Satellite Uplink Link Budget
(Edge of coverage, minimum elevation)

General information			Comments
Satellite altitude	775	km	
Elevation angle to satellite	5	deg	
Satellite angle from nadir	62.5	deg	
Range to earth	2730	km	
Data rate	57.6	kb/s	
Uplink frequency	149.4	MHz	
Uplink:			
Transmit EIRP	40.0	dBW	
Spreading loss	−139.7	dB/m^2	
Pointing loss	0.2	dB	
Atmospheric losses	2.0	dB	
Polarization losses	0.1	dB	GES axial ratio 1.2 dB, S/C 2.0 dB
Multipath fade losses	5.0	dB	
Area of an isotrope	−4.9	dB/m^2	
Power at satellite antenna	−112.0	dBW	
Satellite antenna G/T	−33.3	dBK^{-1}	
Received C/N_0	83.3	dBHz	
Data rate	47.6	dBHz	57.6 kb/s
Received E_b/N_0	35.7	dB	
Ideal E_b/N_0	10.6	dB	BER at $1{:}10^{-6}$
Implementation loss	3.0	dB	
Interference margin	20.0	dB	
Remaining margin	3.1	dB	

SOURCE: Orbcomm 1993.

effects of the ionosphere to be removed. The link budget figures for the 400.1-MHz channel are summarized in Table 13.14. It will be observed from Fig. 13.5 that the satellites carry GPS antennas which enable on-board determinations of the positions of the satellites. This information can then be downloaded on the VHF subscriber channel and used for accurate positioning. Recently Orbcomm announced an agreement with Trimble Navigation to develop hybrid GPS/Orbcomm user equipment for position and navigation purposes (Orbcomm, 1993). The system can also be used for search and rescue, and may well be a strong competitor to the search and rescue service described in Sec. 1.5.

One significant advantage achieved with low earth orbits is that the range is small compared to geostationary satellites (the altitude of the Orbcomm satellites is 775 km compared to 35,876 km for geo-

TABLE 13.11   Satellite-to-Subscriber Downlink Link Budget

(Edge of coverage, minimum elevation)

General information			Comments
Satellite altitude	775	km	
User elevation angle	5	deg	
User data rate	4800	b/s	
Downlink frequency	137.5	MHz	Midrange
Downlink:			
Transmit EIRP	12.5	dBW	
Spreading loss	−139.7	dB/m²	
Atmospheric losses	2.0	dB	
Polarization losses	4.1	dB	S/C 2-dB axial ratio, subscriber linear
Multipath fade losses	5.0	dB	
Satellite pointing loss	0.2	dB	5° off-nadir pointing
Area of an isotrope	−4.2	dB/m²	
Power at user antenna	−143.8	dBW	
Subscriber antenna G/T	−28.6	dBK^{-1}	
Received C/N_0	57.2	dBHz	
Data rate	36.8	dBHz	4.8 kb/s
Received E_b/N_0	20.4	dB	
Ideal E_b/N_0	10.3	dB	BER at 10^{-5}
Implementation loss	9.0	dB	Blockage, implementation and system
Remaining margin	1.1	dB	

SOURCE: Orbcomm 1993.

TABLE 13.12   Subscriber-to-Satellite Uplink Link Budget

(Edge of coverage, minimum elevation)

General information			Comments
Satellite altitude	775	km	
User elevation angle	5	deg	
User data rate	2400	b/s	
Uplink frequency	148.95	MHz	
Uplink budget:			
Transmit EIRP	7.5	dBW	
Spreading loss	−139.7	dB/m²	
Atmospheric losses	2.0	dB	After Ippolito
Polarization losses	4.1	dB	S/C 2-dB axial ratio, subscriber linear
Multipath fade losses	5.0	dB	After Krauss
Satellite pointing loss	0.2	dB	5° off-nadir pointing
Area of an isotrope	−4.9	dB/m²	

TABLE 13.12 Subscriber-to-Satellite Uplink Link Budget (*Continued*)

(Edge of coverage, minimum elevation)

General information			Comments
Power at satellite antenna	−148.5	dBW	
Satellite G/T	−26.0	dBi	
Received P_r/N_0	53.3	dBHz	
Data rate	33.8	dBHz	2.4 kb/s
Received E_b/N_0	19.5	dB	
Ideal E_b/N_0	10.3	dB	BER at 10^{-5}
Implementation loss	2.0	dB	
Required link margin	3.0	dB	
Remaining margin	4.2	dB	

SOURCE: Orbcomm 1993.

TABLE 13.13 Satellite-to-Gateway Earth Station Downlink Link Budget

(Edge of coverage, minimum elevation)

General information			Comments
Satellite altitude	775	km	
Earth station elevation angle to satellite	5	deg	
Data rate	57,600	b/s	
Downlink frequency	137.2	MHz	
Downlink budget:			
Transmit EIRP	6.5	dBW	
Spreading loss	−139.7	dB/m^2	
Pointing loss	0.2	dB	5° off-nadir pointing
Atmospheric losses	2.0	dB	
Polarization losses	0.1	dB	*S/C* 2-dB axial ratio, GES 1.2 dB
Multipath fade losses	5.0	dB	
Area of an isotrope	−4.2	dB/m^2	
Power at satellite antenna	−144.8	dBW	
Gateway antenna G/T	−12.8	dBK^{-1}	
Received C/N_0	71.0	dBHz	
Data rate	47.6	dBHz	57.6 kb/s
Received E_b/N_0	23.4	dB	
Ideal E_b/N_0	10.6	dB	OQPSK at $1:10^{-6}$
Implementation loss	3.0	dB	
Required link margin	3.0	dB	
Interference margin	3.0	dB	
Remaining margin	3.8	dB	

SOURCE: Orbcomm 1993.

TABLE 13.14   UHF-to-Subscriber Downlink Link Budget
(Edge of coverage, minimum elevation)

General information			Comments
Satellite altitude	775	km	
User elevation angle	5	deg	
Downlink frequency	400.1	MHz	
Downlink budget:			
Transmit EIRP	2.5	dBW	
Spreading loss	−139.7	dB/m²	
Atmospheric losses	2.0	dB	
Polarization losses	4.1	dB	*S/C* 2-dB axial ratio, subscriber linear
Multipath fade losses	4.0	dB	
Area of an isotrope	−13.9	dB/m²	
Receiver gain	0.0	dBi	
Antenna-to-receiver losses	−1.0	dB	
Received signal level	−162.5	dBW	
Receiver noise temp.	24.6	dBK⁻¹	(3-dB noise figure)
Receiver loop bandwidth	20.0	dBHz	
Boltzmann's constant	−228.6	dBW/Hz·K	
Receiver noise power	−184.0	dBW	
Received signal-to-noise	21.5	dB	

SOURCE: Orbcomm 1993.

stationary satellites). Thus the free-space loss (FSL) is very much less. Propagation delay is correspondingly reduced, but this is not a significant factor where messaging and data communications, as compared to real-time voice communications, are involved.

The Orbcomm system will provide a capacity of more than 60,000 messages per hour. By using digital packet switching technology, and confining the system to nonvoice, low-speed alphanumeric transmissions, Orbcomm calculates that the service, combined with other LEO systems, will be able to provide for 10,000 to 20,000 subscribers per kilohertz of bandwidth, which is probably unmatched by any other two-way communications service. Although it is a U.S.-based system, because of the global nature of satellite communications, Orbcomm has signed preliminary agreements with companies in Canada, Russia, South Africa, and Nigeria to expand the Orbcomm service (Orbcomm, 1994).

13.8   Iridium

The Iridium concept was originated by engineers at Motorola's Satellite Communications Division in 1987. Originally envisioned as consisting of 77 satellites in low earth orbit, the name Iridium was adopted by analogy with the element iridium, which has 77 orbital electrons. Further studies led to a revised constellation plan requiring 66 satellites (Leopold, 1992). Because of the international character of satellite communications, an international consortium of telecommunications operators and industrial companies named Iridium Inc. was formed to implement and manage the Iridium system. Major investors are located in North America, South America, Japan, China, India, Africa, Europe, the Middle East, and Russia (Iridium, 1994).

The 66 satellites are grouped in six orbital planes, each containing 11 active satellites. The orbits are circular, at a height of 783 km (421.5 nmi). Prograde orbits are used, the inclination being 86°. The 11 satellites in any given plane are uniformly spaced, the nominal spacing being 32.7°. An in-orbit spare is available for each plane at an orbit 130 km (70 nmi) lower in the orbital plane. Some of the other orbital characteristics are listed in Table 13.15.

The satellites travel in corotating planes; that is, they travel up one side of the earth, cross over near the north pole, and travel down the other side. Since there are 11 equispaced satellites in each plane, it will be seen that both sides of the earth are covered continuously. The satellites in adjacent planes travel out of phase, meaning that adjacent planes are rotated by half the satellite spacing relative to one another. This is shown in Fig. 13.6, which presents a view from above the north pole. Collision avoidance is built into the orbital planning, and the closest approach between satellites is 223 km (120 nmi). Satellites in planes 1, 3, and 5 cross the equator in synchronization,

TABLE 13.15   Physical Characteristics of Satellite

Characteristic	Nominal values (3σ range for kilometers and degrees)
Station keeping	2.0 km cross-track
	5.7 km in-track
	4.7 km radial
Antenna pointing accuracy toward earth	1.0° in azimuth
	0.7° in elevation
Attitude stabilization and station-keeping systems	Roll: < 0.5°
	Pitch: < 0.4°
	Yaw: < 0.75°
Electrical energy system	< 1200-W solar array
	< 50 A·h battery
Estimated minimum lifetime of in-orbit satellite	5 years

SOURCE: Motorola 1992.

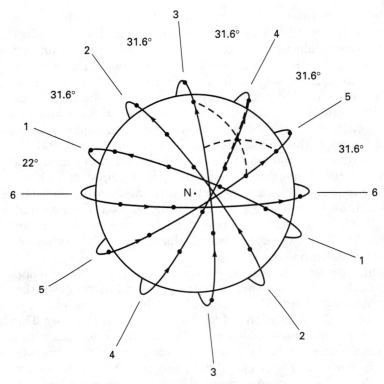

Figure 13.6 A polar view of the Iridium satellite orbits.

while satellites in planes 2, 4, and 6 also cross in synchronization, but out of phase with those in planes 1, 3, and 5.

Although the planes are corotating, the first and last planes must be counterrotating where they are adjacent. This is illustrated in Fig. 13.6. The separation between corotating planes is 31.6°, which allows 22° separation between the first and last planes. The closer separation is needed because earth coverage under the counterrotating "seam" is not as efficient as it is under the corotating seams. Two-way communications links exist between each satellite and its nearest neighbors ahead and front, and to the satellites in the adjacent planes, as shown by the dotted lines in Fig. 13.6.

Figure 13.7 is a system overview. The up/downlinks between subscribers and satellites take place in the L band. A 48-beam antenna pattern is used from each satellite, with each beam under separate control. At the equator, for instance, overlap of patterns will be minimal and all beams may be on, while at high latitudes, considerable overlap occurs, and certain beams will be switched off. Also, in regions where operation is prohibited by the telecommunications administra-

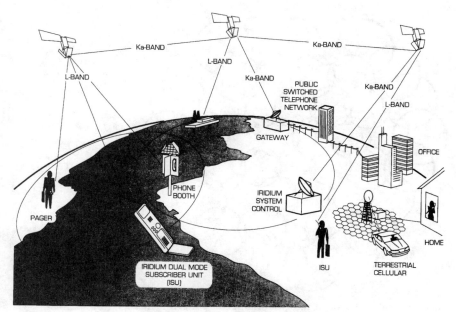

Figure 13.7 Iridium® system overview. (*Courtesy Motorola.*)

tion, the beams can be switched off. The switching of beams is referred to as *cell management*. The beams are similar to the cells encountered in cellular mobile, but with the fundamental difference that the beams move relative to the subscriber, whereas in cellular mobile, the cells are fixed. It must be realized that the satellites are traveling at a much greater velocity than that normally encountered with terrestrial vehicles, which may be considered stationary relative to the satellites. The orbital period is approximately 100 min, and taking an average value of 6371 km for the earth's radius, the surface speed is $2 \times 6371 \times \pi/100$ min ≈ 400 km/min or just over 15,000 mi/h. The 48-cell pattern is shown in Fig. 13.8 and the earth coverage in Fig. 13.9.

Intersatellite links, and the up/downlinks between satellites and gateway stations, operate in the Ka band. The link between the Iridium system control station and the satellites is also in the Ka band. Circular polarization is used on all links (Leopold, 1992). The multiple access utilizes a combination of TDMA and FDMA. Each subscriber unit operates in a burst mode using a single-carrier transmission (Motorola, 1990). The TDMA frame format is shown in Fig. 13.10, and the uplink/downlink frequency plan is shown in Fig. 13.11 (Motorola, 1992).

Figure 13.12 shows an artist's conception of the satellites, and Tables 13.16 to 13.24 (Motorola, 1992) provide details of the link budget calculations. In these tables, *shadowing* refers to the added atten-

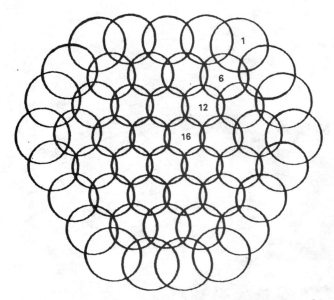

Figure 13.8 48-cell L-band integrated antenna pattern architecture. (*Source: Motorola 1992.*)

Figure 13.9 Motorola beam projections. (*Courtesy Motorola.*)

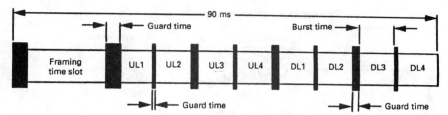

Figure 13.10 TDMA frame format. (*Source: Motorola 1992.*)

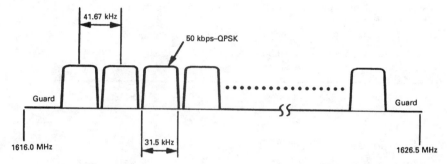

Figure 13.11 L-band uplink/downlink RF frequency plan. (*Source: Motorola 1992.*)

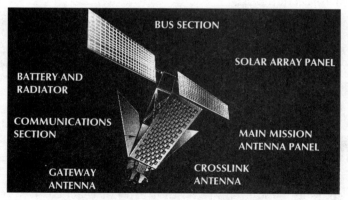

Figure 13.12 Motorola SV deployed configuration. (*Courtesy Motorola.*)

uation resulting from natural vegetation. The following terms are abbreviated in the tables:

  Space vehicle (SV)

  Iridium subscriber unit (ISU)

TABLE 13.16   Link Parameter Summary

	SV–user (ISU)		SV–gateway		SV–SV
	Down	Up	Down	Up	
Multiplexing	TDMA/FDMA		TDM/FDMA		TDM/FDMA
Modulation	QPSK		QPSK		QPSK
Baseband filtering	40% raised-cosine		Filtered		Filtered
FEC rate	Multiple rates		$\frac{1}{2}$	$\frac{1}{2}$	$\frac{1}{2}$
Coded data rate, Mb/s	0.05	0.05	6.25	6.25	25.00
Occupied bandwidth per channel, kHz	31.50	31.50	4375.00	4375.00	17500.00
Center frequency, GHz	1.62125	1.62125	20.00	29.40	23.28
Total bandwidth, MHz	10.50	10.50	100.00	100.00	200.00
Carrier spacing, MHz	0.04167	0.04167	7.50	7.50	25.00

SOURCE: Motorola 1992.

High-power amplifier (HPA)

Edge of coverage (EOC)

Telemetry, tracking, and control (TT&C)

Forward error correction (FEC)

Saturation power flux density (SPFD)

The company stresses that Iridium is not a replacement for existing cellular systems, but rather an extension of wireless telephony. The major advantages listed by the company are

Iridium is less likely to be disabled in disaster situations (earthquakes, fire, floods, and the like), and will therefore be available as an emergency service in the event that terrestrial cellular services are knocked out.

Iridium will be able to provide mobile services to those areas (e.g., remote and sparsely populated areas) that are not reached by terrestrial cellular services.

Iridium will be able to offer more channels with shorter delays, and the ability to connect into worldwide networks for those areas which currently obtain mobile services through geostationary satellites.

Iridium will make it possible to get a service in operation very quickly in areas where no telephone services exist, while terrestrial networks are being installed.

TABLE 13.17 SV–ISU Downlink, with Shadowing
(Representative cells)

	Cell 1	Cell 6	Cell 12	Cell 16
Azimuth angle, deg	32.4	38.3	40.5	60.0
Ground range, km	2215.3	1424.9	957.3	528.8
Nadir angle, deg	61.9	56.4	48.2	33.4
Elevation angle, deg	8.2	20.8	33.2	51.9
Slant range, km	2461.7	1696.2	1278.5	960.0
Space vehicle:				
HPA burst power, W (dBW)	3.5 (5.5)	2.2 (3.5)	1.3 (1.2)	3.0 (4.8)
Transmitter circuit loss, dB	2.1	2.1	2.1	2.1
Eff. EOC antenna gain, dBi	24.3	23.1	22.9	16.8
EIRP, dBW	27.7	24.5	22.0	19.5
Propagation:				
Space loss, dB	164.5	161.3	158.8	156.3
Propagation losses, dB	15.7	15.7	15.7	15.7
Total propagation loss, dB	180.2	177.0	174.5	172.0
Iridium subscriber unit:				
Rcvd signal strength, dBW	−152.5	−152.5	−152.5	−152.5
Antenna gain, dBi	1.0	1.0	1.0	1.0
Signal level, dBW	−151.5	−151.5	−151.5	−151.5
Req. $E_b/(N_0 + I_0)$, dB	5.8	5.8	5.8	5.8
E_b/I_0, dB	18.0	18.0	18.0	18.0
Req. E_b/N_0, dB	6.1	6.1	6.1	6.1
T_s, K	250.0	250.0	250.0	250.0
Signal level req., dBW	−151.5	−151.5	−151.5	−151.5
Link margin, dB	0.0	0.0	0.0	0.0
G/T_s, dBi/K	−23.0	−23.0	−23.0	−23.0
SPFD at ISU, dBW/m²/4 kHz	−135.8	−135.8	−135.8	−135.8

SOURCE: Motorola 1992.

TABLE 13.18   SV–ISU Uplink, with Shadowing
(Representative cells)

	Cell 1	Cell 6	Cell 12	Cell 16
Azimuth angle, deg	32.4	38.3	40.5	60.0
Ground range, km	2215.3	1424.9	957.3	528.8
Nadir angle, deg	61.9	56.4	48.2	33.4
Elevation angle, deg	8.2	20.8	33.2	51.9
Slant range, km	2461.7	1696.2	1278.5	960.0
Iridium subscriber unit:				
HPA burst power, W (dBW)	3.7 (5.7)	3.6 (5.6)	1.9 (2.9)	3.7 (5.7)
Circuit loss, dB	0.7	0.7	0.7	0.7
Antenna gain, dBi	1.0	1.0	1.0	1.0
EIRP, dBW	6.0	5.9	3.2	6.0
Uplink EIRP density (dBW/4 kHz)	−3.0	−3.1	−5.8	−3.0
Propagation:				
Space loss, dB	164.5	161.3	158.8	156.3
Propagation losses, dB	15.7	15.7	15.7	15.7
Total propagation loss, dB	180.2	177.0	174.5	172.0
Space vehicle:				
Received signal strength, dBW	−174.2	−171.1	−171.3	−166.0
Eff. EOC antenna gain, dBi	23.9	22.6	22.8	16.4
Signal level, dBW	−150.3	−148.5	−148.5	−149.6
Req. $E_b/(N_0 + I_0)$, dB	5.8	5.8	5.8	5.8
E_b/I_0, dB	18.0	18.0	18.0	18.0
Req. E_b/N_0, dB	6.1	6.1	6.1	6.1
T_s, K	500.0	500.0	500.0	500.0
Signal level req., dBW	−148.5	−148.5	−148.5	−148.5
Link margin, dB	−1.8	0.0	0.0	−1.1
G/T, dBi/K	−3.1	−4.4	−4.2	−10.6

SOURCE: Motorola 1992.

TABLE 13.19   SV–ISU Downlink, No Shadowing

(Representative cells)

	Cell 1	Cell 6	Cell 12	Cell 16
Azimuth angle, deg	32.4	38.3	40.5	60.0
Ground range, km	2215.3	1424.9	957.3	528.8
Nadir angle, deg	61.9	56.4	48.2	33.4
Elevation angle, deg	8.2	20.8	33.2	51.9
Slant range, km	2461.7	1696.2	1278.5	960.0
Space vehicle:				
HPA burst power, W (dBW)	0.2 (−6.5)	0.1 (−8.5)	0.1 (−10.8)	0.2 (−7.2)
Transmitter circuit loss, dB	2.1	2.1	2.1	2.1
Eff. EOC antenna gain, dBi	24.3	23.1	22.9	16.8
EIRP, dBW	15.7	12.5	10.0	7.5
Propagation:				
Space loss, dB	164.5	161.3	158.8	156.3
Propagation losses, dB	0.7	0.7	0.7	0.7
Total propagation loss, dB	165.2	162.0	159.5	157.0
Iridium subscriber unit:				
Received signal strength, dBW	−149.5	−149.5	−149.5	−149.5
Antenna gain, dBi	1.0	1.0	1.0	1.0
Signal level, dBW	−148.5	−148.5	−148.5	−148.5
Req. $E_b/(N_0 + I_0)$, dB	5.8	5.8	5.8	5.8
E_b/I_0, dB	18.0	18.0	18.0	18.0
Req. E_b/N_0, dB	6.1	6.1	6.1	6.1
T_s, K	250.0	250.0	250.0	250.0
Signal level req., dBW	−151.5	−151.5	−151.5	−151.5
Link margin, dB	3.0	3.0	3.0	3.0
G/T_s, dBi/K	−23.0	−23.0	−23.0	−23.0
SPFD at ISU, dBW/m²/4 kHz	−132.8	−132.8	−132.8	−132.8

SOURCE: Motorola 1992.

TABLE 13.20 SV–ISU Uplink, No Shadowing

(Representative cells)

	Cell 1	Cell 6	Cell 12	Cell 16
Azimuth angle, deg	32.4	38.3	40.5	60.0
Ground range, km	2215.3	1424.9	957.3	528.8
Nadir angle, deg	61.9	56.4	48.2	33.4
Elevation angle, deg	8.2	20.8	33.2	51.9
Slant range, km	2461.7	1696.2	1278.5	960.0
Iridium subscriber unit:				
HPA burst power, W (dBW)	0.4 (−4.5)	0.2 (−6.4)	0.1 (−9.1)	0.3 (−5.2)
Circuit loss, dB	0.7	0.7	0.7	0.7
Antenna gain, dBi	1.0	1.0	1.0	1.0
EIRP, dBW	−4.2	−6.1	−8.8	−4.9
Uplink EIRP density, dBW/4 kHz	−13.2	−15.1	−17.8	−13.9
Propagation:				
Space loss, dB	164.5	161.3	158.8	156.3
Propagation losses, dB	0.7	0.7	0.7	0.7
Total propagation loss, dB	165.2	162.0	159.5	157.0
Space vehicle:				
Received signal strength, dBW	−169.4	−168.1	−168.3	−161.9
Eff. EOC antenna gain, dBi	23.9	22.6	22.8	16.4
Signal level, dBW	−145.5	−145.5	−145.5	−145.5
Req. $E_b/(N_0 + I_0)$, dB	5.8	5.8	5.8	5.8
E_b/I_0, dB	18.0	18.0	18.0	18.0
Req. E_b/N_0, dB	6.1	6.1	6.1	6.1
T_s, K	500.0	500.0	500.0	500.0
Signal level req., dBW	−148.5	−148.5	−148.5	−148.5
Link margin, dB	3.0	3.0	3.0	3.0
G/T, dBi/K	−3.1	−4.4	−4.2	−10.6

SOURCE: Motorola 1992.

TABLE 13.21   SV–Gateway Links

	Downlink, 20.00 GHz		Uplink, 29.40 GHz	
	Rain	Clear	Rain	Clear
Range, km	2326.0	2326.0	2326.0	2326.0
Transmitter:				
Power, dBW	0.0	−9.7	13.0	−11.8
Antenna gain, dB	26.9	26.9	56.3	56.3
Circuit loss, dB	−3.2	−3.2	−1.0	−1.0
Pointing loss, dB	−0.5	−0.5	−0.3	−0.3
EIRP, dBWi	23.2	13.5	68.0	43.2
System:				
Margin, dB	3.2	3.2	2.1	2.1
Space loss, dB	−185.8	−185.8	−189.1	−189.1
Propagation loss, dB	−14.2	−1.5	−30.0	−1.5
Polarization loss, dB	−0.2	−0.2	−0.2	−0.2
Total propagation loss, dB	−201.5	−188.8	−220.2	−191.7
Receiver:				
Received signal strength, dBWi	−180.2	−177.2	−153.4	−149.7
Pointing loss, dB	−0.2	−0.2	−0.8	−0.8
Antenna gain, dB	53.2	53.2	30.1	30.1
Received signal, dBW	−127.2	−124.2	−124.1	−120.4
T_s, K	731.4	731.4	1295.4	1295.4
Noise density, dBW/Hz	−200.0	−200.0	−197.5	−197.5
Noise bandwidth, dBHz	64.9	64.9	64.9	64.9
Noise, dBW	−135.1	−135.1	−132.6	−132.6
Link E_b/N_0, dB	7.9	10.9	8.5	12.2
E_b/I_0, dB	25.0	25.0	16.0	16.0
Comp. $E_b/(N_0 + I_0)$, dB	7.8	10.7	7.8	10.7
Required E_b/N_0, dB	7.7	7.7	7.7	7.7
Excess margin, dB	0.1	3.0	0.1	3.0
SPFD at gateway, dBW/m²/1 MHz	−134.3	−131.3		

SOURCE: Motorola 1992.

TABLE 13.22   SV–SV Intersatellite Links

(Carrier frequency 23.28 GHz)

	E/W	E/W with sun	N/S	N/S with sun
Range, km	4400.0	4400.0	4050.0	4050.0
Transmitter:				
Power, dBW	5.3	5.3	5.3	5.3
Antenna gain, dB	36.7	36.7	36.7	36.7
Circuit loss, dB	−1.8	−1.8	−1.8	−1.8
Pointing loss, dB	−1.8	−1.8	−1.8	−1.8
EIRP, dBWi	38.4	38.4	38.4	38.4
System:				
Margin, dB	1.8	0.0	2.6	0.0
Space loss, dB	−192.7	−192.7	−191.9	−191.9
Polarization loss, dB	0.0	0.0	0.0	0.0
Receiver:				
Received signal strength, dBWi	−156.1	−154.3	−156.1	−153.5
Pointing loss, dB	−1.8	−1.8	−1.8	−1.8
Antenna gain, dB	36.7	36.7	36.7	36.7
Received signal, dBW	−121.2	−119.4	−121.2	−118.6
T_s, K	720.3	1188.3	720.3	1188.3
Noise density, dBW/Hz	−200.0	−197.9	−200.0	−197.9
Noise bandwidth, dBHz	71.0	71.0	71.0	71.0
Noise, dBW	−129.0	−126.9	−129.0	−126.9
Link E_b/N_0, dB	7.8	7.5	7.8	8.3
E_b/I_0, dB	27.0	27.0	27.0	27.0
Comp. $E_b/(N_0 + I_0)$, dB	7.7	7.5	7.7	8.2
Required E_b/N_0, dB	7.7	7.7	7.7	7.7
Excess margin, dB	0.0	−0.2	0.0	0.5

SOURCE: Motorola 1992.

TABLE 13.23   Major Iridium Satellite Characteristics

Stabilization	Three-axis
Mission life	5 years minimum
Station keeping	2.0 km cross-track
	5.7 km in-track
	4.7 km radial-track
Frequency bands	1616–1626.5 MHz
	18.8–20.2 GHz
	27.5–30.0 GHz
	22.55–23.55 GHz
Earth coverage	5.9 million square (statute) miles per satellite
Max. number of L-band uplink channels per satellite	3840
Max. number of L-band downlink channels per satellite	3840
Max. number of intersatellite channels per satellite	About 6000
Max. number of gateway channels per satellite	About 3000
Total occupied bandwidth	10.5 MHz at L band
	200 MHz at Ka band (intersatellite links)
	100 MHz at Ka band (gateway uplink)
	100 MHz at Ka band (gateway downlink)
Polarization:	
L band	Right-hand circular
Ka band (gateway and TT&C links)	Right-hand circular
Ka band (intersatellite links)	Vertical
Transmit EIRP	7.5 to 27.7 dBW at L band
	13.5 to 23.2 dBW at Ka band (gateway)
	38.4 dBW at Ka band (intersatellite)
Satellite G/T	-10.6 to -3.1 dBi/K at L band
	-1.0 dBi/K at Ka band (gateway)
	8.1 dBi/K at Ka band (intersatellite)
Wet mass with reserve	Approximately 700 kg
Orbit	Near polar (six planes) about 86° inclination

SOURCE: Motorola 1992.

TABLE 13.24   Iridium Technical Information

L band (uplink and downlink):	
Frequency	1616–1626.5 MHz (10.5 MHz)
Polarization	Right-hand circular
Center frequency	1.62125 GHz
Channel bandwidth	31.5 kHz
Channel spacing	41.67 kHz
Gateway and TT&C (up/downlink):	
Frequency	27.5–30.0 GHz (100 MHz)
	18.8–20.2 GHz (100 MHz)
Polarization	Right-hand circular
Center frequency (uplink)	29.40 GHz
Center frequency (downlink)	20.0 GHz
Channel bandwidth	4.375 MHz
Channel spacing	7.5 MHz
Intersatellite link:	
Frequency	22.55–23.55 GHz (200 MHz)
Polarization	Vertical
Center frequency	23.28 GHz
Channel bandwidth	17.5 MHz
Channel spacing	25 MHz
Final amplifier output power:	
L band (Cells 1–48)	0.1 to 3.5 W per carrier (burst)
Ka band	
Gateway	0.1 to 1.0 W per channel
Intersatellite	3.4 W per carrier (burst)
Receiving system noise temperature:	
L band	500 K
Ka band	
Gateway	1295 K
Intersatellite	720 K (1188 K with sun)
Gain of each L band channel	(Not a transponder)
Orbital locations:	
Altitude	780 kilometers nominal
Number of planes	Six near-polar planes
Spacing of planes	31.6° nominal (except planes 1 and 6 spaced 22.0° nominal)
Number of satellites per plane	11
Spacing of satellites in plane	32.7°

SOURCE: Motorola 1992.

TABLE 13.24 Iridium Technical Information (*Continued*)

Physical characteristics of satellite	
Station keeping	2.0 km cross-track (3σ)
	5.7 km in-track (3σ)
	4.7 km radial (3σ)
Antenna pointing accuracy toward earth	1.0° in azimuth (3σ)
	0.7° in elevation (3σ)
Estimated minimum lifetime of in-orbit satellite	5 years
Attitude stabilization and station-keeping systems	Roll: <0.5° (3σ)
	Pitch: <0.4° (3σ)
	Yaw: <0.75° (3σ)
Electrical energy system	<1200 W solar array
	<50 A·h battery
Emission limitations (L band)	
Channel spacing	41.67 kHz
Spurious emissions	−30 dB at 1 × channel spacing from carrier
	−60 dB at 2 × channel spacing from carrier

SOURCE: Motorola 1992.

13.9 Problems

13.1. Describe the main differences between high power, medium power, and low power satellite broadcast systems.

13.2. State the main reasons why the Ku band is used for DBS services rather than the C band. Smaller antennas are in use in general for reception of DBS signals in the Ku band compared to C band. Explain why this is so.

13.3. Calculate the rms surface tolerance which must be met for a 12-GHz paraboloidal reflector if the reduction in gain from this source is not to exceed 1 dB.

13.4. Discuss briefly the need for the MSAT system, and explain how this compares with existing terrestrial cellular networks.

13.5. Describe the operation of a typical VSAT system, and list some of the shortcomings of present day VSAT systems.

13.6. Describe the main features of Radarsat. Explain what is meant by a "dawn to dusk" orbit and why the Radarsat follows such an orbit.

13.7. Explain why a minimum of four satellites must be visible at any earth locations utilizing the GPS system for position determination. To what does the term *dilution of position* refer?

13.8. Describe the main features and services offered by the Orbcomm satellite system. How do these services compare with services offered by geostationary satellites and by terrestrial cellular systems?

13.9. Using the range and frequency values given in Table 13.10 calculate the free space loss. Explain how this differs from the spreading loss listed in the table.

13.10. Using the free space loss calculated in Prob. 13.9 calculate the received $[C/N_o]$ value and the received $[E_b/N_o]$ and compare with the values given in Table 13.10.

13.11. Using the range value given in Table 13.10 and the frequency given in Table 13.11 calculate the free space loss, the received $[C/N_o]$ value and the received $[E_b/N_o]$ and compare with the values given in Table 13.11.

13.12. Using the range value given in Table 13.10 and the frequency given in Table 13.12 calculate the free space loss, the received $[C/N_o]$ value, and the received $[E_b/N_o]$ and compare with the values given in Table 13.12.

13.13. Using the range value given in Table 13.10 and the frequency given in Table 13.13 calculate the free space loss, the received $[C/N_o]$ value, and the received $[E_b/N_o]$ and compare with the values given in Table 13.13.

13.14. Describe the main features and services offered by the Iridium satellite system. How do these services compare with services offered by geostationary satellites and by terrestrial cellular systems?

13.15. Using an average value of 6371 km for the earth's radius, calculate the nominal period for a satellite in the Iridium system. Hence determine the velocity of the satellite relative to the earth.

13.16. With reference to Table 13.17 verify the values tabulated for (a) EIRP, (b) received signal level, (c) required E_b/N_o, and (d) the required signal level. Use the coded data rate from Table 13.16.

13.17. With reference to Table 13.18 verify the values tabulated for (a) EIRP, (b) received signal level, (c) required E_b/N_o, and (d) the required signal level. Use the coded data rate from Table 13.16.

13.18. With reference to Table 13.19 verify the values tabulated for (a) EIRP, (b) received signal level, (c) required E_b/N_o, and (d) the required signal level. Use the coded data rate from Table 13.16.

13.19. With reference to Table 13.20 verify the values tabulated for (a) EIRP, (b) received signal level, (c) required E_b/N_o, and (d) the required signal level. Use the coded data rate from Table 13.16.

13.20. With reference to Table 13.21 verify the values tabulated for (a) EIRP, (b) total propagation loss, and (c) received signal level. (d) Calculate also the received $[E_b/N_0]$ value and compare with the tabulated value. Use the coded data rate from Table 13.16.

13.21. With reference to Table 13.22 verify the values tabulated for (a) EIRP, (b) total propagation loss, and (c) received signal level. (d) Calculate also the received $[E_b/N_0]$ value and compare with the tabulated value. Use the coded data rate from Table 13.16.

Answers to Selected Problems

Chapter 2

2.5. 11016 km

2.7. 70.14°

2.9. No. 4 (*a*) 10367.17 n.mi; (*b*) 10580.99 n.mi; (*c*) 2.159.8 n.mi.

No. 5 5948 n.mi

No. 6 (*b*) 2148.5 n.mi; (*c*) 690.6 n.mi.

2.11. 42165 km; 0.7349

2.17. (*a*) 0; (*b*) −3.17 deg/day; (*c*) 14.034 rev/day

2.19. 2446433.165; 2446865.5; 2447215.0; 2447222.1875

2.21. (*a*) Jan 3, 0 hr; (*b*) July 4, 0300 hr; (*c*) Oct 27, 2 hr 55 min 4.2034 sec; (*d*) Jan 3, 7 hr 3 min 57.754 sec

2.23. (*a*) 27027 km; (*b*) 0.7472; (*c*) 12.2761 rad/day; (*d*) 737.0236 min; (*e*) −0.1487 deg/day; (*f*) 7.269×10^{-3} deg/day

2.25. 0.856° W

2.27. (*a*) 54.53 rad/day; (*b*) 234.85°; (*c*) 10,505 km; (*d*) 36.2° S

2.29. 101.9° W

2.31. 249.83°; 84.89°; 290.45°; 290.43°; 6613.87 km

2.33. −0.707 deg/day; −2.502 deg/day; 82.8 rad/day

2.35. −0.205°; 110.063 min

2.37. 0° Latitude, −92.6° longitude

2.41. Azimuth 36.1° and 143.9°; Elevation 48.7°; 0.2 sec

2.43. 131.45°; 36.41°

2.45. 241°; 26.94°

2.51. 6°

2.53. −5°

Chapter 3

3.3. 0.3 dB

3.7. Horizontal 0.0067 dB; vertical 0.0055 dB

3.11. (a) 1.11 dB; (b) 1.17 dB

3.13. 0.006 dB

3.15. 0.015 dB

Chapter 4

4.9. Right hand elliptical

4.13. 59°

4.17. 19°

4.21. PL = 0.11 dB; XPD = 16 dB

4.25. 50.4 dB

4.27. 36.6 dB

Chapter 5

5.1. 26.8 dB

5.3. (1.41, 0.513, 2.598)

5.5. $E_y = 4$ mV; $E_z = -3$ mV

5.7. 2.387 pW/m^2

5.9. 51.8 dB

5.11. 4.6 dB

5.13. 8×10^{-6} m^2

5.19. 21.8 dB

5.25. 0.5 m

5.27. 12.76 m^2; 48.1 dB

5.33. Currents are in-phase

Chapter 8

8.1. $300-3400$ Hz; -14.4 dBm

8.9. $Y = .3R + .59G + .11B$; $Q = .21R - .52G + .31B$; $I = .6R - .28G - .32B$

8.13. 500 Hz

8.17. 152 kHz

8.21. (a) 1.46; (b) 2.32; (c) 7.54

8.23. (a) 12.6 dB; (b) 13.9 dB; (c) 5.45 dB

8.25. (a) 62.4 dB; (b) 68.6 dB

Chapter 9

9.1. (a) $v(t) = A + \dfrac{4A}{\pi}\left[\cos\dfrac{\pi t}{T_b} - \dfrac{1}{3}\cos\dfrac{3\pi t}{T_b} + \dfrac{1}{5}\cos\dfrac{5\pi t}{T_b}\right]$

(b) Same as (a) except for the absence of the d.c. component

(c) $v(t) = \displaystyle\sum_{m=0}^{\infty} \dfrac{2\sqrt{2}A}{(2m+1)\pi}\cos\dfrac{(2m+1)\pi t}{T_b}$

9.5. (a) 11100001, 01100001; (b) 11011111, 01011111; (c) 11010100, 01010100; (d) 10011000, 00011000

9.7. (a) 49.9 dB; (b) 16

9.9. (a) 0.079; (b) 3.87×10^{-6}; (c) approx 0

9.15. 9.51 dB; 0.5×10^{-5}

9.17. 2.118 MHz; 1.058 MHz

9.19. 1.75 dB

9.21. 80 to 120

Chapter 10

10.1. (a) 14.6 dB; (b) 23.6 dBW; (c) 75.6 dBHz; (d) 28.2 dB; (e) 23 dBK

10.3. (a) 42.9 dB; (b) 50.3 dB

10.5. 1.98 m^2

10.7. (a) 4.33 pW/m^2; (b) 193.9 pW

10.9. -98.6 dBW

10.11. 44K

10.13. (a) 8.5 dB; (b) 1740K

10.15. 941K

10.19. 108 dBHz; 32.5 dB

10.21. 6 dB

10.23. (a) 178.6 pW/m^2; (b) -97.5 dBW/m^2

10.25. 45.6 dBW

10.27. 78.1 dBHz

10.29. 84.5 dBW

10.31. 92.1 dBHz; 86.1 dBHz

10.33. 14.6 dB

10.35. 76.1 dBHz

10.37. 14.5 dB

Chapter 11

11.3. For earth station 1, 2.83° and for earth station 2, 2.85°

11.5. 0.5°; 314 km

11.7. 4.5 dB

11.9. 45.4 dB

11.11. 25 dB

11.13. 28.5 dB

11.15. 51.34 dB

11.17. Yes in both cases

11.19. (a) 22.4 dB; (b) 25 dB

Chapter 12

12.11. 26.5 dB

12.13. (a) -146 dBWK^{-1}

12.15. (b)(1) 2.2 MHz; (b)(2) 11.2 MHz

12.21. 0.85

12.23. 0.5

12.25. 0.93; 872

12.27. 0.96; 1794

12.33. 60 Mb/s

12.35. (a) 75.89 dBHz; (b) 49.3 dBW

12.37. 50.5 dBW

12.39. 24 modes

Chapter 13

13.1. 0.65 mm

13.9. 144.6

13.11. 143.9 dB; 57.3 dBHz; 20.5 dB

13.13. 143.9 dB; 71.1 dBHz; 23.5 dB

13.15. 100.4 min; 23,872 km/hr (14,834 mi/hr)

Conic Sections

A *conic section* as the name suggests, is a section taken through a cone. At the intersection of the sectional plane and the surface of the cone, curves having many different shapes are produced, depending on the inclination of the plane, and it is these curves which are referred to generally as conic sections. Although the origin of conic sections lies in solid geometry, the properties are readily expressed in terms of plane geometrical curves. In Fig. B.1*a*, a reference line for conic sections, known as the *directrix,* is shown as *Z-D*. The *axis* for the conic sections is shown as line *Z-Z'*. The axis is perpendicular to the directrix. The point *S* on the axis is called the *focus.* For all conic sections, the focus has the particular property that the ratio of the distance *SP* to distance *PQ* is a constant. *SP* is the distance from the focus to any point *P* on the curve (conic section), and *PQ* is the distance, parallel to the axis, from point *P* to the directrix. The constant ratio is called the *eccentricity,* usually denoted by *e*. Referring to Fig. B.1*a*,

$$e = \frac{SP}{PQ} \tag{B.1}$$

The conic sections are given particular names according to the value of *e,* as shown in the following table:

Curve	Eccentricity e
Ellipse	$e < 1$
Parabola	$e = 1$
Hyperbola	$e > 1$

These curves are encountered in a number of situations. In this book they are used to describe (1) the path of satellites orbiting the earth, (2) the ellipsoidal shape of the earth, and (3) the outline curves for various antenna reflectors.

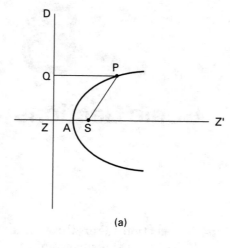

(a)

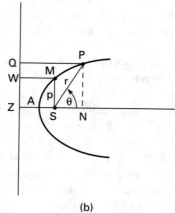

(b)

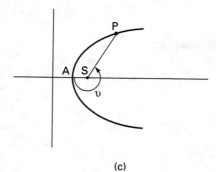

Figure B.1

(c)

A polar equation for conic sections with the pole at the foci can be obtained in terms of the fixed distance p, called the *semilatus rectum*. The polar equation relates the point P to the radius r and the angle θ (Fig. B.1b). From Fig. B.1b,

$$SN = ZN - ZS \tag{B.2}$$

$$= QP - WM$$

$$= \frac{r}{e} - \frac{p}{e}$$

Also,

$$SN = r \cos \theta \tag{B.3}$$

Combining Eqs. (B.2) and (B.3) and simplifying gives the polar equation as

$$r = \frac{p}{1 - e \cos \theta} \tag{B.4}$$

If the angle is measured from SA, shown as v in Fig. B.1c, then, since $v = 180° + \theta$, the polar equation becomes

$$r = \frac{p}{1 + e \cos v} \tag{B.5}$$

The Ellipse

For the ellipse, $e < 1$. Referring to Eq. (B.4), when $\theta = 0°$, $r = p/(1 - e)$. Since $e < 1$, r is positive. At $\theta = 90°$, $r = p$, and at $\theta = 180°$, $r = p/(1 + e)$. Thus r decreases from a maximum of $p/(1 - e)$ to a minimum of $r = p/(1 + e)$, the locus of r describing the closed curve $A'BMA$, Fig. B.2. Also, since $\cos (-\theta) = \cos \theta$, the curve is symmetrical about the axis, and the closed figure results, Fig. B.2a.

The length AA' is known as the *major axis* of the ellipse. The semi-major axis $a = AA'/2$ and e are the parameters normally specified for an ellipse. The semilatus rectum p can be obtained in terms of these two quantities. As already shown, the maximum value for r is $SA' = p/(1 - e)$ and the minimum value is $SA = p/(1 + e)$. Adding these two values gives

$$AA' = AS + SA'$$

$$= \frac{p}{1 + e} + \frac{p}{1 - e}$$

$$= \frac{2p}{1 - e^2}$$

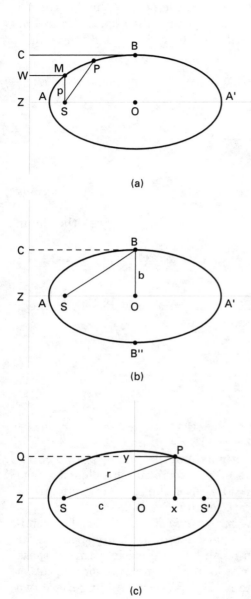

(a)

(b)

(c)

Figure B.2

Since $a = AA'/2$ it follows that

$$p = a(1 - e^2) \tag{B.6}$$

Substituting this into Eq. (B.5) gives

$$r = \frac{a(1 - e^2)}{1 + e \cos v} \tag{B.7}$$

which is Eq. (2.23) of the text.

Equation (B.7) can be rewritten as

$$r = \frac{a(1 + e)(1 - e)}{1 + e \cos v}$$

When $v = 360°$,

$$r = a(1 - e) \tag{B.8}$$

which is Eq. (2.6) of the text.

When $v = 180°$

$$r = (1 + e) \tag{B.9}$$

which is Eq. (2.5) of the text.

Referring again to Fig. B.2a, and denoting the length $SO = c$, it is seen that $AS + c = a$. But as shown above, $AS = p/(1 + e)$ and $p = a(1 - e^2)$. Substituting these for AS and simplifying gives

$$c = ae \tag{B.10}$$

Point O bisects the major axis and length BO is called the *semiminor axis*, denoted by b (BB'' is the minor axis) (Fig. B.2b).

The semiminor axis can be found in terms of a and e as follows. Referring to Fig. B.2b, $2a = SA + SA' = e(AZ + A'Z) = e(2OZ)$ and therefore

$$OZ = \frac{a}{e} \tag{B.11}$$

But $OZ = BC = SB/e$ and therefore $SB = a$. SB is seen to be the radius at B. From the right-angled triangle so formed,

$$a^2 = b^2 + c^2$$

$$= b^2 + (ae)^2$$

From this it follows that

$$b = a\sqrt{1 - e^2} \tag{B.12}$$

and

$$e = \frac{\sqrt{a^2 - b^2}}{a} \tag{B.13}$$

This is Eq. (2.1) of the text.

The equation for an ellipse in rectangular coordinates with origin at the center of the ellipse can be found as follows. Referring to Fig. B.2c, in which O is at the zero origin of the coordinate system:

$$r^2 = (c + x)^2 + y^2$$

But $r = ePQ = e(OZ + x) = e(a/e + x) = a + ex$. Hence

$$(a + ex)^2 = (c + x)^2 + y^2$$

Multiplying this out and simplifying gives

$$a^2(1 - e^2) = x^2(1 - e^2) + y^2$$

Hence

$$1 = \frac{x^2}{a^2} + \frac{y^2}{a^2(1 - e^2)}$$

$$= \frac{x^2}{a^2} + \frac{y^2}{b^2} \tag{B.14}$$

This shows the symmetry of the ellipse, since for a fixed value of y the x^2 term is the same for positive and negative values of x. Also, because of the symmetry there exists a second directrix and focal point to the right of the ellipse. The second focal point is shown as S' in Fig. B.2c. This is positioned at $x = c = ae$ [from Eq. (B.10)].

In the work to follow, y can be expressed in terms of x as

$$y = \frac{b}{a} \sqrt{a^2 - x^2} \tag{B.15}$$

To find the area of an ellipse consider first the area of any segment, Fig. B.3. The area of the strip of width dx is $dA = y\,dx$, and hence the area ranging from $x = 0$ to x is

$$A_x = \int_0^x y\,dx$$

$$= \int_0^x \frac{b}{a} \sqrt{a^2 - x^2}\,dx \tag{B.16}$$

This is a standard integral which has the solution

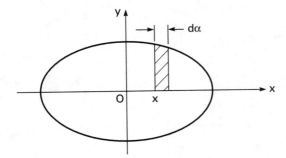

Figure B.3

$$A_x = \frac{xy}{2} + \frac{ab}{2}\,\phi \tag{B.17}$$

where $\phi = \arcsin(x/a)$. In particular, when $x = a$, $\phi = \pi/2$ and the area of the quadrant is

$$A_q = \frac{ab\pi}{4} \tag{B.18}$$

It follows that the total area of the ellipse is:

$$A = 4A_q = \pi ab \tag{B.19}$$

In satellite orbital calculations time is often measured from the instant of perigee passage. Denote the time of perigee passage as T and any instant of time after perigee passage as t. Then the time interval of significance is $t - T$. Let A be the area swept out in this time interval, and let T_p be the periodic time. Then from Kepler's second law:

$$A = \pi ab\,\frac{t - T}{T_p} \tag{B.20}$$

The mean motion is $n = 2\pi/T_p$ and the mean anomaly is $M = n(t - T)$. Combining these with Eq. (B.20) gives:

$$A = \frac{M\,ab}{2} \tag{B.21}$$

The auxiliary circle is the circle of radius a circumscribing the ellipse as shown in Fig. B.4. This also shows the *eccentric anomaly* which is angle E and the true anomaly v (both of which are measured from perigee). The true anomaly is found through the eccentric anomaly. Some relationships of importance which can be seen from the figure are:

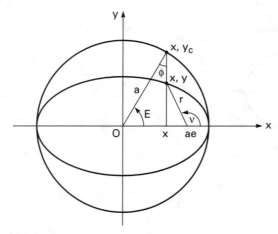

Figure B.4

$$E = \frac{\pi}{2} - \phi \qquad (B.22)$$

$$y_c = a \sin E \qquad (B.23)$$

The equation for the auxiliary circle is $x^2 + y_c^2 = a^2$. Substituting for x^2 from this into Eq. (B.15) and simplifying gives

$$y = \frac{b}{a} y_c \qquad (B.24)$$

Combining this with Eq. (B.23) gives another important relationship:

$$y = b \sin E \qquad (B.25)$$

The area swept out in time $t-T$ can now be found in terms of the individual areas evaluated. Referring to Fig. B.5, this is

$$A = A_q - A_x - A_\Delta$$

$$= \frac{\pi ab}{4} - \left(\frac{xy}{2} + \frac{ab}{2} \phi \right) - \frac{(ae - x)}{2} y$$

$$= \frac{ab}{2} \left(\frac{\pi}{2} - \phi - e \sin E \right)$$

$$= \frac{ab}{2} (E - e \sin E) \qquad (B.26)$$

Comparing this with Eq. (B.21) shows that

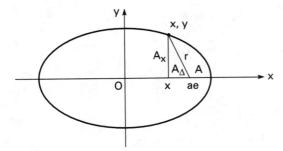

Figure B.5

$$M = E - e \sin E \tag{B.27}$$

This is Kepler's equation given as Eq. (2.27) in the text.

The orbital radius r and the true anomaly v can be found from the eccentric anomaly. Referring to Fig. B.4

$$r \cos (180 - v) = ae - x$$

But $x = a \cos E$ and hence

$$r \cos v = a(\cos E - e) \tag{B.28}$$

Also, from Fig. B.4,

$$r \sin (180 - v) = y$$

and as previously shown, $y = b \sin E$ and $b = a\sqrt{1 - e^2}$. Hence

$$r \sin v = a\sqrt{1 - e^2} \sin E \tag{B.29}$$

Squaring and adding Eqs. (B.29) and (B.28) give

$$r^2 = a^2 (\cos E - e)^2 + a^2 (1 - e^2) \sin^2 E$$

from which

$$r = a(1 - e \cos E) \tag{B.30}$$

This is Eq. (2.30) of the text.

One further piece of manipulation yields a useful result. Combining Eqs. (B.28) and (B.30) gives

$$\cos v = \frac{\cos E - e}{1 - e \cos E}$$

and hence

$$\frac{1-\cos v}{1+\cos v} = \frac{1-\dfrac{\cos E-e}{1-e\cos E}}{1+\dfrac{\cos E-e}{1-e\cos E}}$$

$$= \frac{(1+e)(1-\cos E)}{(1-e)(1+\cos E)}$$

Using the trigonometric identity for any angle α that

$$\tan^2\frac{\alpha}{2} = \frac{1-\cos\alpha}{1+\cos\alpha}$$

yields

$$\tan\frac{v}{2} = \sqrt{\frac{1+e}{1-e}}\,\tan\frac{E}{2} \tag{B.31}$$

This is Eq. (2.29) of the text.

An elliptical reflector has focusing properties which may be derived as follows. From Fig. B.6:

$$OZ = a + AZ$$

$$= a + \frac{AS}{e}$$

But $AS = a(1-e)$ as shown by Eq. (B.8), hence

$$OZ = \frac{a}{e}$$

Referring again to Fig. B.6, $SP = eQP$ and $S'P = PQ'$. Hence

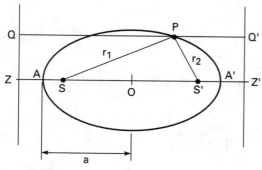

Figure B.6

$$SP + S'P = e(QP + PQ')$$

$$= e(ZZ')$$

$$= e(2 \times OZ)$$

$$= 2a$$

But $SP = r_1$ and $S'P = r_2$ and hence

$$r_1 + r_2 = 2a \qquad \text{(B.32)}$$

This shows that the sum of the focal distances is constant, or in other words, the ray paths from one focus to the other which go via an elliptical reflector are equal in length. Thus electromagnetic radiation emanating from a source placed at one of the foci will have the same propagation time over any reflected path, and therefore the reflected waves from all parts of the reflector will arrive at the other foci in phase. This property is made use of in the Gregorian reflector antenna described in Sec. 5.15.

The Parabola

For the parabola, the eccentricity $e = 1$. Referring to Fig. B.7, since $e = SA/AZ$ by definition it follows that $SA = AZ$. Let $f = SA = AZ$. This is known as the *focal length*. Consider a line LP' drawn parallel to the directrix. The path length from S to P' is $r + PP'$. But $PP' = AL - f - r \cos \theta$, and hence the path length is $r(1 - \cos \theta) + AL - f$.

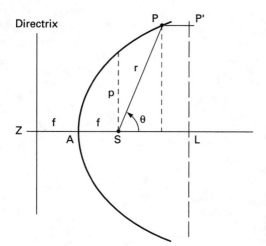

Figure B.7

Substituting for r from Eq. (B.4) with $e = 1$ yields a path length of $p + AL - f$. This shows that the path length of a ray originating from the focus and reflected parallel to the axis (ZZ') is constant. It is this property which results in a parallel beam being radiated when a source is placed at the focus.

It will be seen that when $\theta = 90°$, $r = SP = p$. But $SP = eSZ = 2f$ (since $e = 1$), and therefore $p = 2f$. Thus the radius as given by Eq. (B.4) can be written as

$$r = \frac{2f}{1 - \cos \theta} \qquad (B.33)$$

This is essentially the same as Eq. (5.24) of the text, where ρ is used to replace r, and $\Psi = 180° - \theta$, and use is made of the trigonometric identity $1 + \cos \Psi = 2 \cos^2 (\Psi/2)$. Thus

$$\frac{\rho}{f} = \sec^2 \frac{\Psi}{2} \qquad (B.34)$$

For the situation shown in Fig. B.8, where D is the diameter of a parabolic reflector, Eq. (B.34) gives

$$\frac{f}{\rho_O} = \cos^2 \frac{\Psi_O}{2} \qquad (B.35)$$

But from Fig. B.8 it is also seen that

$$\rho_O = \frac{D}{2 \sin \Psi_O} = \frac{D}{4 \sin \frac{\Psi_O}{2} \cos \frac{\Psi_O}{2}}$$

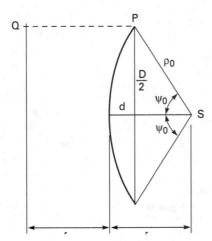

Figure B.8

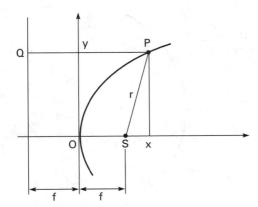

Figure B.9

where use is made of the double angle formula $\sin \Psi_O = 2 \sin (\Psi_O/2) \cos (\Psi_O/2)$. Substituting for ρ_O in Eq. (B.34) and simplifying gives

$$\frac{f}{D} = \frac{1}{4} \cot \frac{\Psi_O}{2} \qquad (B.36)$$

This is Eq. (5.26) of the text.

It is sometimes useful to be able to locate the focal point, knowing the diameter D and the depth d of the dish. From the property of the parabola, $\rho_O = PQ = f + d$. From Fig. B.8 it is also seen that $\rho_O{}^2 = (D/2)^2 + (f - d)^2$. Substituting for ρ_O and simplifying yields

$$f = \frac{D^2}{16d} \qquad (B.37)$$

With the zero origin of the x-y coordinate system at the vertex (point A) of the parabola, the line PQ becomes equal to $x + f$ (Fig. B.9), and $y = r^2 - (x - f)^2$. These two results can be combined to give the equation for the parabola:

$$y^2 = 4fx \qquad (B.38)$$

The Hyperbola

For the hyperbola, $e > 1$. The curve is sketched in Fig. B.10. Equation (B.4) still applies but it will be seen that for $\cos \theta = 1/e$ the radius goes to infinity, in other words the hyperbolic curve does not close on itself in the way that the ellipse does, but lies parallel to the radius at this value of θ. In constructing the ellipse, because e was less than unity, it was possible to find a point A' to the right of S for which the ratio $e = SA'/A'Z = e$ applied in addition to the ratio $e = SA/AZ$. With

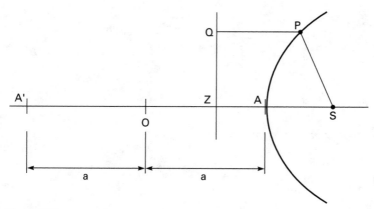

Figure B.10

the hyperbola, because $e > 1$, a point A' to the left of S can be found for which $e = SA'/A'Z$ in addition to $e = SA/AZ$. By setting $2a$ equal to the distance $A'A$ and making the midpoint of $A'A$ equal to the x,y coordinate origin O as shown in Fig. B.10, it is seen that $2\,OS = SA + SA'$. But $SA' = eA'Z$ and $SA = eZA$. Hence

$$2\,OS = e(ZA + A'Z)$$

$$= 2ae$$

Hence

$$OS = ae \qquad\qquad (B.39)$$

Also $SA' - SA = 2a$ and hence

$$2a = e(A'Z - ZA)$$

$$= e2\,OZ$$

Therefore

$$OZ = \frac{a}{e} \qquad\qquad (B.40)$$

Point P on the curve can now be given in terms of the x-y coordinates. Referring to Fig. B.11, S is at point ae and

$$SP^2 = (ae - x)^2 + y^2$$

Also

$$SP = ePQ = e\left(x - \frac{a}{e}\right)$$

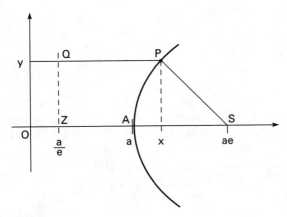

Figure B.11

Combining these two equations and simplifying yields

$$\frac{x^2}{a^2} - \frac{y^2}{b^2} = 1 \tag{B.41}$$

where in this case

$$b^2 = a^2 (e^2 - 1) \tag{B.42}$$

By plotting values it will be seen that a symmetrical curve results as shown in Fig. B.12, and that there is a second focus at point S'. An important property of the hyperbola is that the difference of the two focal distances is a constant. Referring to Fig. B.12, $S'P = ePQ'$ and $SP = ePQ$. Hence

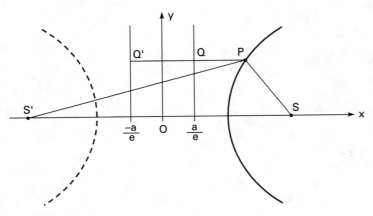

Figure B.12

$$S'P - SP = e(PQ' - PQ)$$

$$= e\left(\frac{2a}{e}\right)$$

$$= 2a \qquad\qquad (B.43)$$

An application of this property is shown in Fig. 5.23a of the text which is redrawn in Fig. B.13. By placing the focus of the parabolic reflector at the focus S of the hyperbola, and the primary source at the focus S', the total path length $S'P + PP''$ is equal to $2a + SP + PP''$ or $2a + SP''$. But as shown previously in Fig. B.7 the focusing properties of the parabola rely on there being a constant path length $SP'' + P''P'$, and adding the constant $2a$ to SP'' does not destroy this property.

The double reflector arrangement can be analyzed in terms of an *equivalent parabola*. The equivalent parabola has the same diameter as the real parabola and is formed by the locus of points obtained at P' which is the intersection of $S'P$ produced to P' and $P''P'$ which is parallel to the x-axis as shown in Fig. B.14. The focal distance of the equivalent parabola is shown as f_e and of the real parabola as f. Looking from the focus S to the real parabola one sees that

$$\frac{h}{d_2} = \frac{y}{f - X_1}$$

Looking to the right from focus S' to the equivalent parabola one sees that

$$\frac{h}{d_1} = \frac{y}{f_e - X_2}$$

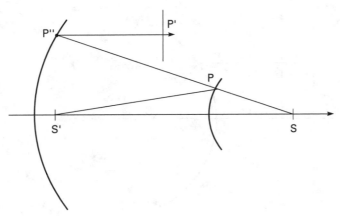

Figure B.13

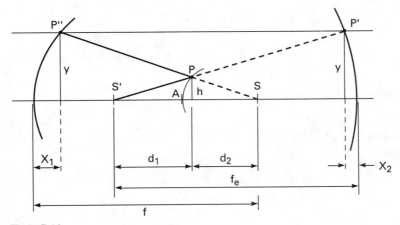

Figure B.14

Hence equating h/y from these two results gives

$$\frac{d_1}{f_e - X_2} = \frac{d_2}{f - X_1}$$

In the limit as h goes to zero, X_1 and X_2 both go to zero and d_1 goes to $S'A$ and d_2 to AS. Hence

$$\frac{S'A}{f_e} = \frac{AS}{f}$$

From Fig. B.12 and Eq. (B. 39)

$$SS' = 2\,OS = 2ae$$

From Fig. B.11

$$AS = ae - a = a(e - 1)$$

From Fig. B.14

$$S'A = SS' - AS = a(e + 1)$$

Hence

$$\frac{a(e + 1)}{f_e} = \frac{a(e - 1)}{f}$$

From which

$$f_e = \frac{e + 1}{e - 1} f \qquad\qquad (B.44)$$

This is Eq. (5.32) of the text.

C

NASA Two-Line Orbital Elements

As stated in Sec. 2.6, prior to August 16, 1994, the two-line orbital elements could be obtained in hard copy in the form of a NASA PREDICTION BULLETIN, as shown in Figure 2.6. These data now have to be obtained through the NASA Bulletin Board Service (BBS), a typical readout being:

1 22969U 94003A 94284.57233250 .00000051 00000-0 10000-3 0 1147

2 22969 82.5601 334.1434 0015195 339.6133 20.4393 13.16724605 34163

The data are presented in basically the same way as in the older Prediction Bulletins. The description of each line follows.

Line Number 1

Name	Description	Units	Example	Field format
LINNO	Line number of element data (always 1 for line 1)	None	1	X
SATNO	Satellite number	None	22969	XXXXX
U	Not applicable	None		X
IDYR	International designator (last two digits of launch year)	Launch yr	94	XX
IDLNO	International designator (launch number of the year)	None	3	XXX
EPYR	Epoch year (last two digits of the year)	Epoch yr	94	XX
EPOCH	Epoch (Julian day and fractional portion of the day)	Day	284.57233250	XXX.XXXXXXXX
NDTO2 or BTERM	First time derivative of the mean motion or ballistic coefficient (depending on the ephemeris type)	Revolutions per day^2 or meters^2 per kilogram	0.00000051	$\pm$.XXXXXXXX[1]
NDDOT 6	Second time derivative of mean motion (field will be blank if NDDOT6 is not applicable)	Revolutions per day^3	00000-0	$\pm$XXXXX-X[2]
BSTAR or AGOM	BSTAR drag term if GP4 general perturbations theory was used. Otherwise will be the radiation pressure coefficient		10000-3	$\pm$XXXXX-X
EPHTYP	Ephemeris type (specifies the ephemeris theory used to produce the elements)	None	0	X
ELNO	Element number	None	1147	XXXX

[1]If NDOT2 is greater than unity, a positive value is assumed without a sign.
[2]Decimal point assumed after the $\pm$ signs.

Line Number 2

Name	Description	Units	Example	Field format
LINNO	Line number of element data (always 2 for line 2)	None	2	X
SATNO	Satellite number	None	22969	XXXXX
II	Inclination	Degrees	82.5601	XXX.XXXX
NODE	Right ascension of the ascending node	Degrees	334.1434	XXX.XXXX
EE	Eccentricity (decimal point assumed)	None	00151950	XXXXXXX
OMEGA	Argument of perigee	Degrees	339.6133	XXX.XXXX
MM	Mean anomaly	Degrees	20.4393	XXX.XXXX
NN	Mean motion	Revolutions per day	13.16724605	XX.XXXXXXXX
REVNO	Revolution number at epoch	Revolutions	34163	XXXXX

Table of Artificial Satellites*

*SOURCE: *ITU Newsletter* 1, 2, and 3, 1995.

Satellite Launchings Notified for Period 9 September to 10 November 1994

Code name and spacecraft description	International number	Country organization (site of launching)	Date	Perigee[1] Apogee[1]	Period[2] Inclination[3]	Frequencies and transmitter power	Observations
STS-64 space shuttle *Discovery*	1994-59-A	United States (Cape Canaveral)	9 Sept.	259 269	89.6 56.9		Carried 12 get away special experiments, an orbit stability experiment and a robot-operated materials processing system. Landed on 20 September 1994
Spartan-1 (Spartan-201)	1994-59-B	United States launched from *STS-64*	13 Sept.	259 269	89.6 56.9		Carried optical instruments to measure solar winds. Was captured back after a few days
Cosmos-2291	1994-60-A	Russia (Baikonur)	21 Sept.				Military spacecraft. *Proton* launcher
Cosmos-2292	1994-61-A	Russia (Plesetsk)	27 Sept.	408 1973	108.6 82.6		Military communications. *Cosmos-3M* launcher
STS-68 space shuttle *Endeavor*	1994-62-A	United States (Cape Canaveral)	30 Sept.	213 226	88.9 57.0		Carried *SIR-C* radar (C- and L-bands), *SAR-X* radar (X-band vertical polarization) and *MAPS* (infrared sensor for measurement of air pollution)
Soyuz-TM20	1994-63-A	Russia (Baikonur)	3 Oct.	228 305	89.8 51.6		Carried a crew of Russian/European cosmonauts as well as experimental apparatus to *Mir-1* orbital complex. Had to be manually docked
Intelsat-703 3640 kg	1994-64-A	United States (Cape Canaveral)	6 Oct.	in geostationary-satellite orbit at 177° E			Twenty-six C-band and ten Ku-band transponders for telephony and television transmission to the Pacific region. *Atlas-2AS* launcher

Name	Designation	Country (Launch site)	Date	Orbital data	Purpose / Notes
Solidaridad-2 Hughes-type *HS 601*: 3-axis stabilized: 2776 kg; 3.4 × 2.8 × 4.5 m (2.5 kW)	1994-65-A	Mexico (Kourou)	8 Oct.	199.3 / 36.431 / 4 / in geostationary-satellite orbit at 113° W over the Pacific Ocean	National communications. Eighteen C-band transponders with a total 35-channel capability in the C-, L-, and Ku-bands. *Ariane-44L* launcher
Thaicom-2 Hughes-type *HS 376 L*; spin-stabilized; 1080 kg; 2.16 × 2.6 × 6.7 m (670 W)	1994-65-B (Kourou)	Thailand	8 Oct.	199.3 / 36.431 / 4 / in geostationary-satellite orbit at 78.5° E over the Indian Ocean	National communications. Ten C-band and two Ku-band transponders. *Ariane-44L* launcher
Okean-01	1994-66-A	Russia/Ukraine (Plesetsk)	11 Oct.	628 / 663 / 82.5 / 97.6	Oceanology and meteorology. *Tsiklon* launcher
Express-1	1994-67-A	Russia (Baikonur)	13 Oct.	in geostationary-satellite orbit	Communications. Twelve-channel capability. *Proton-K* launcher
IRS-P2 3-axis stabilized; 870 kg	1994-68-A	India (Sriharikota)	15 Oct.	804 / 881 / 98.7 / 98.7	Indian Remote Sensing spacecraft. Carried a linear imaging self-scanner camera. *PSLV-D2* launcher
Electro	1994-69-A	Russia (Baikonur)	31 Oct.	in geostationary-satellite orbit at 76° E over the Indian Ocean	Weather survey to enable hurricane, floods, and typhoon warnings. *Proton* launcher
Astra-1D Hughes-type *HS 601*; 3-axis stabilized; 2.9 tonnes	1994-70-A	Luxembourg SES (Kourou)	1 Nov.	in geostationary-satellite orbit at 19.2° E	Carries 24 transponders of which 18 are active to provide direct television and radio broadcast to Europe. *Ariane-42P* launcher
Wind 1250 kg	1994-71-A	United States (Cape Canaveral)	1 Nov.		International Solar-Terrestrial Programme (ISTP). Carries instrument to measure solar wind plasma and magnetic fields. *Delta-2* launcher

Satellite Launchings Notified for Period 9 September to 10 November 1994 (*Continued*)

Code name and spacecraft description	International number	Country organization (site of launching)	Date	Initial orbital data		Frequencies and transmitter power	Observations
				Perigee[1] Apogee[1]	Period[2] Inclination[3]		
Cosmos-2293	1994-72-A	Russia	2 Nov.	412 436	92.7 65		Military communications. *Tsiklon-2* launcher
STS-66 space shuttle *Atlantis*	1994-73-A	United States (Cape Canaveral)	3 Nov.	296 310	90.6 57		Carried *Atlas* laboratory to measure solar emissions in the visible and UV bands, as well as pregnant and non-pregnant rats for biological studies. Landed at Kennedy Space Center on 14 November 1994
CRISTA-SPAS 3.2 tonnes	1994-73-B	Germany launched from *STS-66*	4 Nov.	296 310	90.6 57		Carried spectrophotometers to monitor gasses in the middle atmosphere and lower thermosphere. Was captured back on 12 November 1994
Resurs 1-3	1994-74-A	Russia (Baikonur)	4 Nov.	663.8 691.4	98 98		Natural resources monitoring. *Zenith-2* launcher

1 = km.
2 = min.
3 = degrees.

SOURCE: COSPAR, NASA, specialized press.

Satellite Launchings Notified for Period 11 November to 20 December 1994

Code name and spacecraft description	International number	Country organization (site of launching)	Date	Initial orbital data Perigee[1] Apogee[1]	Period[2] Inclination[3]	Frequencies and transmitter power	Observations
Progress-M25	1994-75-A	Russia (Baikonur)	11 Nov.	342 394	92.4 51.6		Automatic cargo ship. Carried supplies and a *Raduge* capsule. Docked with *Mir-1* orbital complex. *Soyuz-1* launcher
Cosmos-2294 to Cosmos-2296	1995-76-A to 1994-76-C	Russia (Baikonur)	20 Nov.	19 000 19 000	120 71		GLObal NAvigation Satellite System—*GLONASS. Proton-K* launcher
Cosmos-2297	1994-77-A	Russia (Baikonur)	24 Nov.	851 879	102 71		Military communications. *Zenith-2* launcher
Geo-IK	1994-78-A	Russia (Plesetsk)	24 Nov.	1477 1524	116 73.6		Geodesic survey of the Earth. Carried experimental positioning equipment to monitor movements of land and sea transportation. *Tsiklon-3* booster
Orion-1	1994-79-A	Germany (Cape Canaveral)	29 Nov.	35 821 36 022	1438 3		Carries 34 Ku-band transponders for television. *Atlas-24* launcher
DFH-3 2.6 tonnes at launch	1994-80-A	China (Xichang)	29 Nov.				Telecommunications. *Long March-3* launcher
Molnya-1 (88) (Molnya-1 T)	194-81-A	Russia (Plesetsk)	14 Dec.	462 39 155	702 62.4		Telecommunications
Luch 2 tonnes	1994-82-A	Russia (Baikonur)	16 Dec.	35 708 geostationary-satellite orbit at 95° E	1432 2.5		Relays messages from and to *Mir-1* orbital complex and monitors and relays general distress and rescue calls. *Proton* launcher
Cosmos-2298	1994-83-A	Russia (Plesetsk)	20 Dec.	780 804	100 74		Military communications. *Cosmos-3M* rocket

1 = km.
2 = Min.
3 = degrees.
SOURCES: COSPAR, NASA, specialized press.

Satellite Launchings Notified for Period 21 December 1994 to 24 January 1995

Code name and spacecraft description	International number	Country organization (site of launching)	Date	Initial orbital data		Frequencies and transmitter power	Observations
				Perigee[1] Apogee[1]	Period[2] Inclination[3]		
USA-107 (DSP-17)	1994-84-A	United States	22 Dec.				Missile early warning. *Titan-4* launcher
RADIO-ROSTO	1994-85-A	Russia (Baikonur)	26 Dec.	1885 2165	127 64.6		Amateur radio. *SS-19* missile launcher
Cosmos-2299 to Cosmos-2304	1994-86-A to 1994-86-F	Russia (Plesetsk)	26 Dec.	1415 1442	114.2 82.6		Military communications. *Tsiklon-3* launcher
Raduga-32	1994-87-A	Russia (Baikonur)	28 Dec.	in geostationary-satellite orbit at 70° E			Telecommunications. *Proton-K* launcher
Cosmos-2305	1994-88-A	Russia (Baikonur)	29 Dec.	189 306	89.2 64.9		Military communications. *Soyuz-U* rocket launcher
NOAA-14	1994-89-A	United States (Vandenberg Air Force Base)	30 Dec.	845 858	102 98.9		Meteorology. Carries an imaging radiometer, optical sounders to monitor temperature and moisture content of the atmosphere and counters to measure energetic electrons and protons. *Atlas-E* launcher
Intelsat-704 3640 kg at launch	1995-01-A	United States (Cape Canaveral)	10 Jan.	in geostationary-satellite orbit at 66° E over the Indian Ocean			Telecommunications. Twenty-six C-band and 10 Ku-band repeaters for telephony and television transmission to the Pacific region. *Atlas-2AS* launcher
Tsikada	1995-02-A	Russia (Plesetsk)	24 Jan.	982 1031	105 82.9		Maritime navigation. *Cosmos-3M* launcher
Astrid 20 kg	1995-02-B	Sweden (Plesetsk)	24 Jan.	968 1023	105 82.9		Instruments to measure auroral plasma and OFR auroral imaging. *Cosmos-3M* launcher
Faisat 115 kg	1995-02-C	United States (Plesetsk)	24 Jan.	967 1021	105 82.9		Experimental communications. *Cosmos-3M* launcher

[1] = km.
[2] = min.
[3] = degrees

SOURCES: COSPAR, NASA, specialized press.

Illustrating Third-Order Intermodulation Products

The nonlinear voltage transfer characteristic for a TWT can be written as a power series

$$e_o = ae_i + be_i^2 + ce_i^3 + \cdots$$

The third-order term gives rise to the intermodulation products. To illustrate this, let the input be two unmodulated carriers

$$e_i = A \cos \omega_A t + B \cos \omega_B t$$

and let the carrier spacing be $\Delta\omega$ as shown in Fig. E.1. The terms on the right-hand side of the transfer characteristic equation can be interpreted as follows:

First term, ae_i. This gives the desired linear relationship between e_o and e_i.

Second term, be_i^2. With e_i as shown, this term can be expanded into the following components: a dc component; a component at frequency $\Delta\omega$; second harmonic components of the carriers; second harmonic + $\Delta\omega$ components. All these components can be removed by filtering and need not be considered further.

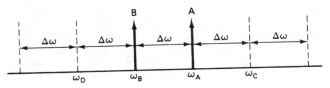

Figure E.1

Third term, ce_i^3. This can be expanded as

$$c(A^3 \cos^3 \omega_A t + B^3 \cos^3 \omega_B t + 3A^2 B \cos^2 \omega_A t \cos \omega_B t + 3AB^2 \cos \omega_A t \cos^2 \omega_B t)$$

The cubed terms in this can be expanded as

$$\cos^3 \omega t = \frac{1}{4}(3 \cos \omega t + \cos 3\omega t)$$

that is, a fundamental component plus a third harmonic component. The third harmonics can be removed by filtering.

The intermodulation products are contained in the cross-product terms $3A^2 B \cos^2 \omega_A t \cos \omega_B t$ and $3AB^2 \cos \omega_A t \cos^2 \omega_B t$. On further expansion, the first of these will be seen to contain a $\cos(2\omega_A - \omega_B)t$ term. With the carriers spaced equally by amount $\Delta \omega$, the $2\omega_A - \omega_B$ frequency is equal to $\omega_A + (\omega_A - \omega_B) = \omega_A + \Delta \omega$. This falls exactly on the adjacent carrier frequency at ω_C as shown in Fig. E.1. Likewise the expansion of the second cross-product term contains a $\cos(2\omega_B - \omega_A)t$ term which yields an intermodulation product at frequency $\omega_B - \Delta \omega$. This falls exactly on the carrier frequency at ω_D.

Acronyms

ACSSB	Amplitude companded single-sideband
AMSC	American Mobile Satellite Consortium
AM	Amplitude modulation
AOS	Acquisition of signal
APM	Amplitude phase modulation
Arabsat	Arabian satellite (communications)
BAPTA	Bearing and power transfer assembly
BOL	Beginning of life
BPSK	Binary phase shift keying
BSS	Broadcasting satellite service
BSU	Burst synchronization unit
CCIR	Comité Consultatif International de Radio
CCITT	Comité Consultatif International de Téléphone et Télégraph
CFM	Companded frequency modulation
Comsat	Communications Satellite (Corporation)
CONUS	Contiguous United States
CPFSK	Continuous phase frequency shift keying
CSSB	Companded single sideband
DAMA	Demand-assignment multiple access
DBS	Direct broadcast satellite
DM	Delta modulation
Domsat	Domestic satellite
DPCM	Differential PCM
DSI	Digital speech interpolation
ECS	European communication satellite

EHF	Extremely high frequency
EIRP	Equivalent isotropically radiated power
EOL	End of life
ERS	Earth resources satellite
ESA	European Space Agency
FCC	Federal Communications Commission
FDM	Frequency-division multiplex
FDMA	Frequency-division multiple access
FEC	Forward error correction
FFSK	Fast FSK
FM	Frequency modulation
FSK	Frequency shift keying
FSS	Fixed satellite service
GHz	Gigahertz
H	Horizontal (polarization)
HEO	Highly elliptical orbit
HPA	High power amplifier
HPBW	Half power beamwidth
Immarsat	International Maritime Satellite (Organization)
Intelsat	International Telecommunications Satellite (Organization)
Intersputnik	International Sputnik (Satellite Communications)
IOT	In orbit testing
ISBN	Integrated Satellite Business Network
ISL	Intersatellite link
ITU	International Telecommunication Union
Landsat	Land survey satellite
LEO	Low earth orbit
LHCP	Left-hand circular polarization
LNA	Low noise amplifier
LOS	Loss of signal
LSAT	Large satellite (European)
MA	Multiple access
MAC	Monitoring alarm and control
MAC	Multiplexed analog compression
Marisat	Marine Satellite (communications)
MHz	Megahertz
MSAT	Mobile Satellite (for mobile communications)
MSK	Minimum shift keying

MUX	Multiplexer
NCS	Network control station
NESS	National Earth Satellite Service
NOAA	National Oceanic and Atmospheric Administration
NTSC	National Television System Committee
PAL	Phase alternation line
PCM	Pulse code modulation
PLL	Phase-locked loop
PSK	Phase shift keying
PSTN	Public Switched Telephone Network
QPSK	Quaternary phase shift keying
Radarsat	Radar satellite (for remote sensing of earth's surface and atmosphere)
RDSS	Radio Determination Satellite Service
RHCP	Right-hand circular polarization
SAR	Synthetic aperture radar
Sarsat	Search and rescue satellite
Satcom	Satellite communications
SBS	Satellite Business Systems
SCPC	Single channel per carrier
SECAM	Sequential Couleur a Mémoire
SFD	Saturation flux density
SHF	Super high frequencies
Spade	Single-channel-per-carrier pulse-code-modulated multiple-access demand-assignment equipment
SPEC	Speech predictive encoded communications
SRB	Synchronization reference burst
SS	Satellite switched
SS	Spread spectrum
SSB	Single sideband
TASI	Time assignment speech interpolation
TDM	Time-division multiplex
TDMA	Time-division multiple access
TDRSS	Tracking data radio satellite system
TELSAT	Telecommunications satellite
TIROS	Television and infrared observational satellite
TMUX	Trans-MUX
TT&C	Telemetry tracking and command
TVRO	TV receive only

TWT	Traveling wave tube
TWTA	Traveling-wave-tube amplifier
UHF	Ultra high frequencies
V	Vertical (polarization)
VHF	Very high frequencies
WARC	World Administrative Radio Conference
XPD	Cross-polarization discrimination

G

Logarithmic Units

Decibels

A power ratio of P_1/P_2 expressed in bels is

$$\log_{10}\left(\frac{P_1}{P_2}\right) \quad \text{bels}$$

The bel was introduced as a logarithmic unit which allowed addition and subtraction to replace multiplication and division. As such, it proved to be inconveniently large, and the decibel, abbreviated dB, is the unit now in widespread use. The same power ratio expressed in decibels is

$$10 \log_{10}\left(\frac{P_1}{P_2}\right) \quad \text{dB}$$

Because the base 10 is understood, it is seldom, if ever, shown explicitly. The abbreviation dB, as shown below, is often modified to indicate a reference level.

Although referred to as a *unit,* the decibel is a dimensionless quantity, the *ratio* of two powers. Power by itself cannot be expressed in decibels. However, by selecting unit power as one of the terms in the ratio, power can be expressed in decibels relative to this. A power expressed in decibels relative to 1 W is shown as dBW. For example, 50 W expressed in decibels relative to 1 W would be equivalent to

$$10 \log \frac{50}{1} \cong 17 \text{ dBW}$$

Another commonly used reference is the milliwatt, and decibels relative to this are shown as dBm. Thus, 50 W relative to 1 mW would be

$$10 \log \frac{50}{10^{-3}} \cong 47 \text{ dBm}$$

By definition, the ratio of two voltages V_1 and V_2 expressed in decibels is

$$20 \log \frac{V_1}{V_2} \text{ dB}$$

The multiplying factor of 20 comes about because power is proportional to voltage squared. Note carefully, however, that the definition of voltage decibels stands on its own, and can be related to the corresponding power ratio only when the ratio of load resistances is also known. Similar arguments apply to current ratios. Also, unit voltage may be selected as a reference, which allows voltage to be expressed in decibels relative to it. For example, 0.5 V expressed in decibels relative to 1 V is

$$20 \log \frac{0.5}{1} \cong -6 \text{ dBV}$$

Another common voltage reference is the microvolt, and 0.5 V expressed in decibels relative to $1 \ \mu V$ is

$$20 \log \frac{0.5}{10^{-6}} \cong 114 \text{ dB}\mu V$$

Decilogs

The decibel concept is extended to allow the ratio of any two like quantities to be expressed in decibel units, these being termed *decilogs*. For example, two temperatures T_1 and T_2 may be expressed as

$$10 \log \frac{T_1}{T_2} \quad \text{decilogs}$$

Just as power and voltage can expressed in decibels relative to the unit quantity, so can other quantities. The name decilog is seldom used however, and common practice is to quote the decibel equivalent for the quantity referred to the selected unit. For example, a temperature of 290 K would be given as

$$10 \log \frac{290}{1 \text{ K}} = 24.6 \text{ dBK}$$

Another example which occurs widely in practice is that of frequency referred to 1 Hz. A bandwidth of 36 MHz is equivalent to

$$10 \log \frac{36 \times 10^6}{1 \text{ Hz}} = 75.56 \text{ dBHz}$$

Apart from voltage and current, all decibel-like quantities are taken as 10 log (.). In this book, square brackets are used to denote decibel quantities, using the basic power definition. A ratio X in decibels is thus

$$[X] \equiv 10 \log X$$

The abbreviation dB is used, with letters added where convenient to signify the reference. For example, a bandwidth $b = 3$ kHz expressed in decibels is

$$[b] = 10 \log 3000 = 34.77 \text{ dBHz}$$

A quantity which occurs frequently in link budget calculations is Boltzmann's constant,

$$k = 1.38 \times 10^{-23} \text{ J/K}$$

Expressed in decibels relative to 1 J/K this is

$$10 \log k = -228.6 \text{ dB}$$

Strictly, the dB units should be written as dBJ/K for decibels relative to 1 joule per kelvin. This is awkward, and it is simply shown as -228.6 dB.

If a voltage ratio V_r is to be expressed in decibels using the square bracket notation, it would be written as $2[V_r]$, since $20 \log V_r = 2[V_r]$. A similar argument applies to current ratios. Thus, it must be kept in mind that the square brackets are identified always with 10 log (.).

The more commonly used decibel-type abbreviations are summarized below:

dBW: decibels relative to one watt

dBm: decibels relative to one milliwatt

dBV: decibels relative to one volt

dBμV: decibels relative to one microvolt

dBK: decibels relative to one kelvin

dBHz: decibels relative to one hertz

dBb/s: decibels relative to one bit per second

The advantage of decibel units is that they can be added directly, even though different reference units may be used. For example, if a power of 34 dBW is transmitted through a circuit which has a loss of 20 dB, the received power would be

$$[P_R] = 34 - 20 = 14 \text{ dBW}$$

In some instances different types of ratios occur, an example being the G/T ratio of a receiving system, described in detail in Sec. 10.6. G is the antenna power gain, and T is the system noise temperature. Now G by itself is dimensionless, so taking the log of this ratio requires that the unit reciprocal temperature, K^{-1} be used. Thus,

$$10 \log \left(\frac{G/T}{1 \text{ K}^{-1}} \right) = 10 \log G - 10 \log \frac{T}{1 \text{ K}} = [G] - [T] \text{ dBK}^{-1}$$

The units are often abbreviated to dB/K, but this must not be interpreted as decibels per kelvin.

The Neper

The *neper* is a logarithmic unit based on natural, or naperian, logarithms. Originally it was defined in relation to currents on a transmission line which decayed according to an exponential law. The current magnitude may decay from a value I_1 to I_2 over some distance x so that the ratio is

$$\frac{I_1}{I_2} = e^{-\alpha x}$$

where α is the attenuation constant. The attenuation is nepers is defined as

$$N = -\ln \frac{I_1}{I_2} = \alpha x$$

The same ratio expressed as an attenuation in decibels would be

$$D = -20 \log \frac{I_1}{I_2}$$

Since the same ratio is involved in both definitions it follows that

$$\frac{I_1}{I_2} = e^{-N} = 10^{-D/20}$$

Thus $D = 20N \log_{10} e = 8.686N$ or 1 neper is equivalent to 8.686 decibels.

Mathcad[1] Notation

Mathcad is a mathematical software package used in problem solving throughout this book. Equations written in Mathcad appear on the screen more or less the way one would write them by hand, which makes them easy to interpret. There are some differences in notation however which need to be explained. Values are assigned to variables by means of a colon.

Typing the statement

$$x:5$$

appears on the screen as

$$x : = 5$$

This is an assignment statement. To find the value of x, one types

$$x =$$

and the screen shows

$$x = 5$$

Suppose a second variable y is typed in as

$$y:x^2$$

This appears on the screen as

$$y : = x^2$$

Typing in

$$y =$$

---

[1]Mathcad is a registered trademark of Mathsoft, Inc.

results in the screen displaying

$$y = 25$$

There are various ways in which a range of values can be assigned to a variable. The statement

$$x:0;9$$

appears on the screen as

$$x := 0 .. 9$$

meaning that x takes on all integer values from 0 to 9. Thus a colon appearing in front of an equal sign means the value or values to the right are being *assigned* to the variable on the left. The normal equal sign gives the value that the variable has, which of course may have changed from the assigned value through computation.

Mathcad has built in units, the default set being SI units. These are best illustrated by means of a simple example. To set a current equal to 3 amperes one types

$$I:3*amp$$

and this appears on the screen as

$$I := 3 \cdot amp$$

Likewise

$$R:10*ohm$$

appears as

$$R := 10 \cdot ohm$$

The statement

$$V:I*R$$

appears as

$$V := I \cdot R$$

Typing

$$V =$$

appears as

$$V = 30 \cdot kg \cdot m^2 \cdot sec^{-2} \cdot coul^{-1} \ \blacksquare$$

This gives the voltage in terms of fundamental units. A units place-holder ■ appears at the end. Placing the cursor on this and typing in the desired units changes these automatically. Thus, setting the cursor on the ■ and typing volt gives

$$V = 30 \cdot \text{volt}$$

The units will change values automatically. Thus, in example 2.1 of the text the mean motion n is defined as

$$n: = \frac{2 \cdot \pi}{1 \cdot \text{day}}$$

This sets n equal to 2π radians per day. Typing

$$n =$$

results in

$$n = 7.272 \cdot 10^{-5} \cdot \text{sec}^{-1} \quad ■$$

Typing rad/sec at the units placeholder results in

$$n = 7.272 \cdot 10^{-5} \cdot \frac{\text{rad}}{\text{sec}}$$

Typing in rad/day results in

$$n = 6.283 \cdot \frac{\text{rad}}{\text{day}}$$

(These values have been rounded off.) Mathcad plus 5.0 is the version used in this book.

References

Abramson, Norman, 1990. "VSAT Data Networks" *Proceedings IEEE,* vol. 78, no. 7, July, pp. 1267–1274.

ADC USAF, 1980. "Model for Propagation of NORAD Element Sets," Spacetrack report no. 3, Aerospace Defense Command, United States Air Force, December.

ADC USAF, 1983. "An Analysis of the Use of Empirical Atmospheric Density Models in Orbital Mechanics," Spacetrack report no. 3, Aerospace Defense Command, United States Air Force, February.

Andrew Antenna, 1985. "1.8 meter 12-GHz Receive-only Earth Station Antenna," bulletin 1206A, Whitby, Ontario, Canada.

Arons, Arnold B., 1965. *Development of Concepts of Physics,* Addison Wesley, Reading, Mass.

Atzeni, C., G. Manes, and L. Masotti, 1975. "Programmable Signal Processing by Analog Chirp-Transformation using SAW Devices," *Ultrasonics Symposium Proceedings,* IEEE.

Balanis, Constantine, 1982. *Antenna Theory Analysis and Design,* Harper & Row, New York.

Bate, Roger R., Donald D. Mueller, and Jerry E. White, 1971. *Fundamentals of Astrodynamics,* Dover, New York.

Baylin, Frank, and Brent Gale, 1991. *Ku-Band Satellite TV,* Baylin Publications.

Bellamy, John, 1982. *Digital Telephony,* John Wiley and Sons, New York.

Bhargara, Vuay K., David Haccoum, Robert Matyas, and Peter P. Nuspl, 1981. *Digital Communications by Satellite.* John Wiley and Sons, New York.

Bischof, I. J., W. B. Day, R. W. Huck, W. T. Kerr, and N. G. Davies, 1981. "Anik-B Program Delivery Pilot Project, A 12-month Performance Assessment," CRC report no. 1349, Dept. of Communications, Ottawa, December.

Brain, D. J., and A. W. Rudge, 1984. "Efficient Satellite Antennas" Electronics and Power, *Journal of the IEE,* January, pp. 51–56.

Brown, Martin P., Jr., ed., 1981. *Compendium of Communication and Broadcast Satellites 1958 to 1980.* IEEE Press.

Brussaard, Gert, and David V. Rogers, 1990. "Propagation Considerations in Satellite Communication Systems," *Proceedings IEEE,* vol. 78, no. 7, July, pp. 1275–1282.

Campanella, S. J., 1983. "Companded Single Sideband (CSSB) AM/FDMA Performance," 0737-2884/83/010025-05 $01.00. © 1983 by John Wiley & Sons Ltd.

CCIR Recommendation 275-2, 1978. "Pre-emphasis Characteristic in Frequency Modulation Radio Relay Systems for Telephony Using Frequency-Division Multiplex," 14th Plenary Assembly, vol. IX, Kyoto.

CCIR Report 338-3, 1978. "Propagation Data Required for Line of Sight Radio Relay Systems," 14th Plenary Assembly, vol. V, Kyoto.

CCIR Report 391-3, 1978. "Radiation Diagrams of Antennae for Earth Stations in the Fixed Satellite Service for Use in Interference Studies and for the Determination of a Design Objective," 14th Plenary Assembly, vol. IV, Kyoto.

CCIR Rec. 523, 1978. "Maximum Permissible Levels of Interference in a Geostationary Satellite Network in the Fixed Satellite Service Using 8-bit PCM Encoded Telephony Caused by Other Networks of This Service," 14th Plenary Assembly, vol. IV, Kyoto.

CCIR Report 263-5, 1982. "Ionospheric Effects upon Earth-Space Propagation," 15th Plenary Assembly, vol. VI, Geneva.

CCIR Recommendation 404-2, 1982. "Frequency Division for Analog Radio Relay Systems for Telephony Using Frequency Division Multiplex," 15th Plenary Assembly, vol. IX, part I, Geneva.

CCIR Recommendation 405-1, 1982. "Pre-emphasis Characteristics for Frequency Modulation Radio Relay Systems for Television," 15th Plenary Assembly, vol. IX, part I, Geneva.

CCIR Report 454-3, 1982. "Method of calculation to determine whether two geostationary-satellite systems require co-ordination," Geneva.

CCIR Report 564-2, 1982. "Propagation Data Required for Space Telecommunication System," Geneva.

CCIR Report 634-2, 1982. "Maximum Interference Protection Ratio for Planning Television Broadcast Systems. Broadcast Satellite Service (Sound and Television)," vols. X and IX, part 2, Geneva.

CCIR Report 708, 1982. "Multiple access and modulation techniques in the fixed satellite service," 14th Plenary Assembly, Doc. 4/5038, Geneva.

CCIR, 1984. *Fixed Services Handbook,* final draft, Geneva.

CCIR Report 719-1, 1982. "Attenuation by Atmospheric Gases," 15th Plenary Assembly, vol. V, Geneva.

CCITT Recommendation G322, 1976. "General Characteristics Recommended for Systems on Symmetric Pair Cables," *International Carrier Analog Systems,* vol. III.2, Geneva.

CCITT G423, 1976. "Interconnection at the Baseband Frequencies of Frequency-Division Multiplex Radio-Relay Systems 1),2)," *International Carrier Analog Systems,* vol. III.2, Geneva.

CCITT Recommendation 567-2, 1986. "Transmission Performance of Television Circuits Designed for Use in International Circuits," vol. XII, Geneva.

Chang, Kai, ed., 1989. *Handbook of Microwave and Optical Components,* vol. 1, John Wiley and Sons, New York.

Chaplin, J., 1992. "Development of satellite TV distribution and broadcasting," *Electronics and Communications,* vol. 4, no. 1, February, pp. 33–41.

Chouinard, G., 1984. "The Implications of Satellite Spacing on TVRO Antennas and DBS Systems," Canadian Satellite User Conference, Ottawa.

Clarke, Arthur C., 1945. "Extraterrestrial Relays," *Wireless World,* vol. 51, October.

Colino, R. R., 1985. "INTELSAT's twentieth anniversary: two decades of innovation in global communications," *Telecommunications Journal,* vol. 52.

Cospas-Sarsat, 1994a. System Data No. 17, February.

Cospas-Sarsat, 1994b. Information Bulletin No. 8, February.

Daly, P., 1993. "Navstar GPS and GLONASS: global satellite navigation systems," *Electronics and Communication Engineering Journal,* December, pp. 349–357.

Dement, D. K., 1984. "United States Direct Broadcast Satellite System Development," *IEEE Communications Magazine,* vol. 22, no. 3, March.

Dixon, Robert C., 1984. *Spread Spectrum Systems,* John Wiley & Sons, New York.

Duffett-Smith, Peter, 1986. *Practical Astronomy with Your Calculator,* Cambridge University Press.

FCC, 1983. "Licensing of Space Stations in the Domestic Fixed-Satellite Service and Related Revisions of Part 25 of the Rules & Regulations," report 83-184 33206, CC Docket 81-184. Federal Communications Commission, Washington, D.C.

FCC, 1994. "Assignment of Orbital Locations to Space Stations in the Domestic Fixed-Satellite Service," information supplied by the Federal Communications Commission, Washington, D.C., Common Carrier Bureau, Domestic Facilities Division.

Franks, L. E., 1980. "Carrier and Bit Synchronization in Data Communication—A Tutorial Review," *IEEE Transactions on Communications,* vol. COM-28, no. 8, August.

Freeman, Roger L., 1981. *Telecommunications Systems Engineering,* John Wiley & Sons, New York.

Fthenakis, Emanuel, 1984. *Manual of Satellite Communications,* McGraw-Hill, New York.

Gagliardi, Robert M., 1991. *Satellite Communications,* 2d ed., Van Nostrand Reinhold.

Glazier, E. V. D., and H. R. L. Lamont, 1958. *The Services Textbook of Radio,* Volume 5, *Transmission and Propagation.* Her Majesty's Stationery Office, London.

Government of Canada, 1983. *Direct-to-Home Satellite Broadcasting for Canada,* Information Services, Department of Communications.

Ha, Tri T., 1990. *Digital Satellite Communications,* 2d ed., McGraw-Hill Publishing Company, New York.

Halliwell, B. J., ed., 1974. *Advanced Communication Systems,* Newnes-Butterworths, London.

Hassanein, Hisham, André Brind'Amour, Karen Bryden, and Robert Deguire, 1989. "Implementation of a 4800 bps Code-excited Predictive Linear Coder on a Single TMS320C25 Chip," Department of Communications, Communications Research Centre, Ottawa.

Hassanein, Hisham, André Brind'Amour, and Karen Bryden, 1992. "A Hybrid Multiband Excitation Coder for Low Bit Rates," Department of Communications, Communications Research Centre, Ottawa.

Hays, R. M., W. R. Shreve, D. T. Bell Jr., L. T. Claiborne, and C. S. Hartmann, 1975. "Surface Wave Transform Adaptable Processor System," *Ultrasonics Symposium Proceedings,* IEEE.

Hays, Ronald M., and Clinton S. Hartmann, 1976. "Surface-Acoustic-Wave Devices for Communications," *Proceedings IEEE,* vol. 64, no. 5, May.

Hogg, David C., and Ta-Shing Chu, 1975. "The Role of Rain in Satellite Communications," *Proc. IEEE,* vol. 63, no. 9, pp. 1308–1331.

Huck, R. W., and J. W. B. Day, 1979. "Experience in Satellite Broadcasting Applications with CTS/Hermes," XIth Inter. TV Symposium, Montreux, 27 May–1 June.

Hughes, C. D., C. Soprano, F. Feliciani, and M. Tomlinson, 1993. "Satellite systems in a VSAT environment," *Electronics and Communications Engineering Journal,* vol. 5, no. 5 October, pp. 285–291.

Hughes TWT and TWTA Handbook, Hughes Aircraft Company, Electron Dynamics Division, Torrance, Calif.

Hwang, Y., 1992. "Satellite Antennas," *Proceedings IEEE,* vol. 80, no. 1, January, pp. 183–193.

Hyndman, John E., 1991. Hughes HS601 Communications Satellite Bus System Design Trades. Hughes Aircraft Company, El Segundo, Ca.

IEEE Proceedings, 1976. Special issue on surface acoustic waves, May.

IEEE Transactions of Communications, 1982. Special issue on spread-spectrum communications, May.

Intelsat 1980. "Interfacing with Digital Terrestrial Facilities," ESS-TDMA-1-8, p. 11.

Intelsat, 1982. "Standard A Performance Characteristics of Earth Stations in the Intelsat IV, IVA and V Systems," BG-28-72E M/6/77.

Intelsat, 1983. "Intelsat TDMA/DSI System Specification (TDMA/DSI Traffic Terminals)," BG-42-65E rev. 2.

Ippolito, Louis J., 1986. *Radiowave Propagation in Satellite Communications,* Van Nostrand Reinhold, New York.

Iridium Today, 1994. Countdown to 98. Realizing the Vision, Fall issue, vol. 1, no. 1, Published by Iridium, Inc., Washington, DC 20005.

ITU, 1985. *Handbook on Satellite Communications (FSS).*

ITU, 1986. *Radio Regulations,* International Telecommunication Union, Geneva.

Jeruchim, M. C., and D. A. Kane, 1970. *Orbit/Spectrum Utilization Study,* Vol. IV, General Electric doc. no. 70SD4293, December 31.

Johnston, E. C., and J. D. Thompson, 1982. "Intelsat VI Communications Payload," IEE Colloquium on the Global Intelsat VI Satellite System, digest no. 1982/76.

Khan, Ahmed S., 1992. *The Telecommunications Fact Book and Illustrated Dictionary,* Delmar Publishers.

Kleusberg, Alfred, and Richard B. Langley, 1990. "The Limitations of GPS," *GPS World,* March/April.

Kummer, W. H., 1992. "Basic Array Theory," *Proceedings IEEE,* vol. 80, no. 1, January, pp. 127–140.

Langley, Richard B., 1991*a.* "Why Is the GPS Signal So Complex?," *GPS World,* May/June.

Langley, Richard B., 1991*b.* The GPS Receiver: An Introduction," *GPS World,* January.

Langley, Richard B., 1991*c.* "The Mathematics of GPS," *GPS World,* July/August.

Leopold, Raymond J., 1992. *The Iridium Communication System,* TUANZ'92, "Communications for Competitive Advantage" Conference and Trade Exhibition, Aotea Centre, Auckland, New Zealand, Aug. 10–12.

Lewis, J. R., 1982. "Satellite Switched TDMA," Colloquium on The Global Intelsat VI Satellite System, digest no. 1982/76 IEE, London.

Lilly, Chris, J., 1990. "Intelsat's new generation," *IEE Review,* vol. 36, no. 3, March.

Lin, S. H., H. J. Bergmann, and M. V. Pursley, 1980. "Rain Attenuation on Earth-satellite Paths—Summary of 10-Year Experiments and Studies," *Bell System Technical Journal,* vol. 59, no. 2, pp. 183–228.

Mahon, J., and J. Wild, 1984. "Commercial launch vehicles and upper stages," *Space Communications and Broadcasting,* vol. 2, pp. 339–362.

Maines, J. D., and Edward G. S. Paige, 1976. "Surface-Acoustic Wave Devices for Signal Processing Applications," *Proceedings IEEE,* vol. 64, no. 5, May.

Martin, James, 1978. *Communications Satellite Systems,* Prentice-Hall, Englewood Cliffs, N.J.

Microwave Filter Company, 1984. "TI and TVROs: A brief troubleshooting guide to suppressing terrestrial interference at 3.7–4.2 GHz TVRO earth stations."

Mattos, Philip, 1992. "GPS," *Electronics World + Wireless World,* December, pp. 982–987.

Mattos, Philip, 1993*a.* "GPS.2: Receiver architecture," *Electronics World + Wireless World,* January, pp. 29–32.

Mattos, Philip, 1993*b.* "GPS.3: The GPS message on the hardware platform," *Electronics World + Wireless World,* February, pp. 146–151.

Mattos, Philip, 1993*c.* "GPS.4: Radio architecture," *Electronics World + Wireless World,* March, pp. 210–216.

Mattos, Philip, 1993*d.* "GPS.5: The software engine," *Electronics World + Wireless World,* April, pp. 296–304.

Mattos, Philip, 1993*e.* "GPS.6: Applications," *Electronics World + Wireless World,* May, pp. 384–389.

Miya, K., ed., 1981. *Satellite Communications Technology,* KDD Engineering and Consulting, Japan.

Morgan, David P., 1985. *Surface-wave Devices for Signal Processing,* Elsevier.

Motorola Satellite Communications Inc., 1990. Application of Motorola Satellite Communications Inc. for IRIDIUM, a low earth orbit satellite system, before the Federal Communications Commission, Washington, D.C., December.

Motorola Satellite Communications Inc., 1992. Minor amendment to application before the FCC to construct and operate a low earth orbit satellite system in the RDSS uplink band File No. 9-DSS-P91 (87) CSS-91-010.

MSAT, 1988. *MSAT, A System and Service Description,* version 4, March 1, Corporate Development Telesat Canada.

MSAT, 1994. *MSAT News,* no. 10, Winter.

NASA 1981. "TIROS-N/NOAA, the 1978–1988 series of polar-orbiting environmental satellites," National Aeronautics and Space Administration, Goddard Space Flight Center, January.

Nudd, G. R., and O. W. Otto, 1975. "Chirp Signal Processing Using Acoustic Surface Wave Filters," *Ultrasonics Symposium Proceedings,* IEEE.

Nuspl, Peter P., and R. de Buda, 1974. "TDMA Synchronization Algorithms," *IEEE EASCON Conference Record,* Washington, D.C.

Olver, A. D., 1992. "Corrugated Horns," *Electronics & Communication Engineering Journal,* February, IEE, London.

Orbcomm, 1993. ORBCOMM Application Amendment and Supplement: file no. 22-DSS-MP-90 (20), December. (Addressed to William F. Caton, Acting secretary, FCC.)

Orbcomm, 1994. News release, June.

Pickholtz, Raymond L., Donald L. Schilling, and Lawrence B. Milstein, 1982. "Theory of Spread Spectrum Communications—A Tutorial," *IEEE Transactions on Communications,* vol. Com-30, no. 5, May.

Pilcher, L. S., 1982. "Overall Design of the Intelsat VI Satellite," Third International Conference on Satellite Systems for Mobile Communications and Navigation, IEE, London.

Pratt, Timothy, and Charles W. Bostian, 1986. *Satellite Communications,* John Wiley & Sons, New York.

Pritchard, Wilbur L., 1984, "The history and future of commercial satellite communications," *IEEE Communications Magazine,* vol. 22, no. 5, May.

Pritchard, W. L., and M. Ogata, 1990. "Satellite Direct Broadcast," *Proceedings IEEE,* vol. 78, no. 7, July, pp. 1116–1140.

Rana, Hamid A., J. McCoskey, and W. Check, 1990. "VSAT Technology, Trends, and Applications," *Proceedings IEEE,* vol. 78, no. 7, July, pp. 1087–1095.

Raney, R. Keith, Anthony P. Luscombe, E. J. Langham, and Shaber Ahmed, 1991. "RADARSAT," *Proceedings IEEE,* vol. 79, no. 6, June, pp. 839–849.

Reinhart, E. E., 1990. "Satellite Broadcasting and Distribution in the United States," *Telecommunication Journal,* vol. 57, no. V1, June, pp. 407–418.

Roddy, Dennis, and John Coolen, 1994. *Electronic Communications,* 4th ed., Prentice Hall, Englewood Cliffs, N.J.

Rosner, Roy D., 1982. *Packet Switching,* Lifetime Learning Publications.

Rudge, A. W., K. Milne, A. D. Olver, and P. Knight, eds., 1982. The Handbook of Antenna Design, vol. 1, Peter Peregrinus Ltd., U.K.

Rusch, W. V. T., 1992. "The Current State of the Reflector Antenna Art—Entering the 1990's," *Proceedings IEEE,* vol. 80, no. 1, January, pp. 113–126.

Sachev, Dharmendra K., Prakash Nadkarni, Pierre Neyret, Leondar R. Dest, Khodadad Betaharon, and William J. English, 1990. "Intelsat V11: A Flexible Spacecraft for the 1990's and Beyond," *Proceedings IEEE,* vol. 78, no. 7, July.

Scales, Walter C., and Richard Swanson, 1984. "Air and sea rescue via satellite systems," *IEEE Spectrum,* March.

Scarcella, T., and R. V. Abbott, 1983. "Orbital Efficiency Through Satellite Digital Switching," *IEEE Communications Magazine,* vol. 21, no. 3, May.

Schwalb, Arthur, 1982*a*. "The TIROS-N/NOAA-G Satellite Series," NOAA technical memorandum NESS 95, Washington, D.C.

Schwalb, Arthur, 1982*b*. "Modified Version of the TIROS-N/NOAA A-G Satellite Series (NOAA E-J)," Advanced TIROS N (ATN) NOAA technical memorandum NESS 116, Washington, D.C.

Scientific Atlanta 1985/6. Satellite Communications Products.

Sciulli, Joseph A., and S. J. Campanella, 1973. "A Speech Predictive Encoding Communication System for Multichannel Telephony," *IEEE Transactions on Communications,* vol. COM-21, no. 7, July.

Sharp, George L., 1983. "Reduced Domestic Satellite Orbital Spacings at 4/6 GHz," FCC/OST R83-2, May.

Sharp, George L., 1984*a*. "Reduced Domestic Satellite Orbit Spacing," AIAA Communications Satellite Systems Conference, Orlando, Florida, March 18–22.

Sharp, George L., 1984*b*. "Revised Gain Function for Adjacent Satellite Interference Program in FCC," OST R83-2, November 30.

Spilker, J. J., 1977. *Digital Communications by Satellite,* Prentice-Hall, Englewood Cliffs, N.J.

Stanley, William D., 1982. *Electronic Communications Systems.* Reston Publishing Co.

Stevenson, S., W. Poley, L. Lekan, and J. Salzman, 1984. "Demand for Satellite-Provided Domestic Communications Services to the Year 2000," NASA tech. memo. 86894, November.

Sweeting, M. N., 1992. "UoSAT microsatellite missions," *Electronic and Communication Engineering Journal,* June, pp. 141–150.

Taub, Herbert, and Donald L. Schilling, 1986. *Principles of Communications Systems,* 2d edition. McGraw-Hill, New York.

Telsat Canada, 1982. *Anik C2 Launch Handbook,* November.

Telesat Canada, 1983. "Telesat, a Technical Description," Business Development Department pub. no. TC83-001.

Telesat Canada. "Design Considerations for Earth Stations to Operate with Anik C3 Satellite."

Thompson, P. T., and E. C. Johnston, 1983. "Intelsat VI. A New Satellite Generation for 1986–2000," *Inter. J. of Satellite Comm.,* vol. 1, pp. 3–14.

Watt, N., 1986, "Multibeam SS-TDMA design considerations related to the Olympus Specialised Services Payload," *IEE Proc.,* vol. 133, pt. F, no. 4, July.

Webber, R. V., J. I. Stickland, and J. J. Schlesak, 1986. "Statistics of attenuation by rain of 13 GHz signals on earth-space paths in Canada," CRC report 1400, Communications Research Centre, Ottawa, April.

Wertz, James R., ed., 1984. *Spacecraft Attitude Determination and Control,* D. Reidel Publishing Co., Holland.

Williamson, Mark, 1994. *IEE Review,* May, pp. 117–120.

Young, Paul H., 1990. *Electronic Communication Techniques,* Merrill Publishing Company.

Index

ABOUT THE AUTHOR

Dennis Roddy, Professor of Electrical Engineering at Lakehead University in Thunder Bay, Ontario, Canada, teaches courses including Satellite Communications, Communications Systems, Optical Communications, and the Analysis and Design of Analog Circuits. His professional background features more than forty years of experience in both industrial and technical education fields. His previous books include *Microwave Technology*, and the original edition of *Satellite Communications*. In addition, he has coauthored several textbooks and published technical papers and reports on a wide variety of topics.